Fog and Fogonomics

Fog and Fogonomics

Challenges and Practices of Fog Computing, Communication, Networking, Strategy, and Economics

Edited by

Yang Yang
Shanghai Institute of Fog Computing Technology (SHIFT)
ShanghaiTech University
Shanghai, China

Jianwei Huang
The Chinese University of Hong Kong
Shenzhen, China

Tao Zhang
National Institute of Standards and Technology (NIST)
Gaithersburg, MD, USA

Joe Weinman
XFORMA LLC
Flanders, NJ, USA

This edition first published 2020
© 2020 John Wiley & Sons, Inc.

The right of Yang Yang, Jianwei Huang, Tao Zhang, and Joe Weinman to be identified as the authors of this work has been asserted in accordance with law.

Registered Office
John Wiley & Sons, Inc., 111 River Street, Hoboken, NJ 07030, USA

Editorial Office
111 River Street, Hoboken, NJ 07030, USA

For details of our global editorial offices, customer services, and more information about Wiley products visit us at www.wiley.com.

Wiley also publishes its books in a variety of electronic formats and by print-on-demand. Some content that appears in standard print versions of this book may not be available in other formats.

Library of Congress Cataloging-in-Publication data applied for

ISBN: 9781119501091

Cover Design: Wiley
Cover Images: Cloud computing © Just_Super/Getty Images,
Network and city background © Busakorn Pongparnit/Getty Images

Set in 10/12pt WarnockPro by SPi Global, Chennai, India

Printed in the United States of America

V10016826_010920

To our families.
– Yang, Jianwei, Tao, and Joe

Contents

List of Contributors

Mohammad Aazam
Carnegie Mellon University (CMU)
USA

Nanxi Chen
Chinese Academy of Sciences
Bio-vision Systems Laboratory
SIMIT
865 Changning Road
200050
Shanghai
China

Shu Chen
IBM Ireland
Watson Client Solution
Dublin
Ireland

Xu Chen
School of Data and Computer Science
Sun Yat-sen University
Guangzhou
China

Mung Chiang
Department of Electrical and
Computer Engineering
Purdue University
West Lafayette, IN
USA

Jaeyoon Chung
Myota Inc.
Malvern, PA
USA

Carnegie Mellon University
University of Colorado Boulder
Boulder, CO
USA

Siobhán Clarke
The University of Dublin
Distributed Systems Group
SCSS
Trinity College Dublin
College Green
Dublin 2
Dublin
Ireland

Abdelouahid Derhab
Center of Excellence in Information
Assurance (CoEIA)
King Saud University
Saudi Arabia

Mohamed Amine Ferrag
LabSTIC Laboratory
Department of Computer Science
Guelma University
Guelma
Algeria

Lin Gao
Department of Electronic and
Information Engineering
Harbin Institute of Technology
Shenzhen
China

Jordi Garcia
Advanced Network Architectures Lab
(CRAAX)
Universitat Politècnica de Catalunya
(UPC)
Vilanova i la Geltrú
Barcelona
Spain

Peter Garraghan
School of Computing and
Communications
Lancaster University
Lancaster
UK

Maria Gorlatova
Department of Electrical Engineering
Princeton University
Princeton, NJ
USA

Sangtae Ha
Department of Computer Science
University of Colorado Boulder
Boulder, CO
USA

Jianwei Huang
School of Science and Engineering
The Chinese University of Hong Kong
Shenzhen
China

Carlee Joe-Wong
Department of Electrical and
Computer Engineering
Carnegie Mellon University (CMU)
Pittsburgh, PA
USA

Fan Li
The University of Dublin
Distributed Systems Group
SCSS
Trinity College Dublin
College Green
Dublin 2, Dublin
Ireland

Tao Lin
School of Computer and
Communication Sciences
École Polytechnique Fédérale de
Lausanne
Lausanne
Switzerland

Zening Liu
School of Information Science and
Technology
ShanghaiTech University
Shanghai
China

George Iosifidis
School of Computer Science and
Statistics
Trinity College Dublin
University of Dublin
Ireland

Yuan-Yao Lou
Graduate Institute of Networking and
Multimedia and Department of
Computer Science and Information
Engineering
National Taiwan University
Taipei City
Taiwan

Leandros Maglaras
School of Computer Science and
Informatics
Cyber Technology Institute
De Montfort University
Leicester
UK

Eva Marín
Advanced Network Architectures Lab
(CRAAX)
Universitat Politècnica de Catalunya
(UPC)
Vilanova i la Geltrú
Barcelona
Spain

Xavi Masip
Advanced Network Architectures Lab
(CRAAX)
Universitat Politècnica de Catalunya
(UPC)
Vilanova i la Geltrú
Barcelona
Spain

David McKee
School of Computing
University of Leeds
Leeds
UK

Mithun Mukherjee
Guangdong Provincial Key
Laboratory of Petrochemical
Equipment Fault Diagnosis
Guangdong University of
Petrochemical Technology
Maoming
China

Ai-Chun Pang
Graduate Institute of Networking and
Multimedia and Department of
Computer Science and Information
Engineering
National Taiwan University
Taipei City
Taiwan

Yichen Ruan
Department of Electrical and
Computer Engineering
Carnegie Mellon University (CMU)
Moffett Field, CA
USA

Sergi Sànchez
Advanced Network Architectures Lab
(CRAAX)
Universitat Politècnica de Catalunya
(UPC)
Vilanova i la Geltrú
Barcelona
Spain

Hamed Shah-Mansouri
Department of Electrical and
Computer Engineering
The University of British Columbia
Vancouver
Canada

Yuan-Yao Shih
Department of Communications
Engineering
National Chung Cheng University
Taipei City
Taiwan

Leandros Tassiulas
Department of Electrical
Engineering, and Institute for
Network Science
Yale University
New Haven, CT
USA

Kunlun Wang
School of Information Science and
Technology
ShanghaiTech University
Shanghai
China

Joe Weinman
XFORMA LLC
Flanders, NJ
USA

Zhenyu Wen
School of Computing
Newcastle University upon Tyne
Newcastle
UK

Gary White
Distributed Systems Group, SCSS,
Trinity College Dublin
The University of Dublin
College Green
Dublin
Dublin 2
Ireland

Vincent W.S. Wong
Department of Electrical and
Computer Engineering
The University of British Columbia
Vancouver
Canada

Jie Xu
School of Computing
University of Leeds
UK

Beijing Advanced Innovation Center
for Big Data and Brain Computing
(BDBC)
Beihang University
Beijing
China

Renyu Yang
School of Computing
University of Leeds
UK

Beijing Advanced Innovation Center
for Big Data and Brain Computing
(BDBC)
Beihang University
Beijing
China

Yang Yang
Shanghai Institute of Fog Computing
Technology (SHIFT)
ShanghaiTech University
Shanghai
China

Tao Zhang
National Institute of Standards and
Technology (NIST)
Gaithersburg, MD
USA

Shuang Zhao
Shanghai Institute of Microsystem
and Information Technology (SIMIT)
Chinese Academy of Sciences
China

Liang Zheng
Department of Electrical Engineering
Princeton University
Princeton, NJ
USA

Zhi Zhou
School of Data and Computer Science
Sun Yat-sen University
Guangzhou
China

Preface

In the eternal dance driven by the evolution of technology and its applications, computing infrastructure has evolved through numerous waves, from the mainframe, to the minicomputer, to the personal computer, client-server, the smartphone, the cloud, and the edge. Whereas the cloud typically is viewed as pooled, centralized resources and the edge comprises the distributed resources that connect to endpoint devices and things, the fog, which is the latest wave, spans the cloud to device continuum.

To understand the fog, it helps to first understand the cloud. Cloud computing has a variety of definitions, ranging from those of standards bodies, to axiomatic and theoretical frameworks, to various vendor and analyst marketing and positioning statements. It typically is viewed as processing, storage, network, platform, software, and services resources that are available to multiple customers and various workload types. These resources are available "for rent" under a variety of pricing models, such as by the hour, by the minute, by the transaction, by the user, and so forth. Further variations include freemium models, discounts for advance reservation and purchase, for sustained flat use, and dynamic pricing. While some analysts define the cloud as having these resources accessed over the (public) Internet, there is no reason that other networking technologies cannot be used as well, ranging from cellular wireless radio access networks to interconnection facilities to dense wave division multiplexing and a variety of other public and private networks.

In any event, the reality of the cloud is that the major cloud providers have each built dozens of large hyper-scale facilities packed with thousands, or even hundreds of thousands of servers, whose capacity and services are accessible on demand and with pay-per-use charging by a wide variety of customers. This "short-term rental" consumption and business model exists in many other industries beyond cloud computing, e.g. overnight stays in hotels for a per-night fee; cars rentals with a daily-rate; airline, train, and bus ticket for each usage; dining at restaurants and cafés. It even exists in places that we do not normally consider: a bank loan is a means of renting capital by the day or month, where the pay-per-use fee is called the interest rate.

Cloud computing use is still growing at astronomical rates, due to the many advantages that it offers. Clouds gain their strength in large part through their consolidation into large masses of resources. This enables cost-effective dynamic allocation of resources to customers on demand and with a pay-per-use charging model. Large hotels can offer rooms for rent at attractive rates because when one convention leaves, another one begins checking in, and the remaining breakage is rented out to other people. Rental car agencies have thousands of customers; when some are returning cars, others are driving them, and still others are arriving at the counters to begin their rentals. In addition to economies of scale, these demand smoothing effects through statistical multiplexing of multiple diverse customer workloads help generate a compelling customer value proposition. They enable elasticity for many workloads, and smoothing enables higher utilization than if the varying workloads were partitioned into smaller silos. Higher utilization reduces wasted resources, lowering the unit cost of each resource.

However, this main advantage of the cloud – consolidated resources – is also its main weakness. Hyper-scale size and centralized pooled resources mean that computing and storage are located far from their actual use in factories, automobiles, smartphones, wearables, irrigation sensors, and the like. Moreover, in stark contrast to the days when computers were housed in temples and only acolytes could tend to them, computing has become pervasive, ubiquitous, low power, and cheap. Rather than the alleged prognostication from decades ago that there was a world market for "maybe five computers," there are tens of billions of intelligent devices distributed in the physical world. It is clear that sooner or later, we will have hundreds of billions – or even a trillion – smart, connected, digital devices. It is an easy calculation to make. There are seven billion people in the world, so it only takes 15 devices per person, on average, to reach 100 billion globally. In the developed world, it is not unusual for an individual to have 4 or 5 video surveillance cameras, a few smart speakers, a laptop, a desktop, a tablet, a smartphone, some smart TVs, a fitness tracker, and a few Wi-Fi lightbulbs or outlets. To this basic observation one can add three main insights.

First, the global economy is developing even as the price of technology is plummeting, suggesting that every individual will be able to own multiple such devices.

Second, ever more devices are becoming smart and connected. For example, the smart voice-activated microwave has been introduced by Amazon; soon it will be virtually impossible to buy an object that is not smart and connected.

Third, these calculations often undercount the number of devices out there. Because in addition to consumer devices with dedicated ownership by an individual or household, there will be additional tens and hundreds of billions of devices such as manufacturing robots and traffic lights and retail point-of-sale systems and hospital wheelchair tracking systems and autonomous delivery vehicles. A trillion connected devices can be deployed if every individual has 60 or seventy devices – not unlikely once you start adding in light bulbs and outlets and nonconsumer device counts make up the other half-trillion.

These massive resource-limited devices with various functionalities and capabilities, when they are deployed and connected, contribute to the future Internet of Things (IoT) to enable different intelligent applications and services, such as environment monitoring, autonomous driving, city management, and medicine and health care. Moreover, emerging wireless capabilities, as embodied in 5G, reduce latency from tens of milliseconds to single digits. To fully take advantage of these capabilities requires processing and storage resources in proximity to the device. There is absolutely no way that the optimal system architecture in such a situation would be to interconnect all these devices across a dumb wide area network to a remote consolidated facility, i.e. the cloud. Instead, multiple layers of processing and storage are needed to bring order, collaboration, intelligence, and solutions out of what otherwise would be a random chaos of devices.

This is the fog.

A number of synonyms and related concepts with nuanced differences exist, such as edge computing, mobile edge computing, osmotic computing, pervasive computing, ubiquitous computing, mini-cloud, cloudlets, and so on and so forth.

And, various bodies have proposed various definitions. The OpenFog Consortium defines fog computing as "a system-level horizontal architecture that distributes resources and services of computing, storage, control and networking anywhere along the continuum from Cloud to Things." The US National Institute of Standards and Technology similarly defines it as a "horizontal, physical or virtual resource paradigm that resides between smart end-devices and traditional cloud or data centers. This paradigm supports vertically-isolated, latency-sensitive applications by providing ubiquitous, scalable, layered, federated, and distributed computing, storage, and network connectivity."

In other words, the fog is simply multiple interconnected layers of computing along the continuum from cloud to endpoints such as user devices and things. This may include racks or microcells in server closets, residential gateways, factory control systems, and the like.

Whereas clouds are hyper-scale, fog nodes may be intermediate size, or even miniature. Whereas clouds rely on multiple customers and workloads, fog nodes may be dedicated to one customer, and even one use. Whereas clouds have state of the art power distribution architectures including multiple grids with diverse access, generators and/or fuel cells, or hydrothermal energy, fog nodes may be powered by batteries or even energy scavenging. Whereas clouds use advanced thermal management strategies including hot-cold aisles, water cooling, airflow simulation and optimization, fog nodes may be cooled by the environmental ambient. Whereas clouds are built in walled data centers, fog nodes may be in homes, factories, agricultural fields, or vineyards. Whereas clouds have fixed street addresses, fog nodes may be mobile. Whereas clouds are engineered for uptime and five nines connectivity, fog nodes may be only intermittently powered, available, within a coverage area, or functional. Whereas clouds are offered by a specific vendor, fog solutions are inherently heterogeneous ecosystems.

Perhaps this is why fog is likely to have an impact across many domains – the economy, technology, standards, market disruption, society and culture, and innovation – on par with cloud computing's impact.

Of course, similar to how cloud's advantages are their weaknesses, fog's advantages can also be its weaknesses. The strength of mobility can lead to intermittent connectivity, which increases the challenges of reliable message passing. Low latency to endpoints means high latency for massive databases, which can be in the cloud. Small footprints can mean an inability to process massive compute jobs. Heterogeneity can create robustness by largely eliminating systemic failures due to design flaws; it can also create a nightmare for monitoring, management, and root cause analysis. This book will document, explore, and quantify many of these challenges and identify and propose solutions and promising directions for future research.

We, the editors, sincerely hope that this collection of insights from the world's leading fog experts and researchers helps you in your journey to the fog.

Shanghai, China, 27 May 2019

Yang Yang
Jianwei Huang
Tao Zhang
Joe Weinman

1

Fog Computing and Fogonomics

Yang Yang[1], Jianwei Huang[2], Tao Zhang[3], and Joe Weinman[4]

[1] Shanghai Institute of Fog Computing Technology (SHIFT), ShanghaiTech University, Shanghai, China
[2] School of Science and Engineering, The Chinese University of Hong Kong, Shenzhen, China
[3] National Institute of Standards and Technology (NIST), Gaithersburg, MD, USA
[4] XFORMA LLC, Flanders, NJ, USA

As a new computing paradigm, fog computing serves as the bridge that connects centralized clouds and distributed edges of the network and plays the crucial role in managing and coordinating multitier computing resources at the cloud, in the network, at the edge, and on the things (devices). In other words, fog computing provides a new architecture that spans along the cloud-to-things continuum, thus effectively pooling dispersive computing resources at global, regional, local, and device levels to quickly meet various service requirements. Together with the edge, fog computing ensures timely data processing, situation analysis, and decision-making at the locations close to where the data are generated and should be used. Together with the cloud, fog computing supports more intelligent applications and sophisticated services in different industrial verticals and scenarios, such as cross-domain data analysis, pattern recognition, and behavior prediction. Some infrastructure challenges and constraints in communication bandwidth, network connectivity, and service latency can be successfully addressed by fog computing, since it makes computing resources in any networks more accessible, flexible, efficient, and cost-effective. It is no doubt that fog computing will not only empower end users by enabling intelligent services in their neighborhoods but also more importantly, deliver a broad variety of benefits to business, consumers, governments, and societies. This book aims at providing a state-of-the-art review and analysis of key opportunities and challenges of fog computing in different application scenarios and business models.

The following three chapters address different technical and economic issues in collaborative fog and cloud scenarios. Specifically, Chapter 2 introduces the hybrid fog–cloud scenario that combines the whole set of resources from the edge up to the cloud, describing the challenges that need to be addressed to

Fog and Fogonomics: Challenges and Practices of Fog Computing, Communication, Networking, Strategy, and Economics, First Edition. Edited by Yang Yang, Jianwei Huang, Tao Zhang, and Joe Weinman.
© 2020 John Wiley & Sons, Inc. Published 2020 by John Wiley & Sons, Inc.

enable realistic management solutions, as well as a review of the current efforts. The authors propose an architectural solution called Fog-to-Cloud (F2C) as a candidate to efficiently manage the set of resources in the IoT-fog-cloud stack. Such an architectural solution is conceptually supported by a service and technology agnostic software solution, which is discussed thoroughly in this chapter in comparison to other existing initiatives. The proposed F2C architecture has two key advantages: (i) it is open and secure by design, easily adoptable by any system environments through distinct software suites and (ii) it has an inherent collaborative model that supports multiple users to optimize resources utilization and services execution. Finally, the authors analyze main challenges for building a stable, scalable, and optimized solution, from both the resource and service perspectives, with special attention to how data must be managed.

In Chapter 3, the authors give an overview of fog computing and highlight the challenges due to the tremendous growth of various Internet of Things (IoT) systems and applications in recent years. They propose a mechanism to efficiently allocate the computing resources in the cloud and fog to different IoT users, in order to maximize their quality of experience (QoE), i.e. less energy consumption and computation delay. The competition among multiple users is modeled as a potential game to determine the computation offloading decisions. The existence of a pure Nash equilibrium (NE) is proven for this game, and it is shown that the equilibrium efficiency loss due to the strategic behavior of users is bounded. A best response strategy algorithm is then developed to obtain an NE of the computation offloading game. Numerical results reveal that the proposed mechanism significantly enhances the overall QoE, and in particular, 18% more users can benefit from computing services than the existing offloading mechanism. The results also demonstrate that the proposed mechanism is promising to enable low-latency computing services for delay-sensitive IoT applications.

In Chapter 4, the authors examine the pricing and performance trade-offs in data analytics. First, they introduce different types of computing devices employed in fog and cloud scenarios, review the current pricing techniques in use, and discuss their implications for performance criteria like accuracy and latency. Then, a data analytics case is studied under a testbed of temperature sensors, where the temperature readings can be analyzed either at local Raspberry Pis or on a cloud server. Local analysis can reduce the communication overhead as raw data are no longer sent to the cloud server, but it lengthens the computation time as Raspberry Pis have less computing capacity than cloud servers. Thus, it is not immediately clear whether fog-based or cloud-based analysis leads to a lower overall completion time; indeed, a hybrid algorithm that can utilize both types of resources in parallel will likely minimize the completion time. However, the choice between a fog-based, cloud-based, or hybrid algorithm also induces different monetary costs (including both computation and data transmission costs) and may lead to different levels of accuracy, since the local analysis involves analyzing only subsets of data and later combining their results due to the Raspberry Pis' limited computing capacity. The authors examine these trade-offs for a simple linear regression scenario and show that there is a threshold number of samples above which a hybrid algorithm is preferred to the cloud-based one.

In Chapter 5, the authors outline a number of qualitative and quantitative arguments and frameworks to help rationally assess the economic benefits and trade-offs between different approaches. For example, resource consolidation tends to increase latency to and from distributed edge and fog services. On the other hand, it tends to reduce latency to cloud-based data and services. The statistics of independent, identically distributed workload demands can benefit from aggregation: multiple independent varying workloads tend to "cancel" each other out, leading to a precisely quantifiable smoothing effect that boosts utilization for a given resource level, which in turn reduces the weighted unit cost of resources. In short, there are many quantifiable characteristics of the fog, which can be evaluated in light of alternative architectures. Ultimately this illustrates that there is no "perfect" solution, as trade-offs need to be quantified and assessed in light of specific application requirements.

In Chapter 6, the authors analyze the design challenges of incentive mechanisms for encouraging user engagements in user-provided infrastructures (UPIs). Motivated by novel business models in network sharing solutions, they focus on mobile UPIs, where the energy consumption and data usage costs are critical, while storage and computation resources are limited. Hence, these parameters have large impacts on users' decisions for requesting/offering their resources from/to UPIs. This chapter reviews a set of incentive schemes that have been proposed for such UPIs, leveraging cooperative game theory, bargaining theory, and auctions. The authors shed light on the attained equilibriums, and study the efficiency and sensitivity on various system parameters. Furthermore, the impact of the network graph on the collaboration benefits in UPI systems is modeled and analyzed, and whether local user interactions achieve system-wide efficient sharing equilibriums is explored. Finally, key bottleneck issues are discussed in order to unleash the full potential of UPIs in fog computing.

In Chapter 7, the authors introduce a Fog-based Service Enablement Architecture (FogSEA), which is a light-weight, decentralized service enablement model. It supports fog services sharing at network edges by adopting a hierarchical management strategy and underpins cross-domain IoT applications based on a semantic-based overlay network. They also propose the Semantic Data Dependency Overlay Network (SeDDON) network, which maintains the semantic information about available microservices. SeDDON aims to reduce the traffic cost and the response time during service discovery. FogSEA produces less traffic and takes less time to return an execution result, comparing to the baseline approach. Generally, traffic increases as more microservices join the network. SeDDON creation needs to send less messages at varying connectivity densities and microservice numbers. The main reason is that SeDDON allows microservices to advertise their services only once when they join the network, and only the microservice that detects the new node as a reverse-dependence neighbor needs to reply.

In Chapter 8, the authors firstly discuss new characteristics and open challenges of realizing fog orchestration for IoT services before summarizing the fundamental requirements. Then, they propose a software-defined orchestration architecture that decouples software-based control policies from the dependencies and operations of heterogeneous hardware. This design can intelligently compose and orchestrate thousands of heterogeneous fog appliances. Specifically, a resource filteringbased resource assignment mechanism is developed to optimize the resource utilization and fair resource sharing among multitenant IoT applications. Additionally, a component selection and placement mechanism is adopted for containerized IoT microservices to minimize the latency by harnessing the network uncertainty and security while considering different application requirements and appliance capabilities. Finally, a fog simulation platform is presented to evaluate the aforementioned procedures by modeling the entities, their attributes, and actions. The practical experiences show that the proposed parallelized orchestrator can reduce the execution time by 50% with at least 30% higher orchestration quality.

In Chapter 9, the authors focus on the problem of reliable Quality of Service (QoS) – aware service choreography within a fog environment where service providers may be unreliable. A distributed QoS optimized adaptive system is proposed to help users in selecting the best available service based on its reputation and to monitor the run-time performance of the service according to the predetermined Service Level Agreement (SLA). A service adaptation model is described to keep the system up with an expected run-time QoS when the SLA is violated. In addition, a performance validation mechanism is developed for the fog environment, which adopts a monitoring and negotiation component to enable the reputation system.

In Chapter 10, the authors propose a typical fog network consisting of multiple fog nodes (FNs), wherein some task nodes (TNs) have heavy computation tasks, while some helper nodes (HNs) have spare resources for sharing with their neighboring nodes. To minimize the delay of every task in such a fog network, a noncooperative game is formulated and investigated to model the competition among TNs for the communication resources and computation capabilities of HNs. Then, a comprehensive analytical model that considers circuit, computation, offloading energy consumptions is developed for accurately evaluating the overall energy efficiency. With this model, the trade-off relationship between performance gains and energy costs in collaborative task offloading is investigated. A novel delay energy balanced task scheduling (DEBTS) algorithm is proposed to minimize the overall energy consumption while reducing average service delay and delay jitter. Further, extensive simulation results show DEBTS can offer much better delay-energy performance in task scheduling challenges.

In Chapter 11, the authors explore both noncooperative and cooperative perspectives of resource sharing issues in multiuser fog networks. On one hand, for the noncooperative distributed computation offloading scenario, the authors develop a game theoretic mechanism with fast convergence property and good performance guarantee. On the other hand, for the cooperation-based centralized computation offloading scenario, the authors devise a holistic dynamic scheduling framework for collaborative computation offloading, by taking into account a variety of system factors including resource heterogeneity and energy efficiency. Extensive performance evaluations demonstrate that the proposed competitive and cooperative computation offloading schemes can achieve superior performance gains over the existing approaches.

In Chapter 12, the authors design and implement an elastic fog storage solution that is fully client-centric, allowing to handle variable availability and possible untrustworthiness at different remote storage locations. The availability, security, and storage efficiency are ensured by employing data deduplication and erasure coding to guarantee a user's ability to access his or her files. By using the FUSE library, a prototype with proper POSIX interfaces is developed and implemented to study the feasibility and practicality issues, such as reusing file statistic information in order to avoid the metadata management overhead from a database system. The proposed method is evaluated by Amazon S3 as a cloud server and five edge/thing resources, and our solution outperforms cloud-only solutions and is robust to edge node failures, seamlessly integrating multiple types of resources to store data. Other fog-based applications can take advantage of this service as a data storage platform.

In Chapter 13, the authors propose a system design of Virtual Local-Hub (VLH) to effectively communicate with ubiquitous wearable devices, thus extending connection ranges and reducing response time. The proposed system deploys wearable services at edge devices and modifies the system behavior of wearable devices. Consequently, wearable devices can be served at the edge of the network without data traveling via the Internet. Most importantly, the system modifications on wearable devices are transparent to both users and application developers, so that the existing applications can fit into the system naturally without any modifications. Due to the limited computing capacity of edge devices, the execution environment needs to be light-weight. Thus, the system enables remote sharing of common and native function modules on edge devices. By using off-the-shelf hardware, a testbed is developed to conduct extensive experiments. The results show that the execution time of wearable services can be reduced by up to 60% with a low system overhead.

In Chapter 14, the authors present an overview of the primary security and privacy issues in fog computing and survey the state-of-the-art solutions that deal with the corresponding challenges. Then, they discuss major attacks in fog-based IoT applications and provide a side-by-side comparison of the state-of-the-art methods toward secure and privacy-preserving fog-based IoT applications. This chapter summarizes all up-to-date research contributions and outlines future research directions that researcher can follow in order to address different security and privacy preservation challenges in fog computing.

We hope you enjoy reading this book on both technical and economic issues of fog computing. More importantly, we will be very happy if some chapters could inspire you to generate new ideas, solutions, and contributions to this exciting research area.

2

Collaborative Mechanism for Hybrid Fog-Cloud Scenarios

Xavi Masip, Eva Marín, Jordi Garcia, and Sergi Sànchez

Advanced Network Architectures Lab (CRAAX), Universitat Politècnica de Catalunya (UPC), Vilanova i la Geltrú, Barcelona, Spain

The collaborative scenario is introduced in Section 2.1, paying special attention to the architectural models as well as to the challenges posed when considering the fog–cloud collaborative model enriched with new strategies such as innovative resource sharing. Then, next section briefly introduces fog-to-cloud (F2C), as one of the proposed architectural contributions in this hybrid scenario, showing its main benefits in different verticals, also emphasizing some open questions remaining unsolved yet. Section 2.3 describes main F2C challenges (could be also applied to any F2C-like architecture), split into three domains, research, industry, and business, in order to illustrate what the issues are on a large view for a successful deployment. Ongoing work done in well-established fora is introduced next in Section 2.4, with the aim of providing users with pointers to main active efforts in the area. Certainly, as of today, this is a very active area; hence, many other relevant works are simultaneously running out there, as it can be seen after a light pass on the recent programs of highly reputed conferences or highly impacting journals. However, it is not the aim of this chapter to report all contributions dealing with all foreseen challenges, rather to point out the reader to the most active repositories. Section 2.5 addresses the insights of the data management in the Internet of Things (IoT)-fog-cloud scenario, emphasizing challenges, as well as the benefits an F2C-like architecture may bring to sort them out. Finally, this chapter ends in Section 2.6 summarizing what the near future is expected to bring in and opening some additional questions for further discussions.

2.1 The Collaborative Scenario

It is widely accepted that fog computing, as a recently coined computing paradigm, is nowadays driving and will definitely drive, many opportunities

Fog and Fogonomics: Challenges and Practices of Fog Computing, Communication, Networking, Strategy, and Economics, First Edition. Edited by Yang Yang, Jianwei Huang, Tao Zhang, and Joe Weinman.
© 2020 John Wiley & Sons, Inc. Published 2020 by John Wiley & Sons, Inc.

in the business sector for developers and also for service and infrastructure providers, as well as many research avenues for the scientific community. However, beyond the characteristics explicitly inherent to fog computing, all undoubtedly bringing many benefits leading to an optimal service execution, a key contribution from a successful fog computing deployment falls into the novel opportunities brought in by its interaction with cloud computing. The central aim of this chapter is to roll out the insights into such a novel collaborative scenario, emphasizing the expected benefits, not only for users as services consumers but also for those within the whole community willing to actively participate in the new roles envisioned for this new paradigm.

Recognized the fact that cloud computing (and certainly the relevance and added value of its products) has been instrumental in facilitating a wider deployment of the so-called IoT services and also in empowering society toward a wide utilization of Internet services in general, some specific aspects of its deployment make the need to highlight several noteworthy constraints. Indeed, although cloud computing is the major and widely adopted commodity [1] designed to addressing the ever increasing demand for computing and processing information, conceptually supported by its massive storage and huge processing capabilities, cloud computing presents well-known limitations to meet the specific demands of IoT services requiring low latency, which can neither be overlooked in near-future IoT deployments nor easily addressed with current network transport technologies. Additionally, beyond the added delay, the long distance from the edge device – where data are collected, and services are requested – to the far datacenters is adding non-negligible issues notably impacting on key performance aspects. For example, the need to convey huge volumes of data from the edge up to the cloud, significantly overloads the network with traffic that, if handled at the edge would not require to be forwarded up to the cloud. Also important, the huge gap between the edge and the cloud drives the need for considering specific security aspects that might be also avoided if staying local at the edge.

Fortunately, fog computing comes up leveraging the capabilities devices located at the edge of the network, such as smart vehicles or 5G mobile phones, bring in to enabling service execution closer to IoT users. Thus, the overall service response time – suitable for real-time services – may be substantially reduced, while simultaneously removing the need to forward traffic through the core network, and some of the cloud security gaps, all in all with a notable impact on the energy consumption [2]. Nevertheless, as a novel technology, some major issues surround fog computing, which if unaddressed may hinder its real deployment and exploitation as well as limit its applicability. Two major fog characteristics must be undeniably considered: (i) the fog storage and processing capabilities are limited when compared to cloud and (ii) the resources volatility, inherent to the mobility and energy constraints of fog devices, might cause undesired service disruptions. These challenges may, with no doubt hinder the adoption of fog computing by the potential users, be it traditional datacenter operators, ISPs, or new actors such as smart city managers or smart services clients. For the sake of literature review, and assuming fog computing is in its infancy, a huge bunch of research contributions are being published in the fog arena, from specific solutions to particular problems to wide surveys highlighting the main fog research avenues, see for example [3].

It is also worth emphasizing that the aforementioned limitations on fog and cloud computing came to stay, for long. In other words, the specific nature of the IoT scenario and the envisioned IoT evolution are both driving toward a more dynamic scenario – mobility becomes an intrinsic attribute in many devices, where heterogeneity at many levels – hardware, network technologies, interfaces, programming models, etc. – , is a key characteristic, and even worse, with no foreseen boundary on that evolution. In fact, nowadays, there is no limit foreseen for IoT, either on the amount of devices to be deployed or on the services to be offered. Many reports from well-reputed consultant companies worldwide foresee an impressive growth in all IoT-related aspects what not only shows the good health these technologies have, the extraordinary impact they will have on the wide society, but also the demand for a persistent endeavor for the research and industrial sectors.

Thus, it seems reasonable to conclude that although cloud computing as a high-performance computing paradigm, already has a solid footprint in nowadays life for many users/services all over the world, a great potential is envisioned when putting together the benefits of cloud computing along with the innovative scenario defined in fog computing. In fact, cloud computing and fog computing are not competing, but collaborating toward a novel hybrid scenario where services execution many benefit from the whole set of advantages brought by them both.

Designed not to compete but to complement cloud computing, fog computing when combined with cloud computing paves the way for a novel, enriched scenario, where services execution may benefit from resources continuity from the edge to the cloud. From a resources perspective, this combined scenario requires resource continuity when executing a service, whereby the assumption is that the selection of resources for service execution remains independent of their physical location. Table 2.1, extending the data cited in [4] – considering different computational layers from the edge up to the cloud – , shows how the different layers can allocate different devices as well as the corresponding relevant features of each of them, including application examples. As it can be seen, an appropriate resources categorization and selection is needed to help optimize service execution, while simultaneously alleviating combined problems of security, resource efficiency, network overloading, etc.

From a formal perspective, it is also worth devoting some efforts to converge on a widely accepted set of terms so as to avoid misleading information. Aligned to this effort, the OpenFog Consortium (OFC) and the Edge Computing Consortium (ECC) are doing a great task in setting a preliminary glossary of terms that may align the whole community into the same wording

Table 2.1 Resource continuity possibilities in a layered architecture (from [4]).

		Resource continuity from edge to cloud			
		Fog			Cloud
	Edge devices	Basic/ aggregation nodes	Intermediate nodes	Cloud
Device	Sensor, actuator, wearables	Car, phone, computer	Smart building, cluster of devices	Datacenter
Features	Response time	Milliseconds	Subseconds, seconds	Seconds, minutes	Minutes, weeks, days
	Application examples	M2M com- munication haptics	Dependable services (e-health)	Visualizations, simple analytics	Big data analytics statistics
	How long IoT data are stored	Transient	Minutes, hours	Days, weeks	Months, years
	Geographic coverage	Device	Connected devices	Area, cluster	Global

(cf. [5, 6], respectively). For example, although many contributions are equally referring to edge computing or fog computing, the OFC in [6] considers fog computing as "a superset of edge computing," also defining the concept behind edge computing as the scenario where "applications, data and processing are placed at the logical extremes of a network rather than centralizing them." Another interesting discussion focuses on what a fog node should be, including terms such as cloudlets or mini DC, or even the need to consider virtual instances (cf. [7]).

All in all, the envisioned scenario can be seen as a collaborative IoT-fog-cloud context (also referred to as IoT continuum), distributed into a set of layers, each gathering distinct devices (resources) according to their characteristics. Figure 2.1 shows an illustrative representation of the whole hybrid fog–cloud scenario, including the IoT devices at the bottom, the cloud datacenter at top, and some smart elements, i.e. fog nodes, with capacity enough to execute some data processing located in between, setting the fog. These fog nodes are respon-sible for a preliminary data filtering and processing, properly customized to the nodes capacities and characteristics, limiting the scope of far cloud datacenters to specific needs not covered by the set of fog nodes. Finally, as also shown in Figure 2.1, since all components in Figure 2.1 are to be connected, network technologies deployed to guarantee that such a connectivity plays a significant role as well. Certainly, the advent of new communication technologies, such as

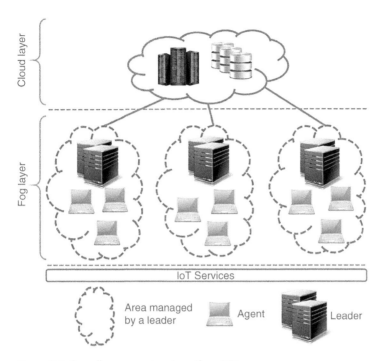

Figure 2.1 Overall resources topology (from [8]).

5G or LoRa, endows the whole scenario with capacities not foreseen yet, that certainly may, with no doubt drive outstanding innovations impacting on the whole society daily activities.

The next subsections go deep into the envisioned collaborative hybrid scenario, first introducing a zoom-out view of the F2C model, setting some interesting discussions on what a fog node should be or strategies to deploy the envisioned F2C model, and later introducing a preliminary high-level architecture best suiting main scenario demands, including the key architectural blocks and the main concepts toward a successful resources sharing strategy.

2.1.1 The F2C Model

The collaborative scenario set when putting together fog and cloud resources may be graphically depicted in terms of several layers, as shown in Figure 2.1. From a service execution perspective, and aiming at using the resources best suiting individual service demands, the scenario in Figure 2.1 can be also mapped into a stack of resources, setting a hierarchical resources pyramid, as shown in Figure 2.2, filling the gap known as the IoT continuum.

After a careful analysis of Figure 2.2, some conclusions may be inferred. For example, it is pretty obvious that the higher in the hierarchy, the higher the

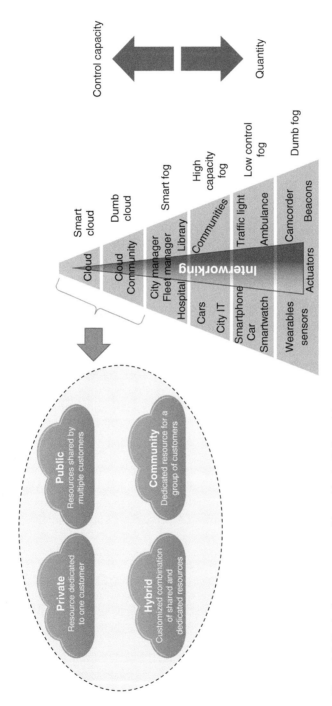

Figure 2.2 Stack of resources as envisioned in the F2C model.

amount of resources, what is evident since datacenters at cloud are endowed with the highest capacities. It is also patent that the lower in the hierarchy, the higher the amount of devices and the lower the control on them as well. This assessment is also unmistakable, since it is a reality the fact that the number of devices grows when getting closer to the edge. But, the interesting discussion comes in between. Indeed, Figure 2.2 does not simply show a single fog layer but many of them instead. These different fog layers are built by categorizing existing devices according to a specific policy, what is actually the main rationale for fog-to-cloud (F2C) [9], positioned to be one of the active approaches intended to solve the IoT continuum challenges. Indeed, the rationale behind the F2C concept is to propose a management architecture to efficiently and properly manage the whole set of resources from the edge up to the cloud aimed to an optimized service execution while simultaneously guaranteeing an efficient resources utilization.

Before going deep into the specific characteristics envisioned for such a management architecture, several specific issues must be clarified – notice that specific challenges are handled in Section 2.3.

2.1.1.1 The Layering Architecture

Certainly, the number of layers will have notable effects on the global control and management. While a low number of layers, i.e. IoT, fog, and cloud, will represent the root for the F2C concept, the main benefits will come when considering additional layers, thus categorizing the resources into distinct layers and hence easing a correct resources selection with no need to go to cloud as well as paving the way for a proper control and management architecture. However, in order not to add unnecessary control overhead due to an excessive number of layers, the granularity to be used in the layering policy must meet a balanced trade-off that certainly will not be an easy task in scenarios where mobility is a key characteristic.

Aligned to the layering model, it is worth paying attention to the fog nodes distribution. Indeed, the different nodes distributed among the distinct fog layers will be endowed with different characteristics and capacities, all controlled through a sort of policies and strategies, all in all configuring the control system. Many approaches may be adopted to set a control architecture, although the right solution should consider how the global topology will look. In the envisioned scenario, with plenty of devices at the edge, a solid, stable, but also proprietary infrastructure at cloud, and finally, many yet unknown fog nodes, it seems reasonable not to go for a centralized approach, rather for a decentralized or distributed approach. The F2C approach proposes a decentralized, hierarchical control architecture, as shown in Figure 2.3, intended to minimize the control overhead while maximizing the resources utilization and services performance.

Another relevant aspect in the layering model refers to the horizontal devices distribution. In order to facilitate a proper control of all devices on all layers and

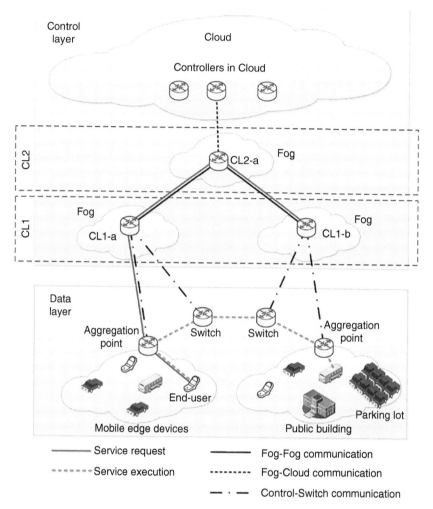

Figure 2.3 F2C hierarchical control architecture (from [10]).

assuming the different range of capacities on each layer, a sort of grouping must be also designed. To that end, F2C proposes the definition of horizontal areas that cover many layers and devices, as shown in Figure 2.3 as well. Although a policy must also be designed for such a classification, it looks reasonable to include devices connectivity as a key parameter to be considered.

2.1.1.2 The Fog Node

Although it may seem unnecessary, the definition of what a fog node is looks not to be easy. Many issues may be highlighted. For example, do we refer to specific hardware? (similar to a router), or do we mean specific software? (similar to Network Functions Virtualization – NFV), is there any specific set of requirements to meet for a device to become a fog node?, if so, who should be the responsible for such a definition?, shall mobility be considered in a fog node?, and certainly, many other. Actually, the concept of "node" may look too old and that may be the fact driving some scientists to refuse the need to work on defining what a fog node should be at the early fog age.

Notwithstanding, current trends are clearly pushing for setting what a fog node should be, as presented in the glossary of terms from the OFC in [6]. Even some industrial efforts are aligned to that trend. See, for example, Nebbiolo Technologies proposing a fogNode to be a flexible hardware architecture defined as "A modular computer for advanced edge computing and secure data storage with a variety of network interfaces for broad IoT connectivity."

A factor that is key in the design and implementation of a fog node refers to its elasticity. The subjacent idea for a fog node focuses on granting a fog node to be agnostic of services and technologies and thus to have high chances to get adapted to the specific needs anytime. This concept is somehow related to the modularity or flexibility introduced in the industrial definition above and also to the definition by the OFC or the one illustrated in [7]. In fact, OFC refers to a fog node as:

> The physical and logical network element that implements fog computing services that allow it to interoperate with other fog nodes. It is somewhat analogous to a server in cloud computing. Fog nodes may be physical, logical, or virtual fog nodes and may be nested (e.g. a virtual fog node on a physical fog node).

And also to a Fog Node Cluster as:

> Commonly referred to as logical fog node, this represents a group of nodes that are managed and orchestrated as a single logical entity in the fog.

Both definitions are supported by the logical concept previously introduced by [7] as:

> Fog nodes are distributed fog-computing entities enabling the deployment of fog services and formed by at least one or more physical devices with processing and sensing capabilities (e.g. computer, mobile phone, smart edge device, car, temperature sensors, etc.) All physical devices of a fog node are connected by different network technologies (wired and wireless) and aggregated and abstracted to be viewed as one single logical entity, that is the fog node, able to seamlessly execute distributed services, as it were on a single device.

However, how various features of edge devices can be presented as logical instances in a fog node is yet an open question. Also, what the computing entity in the whole system is and its locality where these abstractions are created and managed are both open to discussion. For instance, one of the physical devices building the fog node (preferably the one with higher computing capacity) can be made responsible for deploying the abstraction strategy, similar to the concept of cluster head, while also granting communication between other fogs and the cloud. In a more appropriate parlance of today's systems, this could be referred to as the fog node controller. The resource discovery is another open challenge, whereby various IT capacities (CPU, memory, storage) of a fog node can be presented in form of few virtualized computing unities. The same analogy applies with the sensors forming part of a fog node and the network connecting the different fog node devices, assuming that according to the logical concept, all resources managed by a fog node should be abstracted, not only the IT resources but also the sensing and network resources.

Therefore, although the adoption of such a logical concept seems reasonable to build a lasting definition, the main objective must be to provide an unblemished answer to the question of what a fog node should be in the context of a holistic, combined edge, fog, and cloud computing ecosystem, where the key role of a fog node is presenting to higher layers an abstracted and virtualized view of the underlying fog resources and the networks connecting them. For the sake of illustration, Figure 2.4a (extracted from [7]) shows a tentative set of physical devices as well as the physical network forming the fog node, while Figure 2.4b shows a potential abstraction of these physical resources, in form of virtual machines (VMs), virtual sensors (VSs), and possible virtual networks, as seen by higher layers in the combined F2C architecture, setting all together a preliminary approach to a candidate Fog node architecture, including a FAN (Fog Area Network) controller, as well as two modules, the IT abstraction and the Wireless Sensor Network (WSN) controller. Figure 2.4a also illustrates the fact that fog node devices can be physically connected between them, using different network technologies, such as 3G/4G, LTE, Ethernet, Wi-Fi, Bluetooth, etc., while the FAN controller would take over the network virtualization (virtual networks VN1 and VN2).

Finally, it also looks reasonable to adopt the fact that the logical concept, the inferred abstraction model and the consequent integration with the cloud will not only ease the fog computing deployment but also fundamentally change the cloud systems as we know them today, toward a more distributed and more decentralized operation, with all the qualities of the today's datacenter–based service provisioning.

2.1.1.3 F2C as a Service

Once the main objective for F2C is introduced, and before going deep into the key foundational blocks and performance, a key issue to be discussed

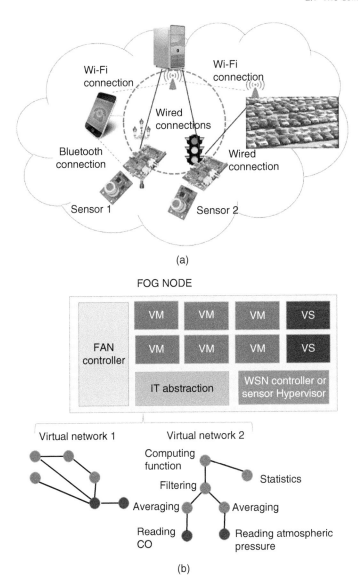

(a)

(b)

Figure 2.4 Fog node proposal (from [7]). (a) Physical devices forming a fog node. (b) Fog node.

refers to how F2C is offered to users. Certainly, this is a critical point since highly impacting on potential business models and also on a wide adoption for potential users.

There are many aspects to be considered when looking for the optimal F2C scenario. With no specific order, the first deals with the highly virtualized scenario F2C will be engaged into. In fact, from the edge up to the cloud, virtualization is a key characteristic bringing value to facilitate deployment, optimal resource utilization, transparency and agnosticism as well as easy technologies integration; hence, should F2C be considered as a virtual entity? The second highlights the envisioned concept for a fog node that may drive into individual solutions be it for hardware or software; hence, shall F2C be supported by a global solution or rather accept proprietary developments? A third aspect may focus on how users access to an F2C system; hence, shall users install specific software in their devices? The three aspects highlighted previously are just a first approach to pragmatic issues to be considered for a wide F2C deployment.

Although there is no clear trend on how such a kind of hybrid fog–cloud systems may be deployed, two main trends may be considered. The first trend would follow the traditional networking model, where specific devices are deployed setting the concept of fog network, where the IoT devices are connected to. In this scenario, potential users would not require any specific installation but just "using" the existing infrastructure and offered capacities. On the other hand, the second trend would be service-oriented, thus offering F2C as a Service. In this scenario, potential users should install the F2C service that will "connect" users to the F2C system, thus enabling them to execute the set of F2C apps included in a future F2C portfolio.

Certainly, several pros and cons may be linked to each option, and an attractive research avenue may focus on looking for the trend best suiting the scenario demands. For example, would it make sense to have different suits of an existing F2C service customized to different services/users/technologies?, or what would be the effort on developing interfaces to guarantee wider "connectivity"?

Moving to real developments, in the industrial sector, we may highlight the effort done by Nebbiolo in designing a fog node, a fog OS and a fog management system, all together pretty close to the first trend explained previously. Differently, the mF2C project [11] follows the second approach, pushing for a sort of app that should be downloaded and installed to run the F2C system. The "running" concept in this case, means the capacity to belong or to join an F2C scenario. The envisioned scenario is very simple, that is devices are grouped into two categories, with and without capacity to support the mF2C requirements. These requirements should be the minimum demands for a device to provide fog features, e.g. storage and processing (in a first approach). The F2C app will be deployed in those devices with enough capacity to support the app, thus abstracting the services to be executed from the real infrastructure. Finally, it is also worth mentioning that since this is yet an ongoing initiative, many challenges remain unsolved.

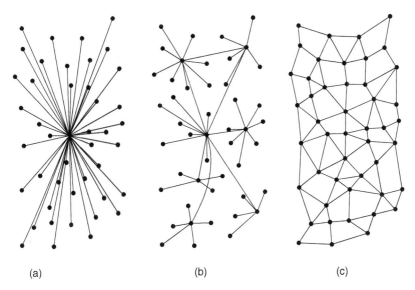

Figure 2.5 Control architectures. (a) Centralized. (b) Decentralized. (c) Distributed.

2.1.2 The F2C Control Architecture

This section introduces a potential preliminary control architecture for a coordinated and efficient resources management in hybrid fog and cloud scenarios, highlighting key pros and cons, challenges for a successful deployment.

A key aspect to be considered when defining a control architecture refers to the topology, i.e. should the control rely on a centralized, decentralized, or distributed architecture? Figure 2.5 presents the three models applied to a combined fog–cloud scenario. Certainly, several benefits but also strong limitations may be associated to each approach, so that a wrong or nonproper selection may hinder the adoption of the whole system. Table 2.2 also summarizes main characteristics for each approach. Considering what the main requirements for a combined IoT-fog-cloud scenario may be (high scalability, high reliability, easy maintenance and settings as well as the capacity to support a continuous evolution on the technology), after carefully observing Table 2.2, we may conclude that the most suitable approach would be a decentralized approach, that basically responds moderately to the envisioned requirements. In fact, the low scalability and reliability inherent to a centralized approach and the low capacity to evolve toward new technologies and to ease the whole management are all making centralized and distributed architectures not to be the proper ones in the envisioned scenario.

Table 2.2 Control architectures characteristics.

Characteristic	Centralized	Decentralized	Distributed
Reliability	L	M	H
Maintenance	H	M	L
Stability	L	M	H
Scalability	L	M	H
Settings	H	M	L
Evolution	H	M	L

L: low; M: moderate; H: high.

While a decentralized approach for control may be the proper architecture, the interoperation model among the different devices must also be defined. A not new, widely tested and largely used strategy is a hierarchical approach, leveraging the hierarchical concept already applied in many sectors and domains (cf. [12] for an earlier example in the optical networks). In a hierarchical architecture, control data forwarding is well defined with no option for misleading information. However, hierarchical architectures may also suffer from a bad design, what may drive to nonoptimal communication among the different components or to components isolation. Nevertheless, we clearly see a hierarchical approach as the best option for control, while a flat architecture should provide the best results for data communication. Notice that one of the harmful effects of a hierarchical architecture, which refers to the need to move the high-level control decision process to a centralized entity, is solved in the envisioned decentralized approach, where in addition, this highest level will be located at cloud.

Next, a tentative hierarchical architecture for the envisioned IoT-fog-cloud scenario is presented (also illustrated in a smart city scenario) emphasizing key functional blocks, control data policies, and the main characteristics and challenges brought in by considering resources sharing.

2.1.2.1 Hierarchical Architecture

We present the envisioned coordinated fog and cloud scenario in the layered architecture shown in Figure 2.6, where resources are allocated to different layers depending on their capacities and features, as earlier presented in Table 2.1. Figure 2.1 illustrates a particular case considering four layers, as proposed in [13], with traditional cloud as the top layer and edge devices at the bottom. In this approach, all fog devices need to be mapped into three layers, which we refer to as edge, basic/aggregation and smart, similar to Table 2.1. As shown in Figure 2.6, the higher layer, the higher the capacities, the control, with implications on the comparably lower number of devices and, perhaps, higher security and a lower privacy (e.g. in cloud).

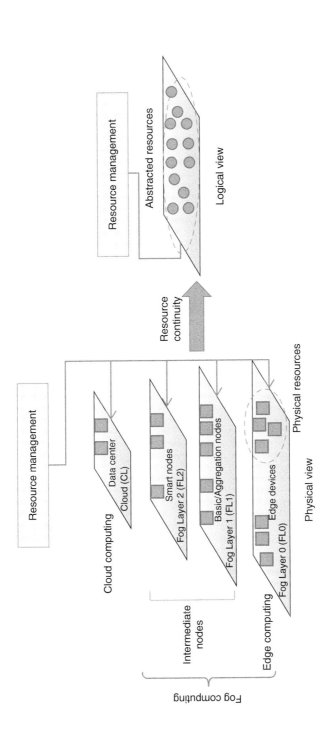

Figure 2.6 The abstraction model filling in the resource continuum (from [13]).

In regard to the resource continuity concept, the main idea is to abstract the layered resources when allocating resources. We note that the physical resources can be first categorized. The categorization has the goal to advertise the resources from the edge to the cloud, as an abstracted entity distributed within one unique layer. As we can see in Figure 2.6, the control view consisting of four layers (FL0, FL1, FL2, and CL) is mapped into a single logical resources view. Individual resources are abstracted in order to be coherently managed by a Resource Management function, part of the envisioned control and management plane, responsible for handling devices mobility, monitoring, and resources categorization.

The resources abstraction can be implemented through virtualization. What is important here is to provide a coherent view of the whole set of available resources, from the edge to the cloud. For instance, if virtual machines are used to manage datacenters at cloud, virtual machines should be also used to abstract edge devices, otherwise the management strategy must support different abstraction strategies and some sort of adaptation, such as between containers and virtual machines. For any architecture to deliver on its promise on resource continuity, the following requirements must be met: (i) coherent view of the whole set of resources to optimize service requirements, (ii) unique control and management plane managing all resources, independently of their location or ownership, (iii) resources categorization and classification to efficiently tailor resource selection to service demands, (iv) resource selection depending on resources capacities, and (v) efficient usage of resources.

We now present an example of a specific implementation of the proposed architecture, on an example of the so-called Smart city. Figure 2.7 illustrates an example of a total of three layers – two layers are defined as Fog Layers (Layers 1 and 2) and one layer as a Cloud Layer (Layer 3), – aggregating the existing computing, networking, and storage resources. In this example, simple actuators and sensors can be in the lowest layers, while "smarter" devices, such as user phones can build layers above. We assume devices are grouped into fog domains (hereinafter referred to as "Fogs," see Figure 2.7), according to a certain policy, depending for example on real connectivity, proximity, capacity, or some business goals. There are three different clouds in Layer 3, with no particular restrictions on whether they are public, private, or hybrid clouds, – notice that interoperability between clouds is considered an innate feature.

There are two different fogs (fog domains) in Layer 2 (fog 1.2 and fog 2.2) and four in Layer 1 (fog 1.1, fog 2.1, fog 3.1, and fog 4.1). We here propose a distributed management architecture that includes two main management artifacts, namely Areas and Control Elements. An Area consists of a group of fogs located at different layers, facilitating vertical resources coordination among different layers. It is yet open to research to go into the details of Areas configuration, their definition or the interaction mechanisms between them. Control Elements, on the other hand, are responsible for providing the different

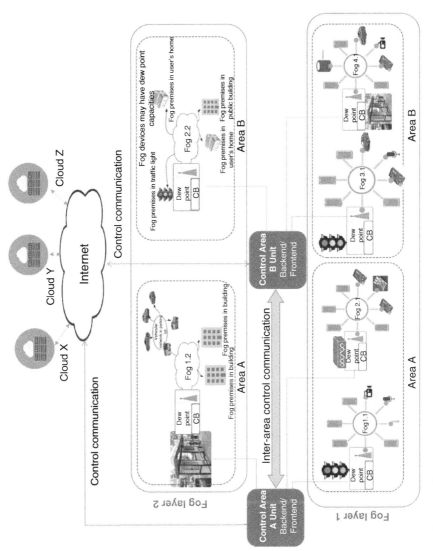

Figure 2.7 Smart city example.

management tasks required in each corresponding domain. The definition of the Control Elements is yet ongoing work, although it seems reasonable to design an element responsible for handling all control and management functionalities for a specific fog (similar to the fog controller introduced in Section 2.1.1), and another one responsible for the management of a specific Area. Certainly, the elements location and the proper communication among them all are nowadays constituting an attractive research avenue as well.

Having set the main components for the proposed management architecture, we now briefly introduce the basic set of functional blocks that would be required to develop the proper management functionalities.

2.1.2.2 Main Functional Blocks

Although some trends are currently investing efforts on the design of a management architecture for the envisioned IoT-fog-cloud scenario, there are some aspects that must be met on whatever the design is. For the sake of understanding and in order to also highlight existing advances done within one of the active trends, next we present the main blocks as defined by the mF2C project [11].

The mF2C project envisions a hierarchical architecture where all nodes install the mF2C app to be F2C capable, i.e. nodes are endowed with the required management and control functionalities to support an F2C system. This approach is highly beneficial rooted on both, it avoids the need to deploy specific hardware to support the set of foreseen F2C services, and it facilitates the collaborative model also envisioned in F2C systems. Indeed, assuming that any device with enough capacities to install the mF2C app must be able to support the defined mF2C features, the project identifies a device embedding the mF2C app as an Agent, so in a hierarchical architecture, Agents are distributed and led by an Agent that becomes a Leader (playing one of the responsibilities linked to the control elements foreseen before). The Leader may be defined as an Agent that performs the required control actions in the F2C hierarchical architecture. Certainly any node with the Agent may become a Leader, and surely some policies should be defined to that end. The hierarchical structure in terms of Agents and Leaders is shown in Figure 2.8, where distinct clusters (or fog domains) are set, each governed by a Leader. The policies to define clusters, to define how and when a node joins a cluster, and all aspects related to mobility and migration, are yet unsolved issues.

Thus, the Agent is the software implementing all expected F2C functionalities. From a design perspective, the Agent is split into two main modules, the Platform Manager (PM) and the Agent Controller (AC). The first, PM, is responsible for guaranteeing the system to be smart, i.e. it implements all decision-making systems and processes. The second, AC, addresses all actions required by the PM at the local level. For example, the PM looks for the set of resources to execute a particular service among the whole stack of resources, and once selected, the AC is responsible for allocating the resources in each individual device.

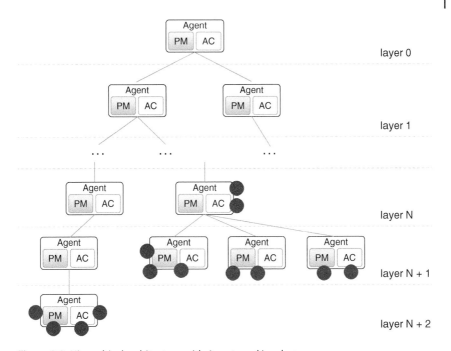

Figure 2.8 Hierarchical architecture with Agents and Leaders.

As defined within the mF2C project, the PM should include blocks to provide the following three main functionalities:

- Service orchestration: Including a life cycle manager, a topology database, an Service Level Agreement (SLA) management, and a resource selection mechanism.
- Runtime: Including a task management, a task scheduling, and policies.
- Monitoring: Including a component dealing with analytics and another one with intelligent instrumentation.

The AC, on the other hand, will include the following main blocks:

- Service management: Including service categorization, services to resources mapping, and Quality of Service (QoS) provisioning.
- Resource management: Including resources discovery, identification and categorization, and monitoring and policies.
- User management: Including users' profiling, sharing models, and assessment.

2.1.2.3 Managing Control Data

An important aspect in the control architecture refers to how control data is handled. The approach designed in the mF2C project considers a unique database shared by both the PM and the AC. Aligned to the proposed hierarchical view, the information to be stored in the database must include resources for an agent node and aggregated cluster resources for the leader. Certainly, a proper updating strategy must be identified to guarantee an accurate picture of the different component status and also a proper strategy to aggregate the information. Moreover, the Leader must embed solutions for reliability in case the Leader is down (back-up leaders are proposed in mF2C to address this issue).

Figure 2.9 depicts a simple, illustrative example showing the information to be stored on each individual agent as well as the information to be aggregated in the leader. Indeed, the database in each agent contains its local information, periodically copied or summarized, according to a certain policy in the same device, and the aggregated data, denoted as AGGR in Figure 2.9, is also periodically synchronized with the local data of its leader. In this particular example, the aggregation is carried out by the addition of the quantitative characteristics and the union of the qualitative characteristics of the devices managed by the leader.

2.1.2.4 Sharing Resources

This subsection aims at introducing what and how sharing, as a concept, when applied to any kind of resource may bring to the proposed collaborative model. Undoubtedly, a promising future is foreseen for resource sharing, but many distinct challenges come up as well. Nowadays, it is widely accepted that models based on collaborative actions become a key instrument in the whole society development. Indeed, the concept of benefitting from "infrastructure" already deployed to implement additional "services," unquestionably looks very attractive from an economical perspective but also for efficient resource utilization. Several examples have currently gained their momentum, such as Uber in transportation or AirBnB in housing. These are just two examples of services where users voluntarily share their resources. There are other scenarios where apps utilize users' resources in a nontransparent way (such as Skype with super nodes).

From a different perspective, and thus in a different scenario, the collaborative model may be also seen as the different roles a user may play when executing a service. For example in the news context, users commonly play as clients consuming the news. However, a new trend recently came up describing users as prosumers, thus acting as consumers but also producers. Certainly, this is a very interesting approach for the IoT world where many services may depend and also rely on information to be collected from the user. In short, we may with no doubt argue that there is a clear trend toward a largely collaborative model, leveraging infrastructure but also data sharing.

However, such a collaborative scenario poses many challenges not easy to solve. It is apparent that collaborative, as a concept, means collaboration among different actors, thus potentially including communications, interaction,

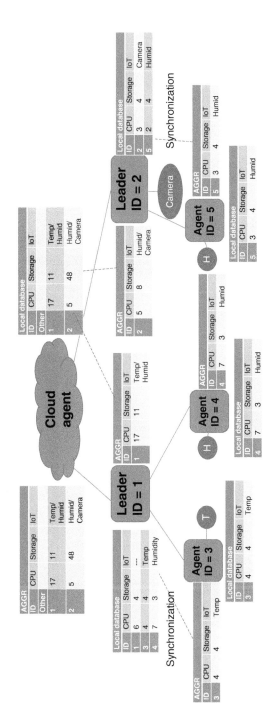

Figure 2.9 Aggregation strategy in mF2C.

interoperation, and even resources sharing or information disclosure. Although there is a nonavoidable trend toward openness, many obstacles may hinder a complete adoption of such an envisioned collaborative scenario, mostly rooted on the aforementioned issues. For example, in a combined fog and cloud scenario, collaboration may refer to cloud and fog provider's interoperation. However, a key issue here is to define what a fog provider is and how and what kind of agreements may be set between them. It is also noticeable to highlight the different business models that may be applied. While cloud is mainly centered on a profit approach, the fog model may support a nonprofit approach where public organizations invest to enrich and improve the set of services to be offered to citizens. And even worse, a user may also become a fog provider when voluntarily sharing resources. How users playing that fog role interact with traditional cloud providers or well-defined fog providers is also a challenge. Another example may consider the scenario where users' devices (cars, smart phones, etc.) are configured to "negotiate," intended to set a cluster of devices to be offered to others. This scenario is adding a nondesirable component of uncertainty due to the inherent mobility linked to users' devices.

As a conclusion, although the development of a potential collaborative model may bring some relevant benefits in terms of resource utilization and service performance, its real applicability is yet constrained by many significant issues, some of them not easy to solve.

2.2 Benefits and Applicability

Recognized the benefits brought by cloud computing, it is with no doubt that the addition of fog computing may only drive to substantially increasing such benefits. Several works have already been published identifying the potential advantages of deploying fog computing in many different sectors. In short, any service or application requiring low latency is a strong candidate to benefit from fog computing. That said, the envisioned scenario where cloud and fog resources are smartly managed to get the most out of both becomes an attractive scenario to ease the development of innovative IoT applications in many sectors.

Indeed, when combining the characteristics inherent to cloud and fog computing, applications may benefit from the high processing capacities at cloud and the immediate react capacities inherent to fog infrastructure, what for example facilitates the deployment of decision-making systems enriched with historical analytics. For example, predictive models can be running at cloud, processing large amount of previous results stored at cloud, and feeding fog resources when needed. In another direction, a smart management system may properly execute applications in parallel, deploying resources at different layers and splitting the application into parallel tasks, as previously done many years ago in high-performance computing. In short, it is a fact that the combination of both paradigms will bring a large set of advantages that must fuel the deployment of innovative services.

Just for illustrative purposes, we can mention recent contributions in the fog arena that while benefitting from fog concepts might also make the most out of the hybrid fog–cloud scenario: [14–16] in e-health; [17–20] in smart cities; [21–24] in energy efficiency; [2, 25, 26] in IoT; [27] in industrial environments; [28] in radio access networks (RANs); as well as [29–31] in vehicular networks; and finally, preliminary contributions in the e-health area leveraging the F2C model [32]. Indeed, this set of efforts have turned into a notable record of scientific publications or even industrial-led forums (such as the OFC or the Open Edge) putting together the industrial and academic sectors to discuss on new ideas and opportunities.

Beyond the traditional verticals, new opportunities have recently come up, promoting the utilization of the F2C scenario in innovative applications. A clear example is AR/VR. The large computing capacities at cloud may be notably enriched with fog resources to provide a much faster response to AR/VR apps. Another example refers to video chat services, where the cloud infrastructure may be used only for analytics or specific connections.

However, the applicability of the envisioned model depends on a key pillar. Indeed, F2C or any architecture aimed at managing the whole set of resources from the edge up to the cloud must rely on the fog infrastructure to execute the applications. In fact, this is a tricky sentence, since as of today, it is not yet clear what a fog provider should be, or even worse if there is a need for such a fog provider. We have previously discussed on the term fog node, so extending the discussion toward a fog provider looks evident. What is extremely clear is that a large set of infrastructure is nowadays deployed – think for example on all smart systems and devices deployed in a city, including cars, bus, or train stations or even municipal buildings. However, what is not yet clear is how potential F2C clients may use this infrastructure. Or even wider, may users also share their resources to become some sort of fog providers thus contributing to build the overall fog infrastructure? Many questions in this regard are yet open for discussion, all pretty related to the potential business models to be deployed. What is undoubtedly evident is that a new resources sharing model is coming and that traditional providers (telcos, cloud, services), but also new actors (municipalities, governments, etc.), will have a key role in the deployment of such a new computing paradigm.

2.3 The Challenges

The envisioned scenario putting together a wide set of heterogeneous resources at the edge with a large and powerful infrastructure at cloud opens many different challenges requiring substantial efforts from the scientific and industrial communities. This section aims at illustrating the main set of open issues where activity is highly required. To that end, three different domains are proposed, research, industry, and business, aimed at differentiating and also highlighting missing aspects. Certainly, a real and successful F2C (or any F2C-like approach) deployment will mandatory require notable advances in the whole set of challenges.

2.3.1 Research Challenges

The hybrid IoT-fog-cloud scenario puts together a set of highly attractive benefits, as discussed in the previous section, whose real deployment will strongly depend on a successful research activity aimed at overcoming several key research challenges. In this section, main research challenges are pointed out.

2.3.1.1 What a Resource is

The main benefit of the proposed coordinated management of the stack of resources from the edge up to the cloud is the capacity to select those resources best suiting the services demands. Certainly, a resource could be defined in terms of many different attributes in a device, for example memory or processing capacities, but a resource can also be a sensor or an actuator playing a specific functionality. Moreover, it must be also highlighted the fact that a resource can be a physical device but also a virtualized entity, as traditionally done in cloud and thus envisioned in fog. However, beyond identifying resources (the categorization pointed out in the first bullet below), it is also critical to identify what a resource must have to be part of an F2C system. Is there any specific demand for a resource to join the system? Shall physical and virtual entities be similarly managed? Which is the virtualization strategy best suiting the F2C scenario? These are just some of the key aspects that must be solved to move forward in the design of an F2C system.

2.3.1.2 Categorization

The envisioned scenario is built upon a stack of resources, from the edge up to the cloud, putting together an enormous diversity of devices, what unquestionably makes a coordinated resources management pretty challenging. Recognized the fact that an efficient management of such a set of heterogeneous devices is a crucial challenge for any IoT computing platform to succeed, it is essential to know what the resources characteristics and attributes are. Moreover, such a resources characteristics assessment will also facilitate an optimal service execution, an optimal resources utilization and therefore, will help to build an effective management solution. However, such a successful management will not only depend on the resources categorization but also on what the services to be executed are. Certainly, this means not only resources to be classified but also services. Indeed, an optimal performance will be possible only when services are executed on those resources best suiting the services needs.

Thus, to that end, it is mandatory to categorize both resources, setting a sort of resources catalogue, and services, describing its main characteristics. On the services side, some part of the work is expected to be done by the service developer itself, providing some service related information that should be later used to fine tuning the final service description. On the resources side, apparently, to build such a description, it is necessary to identify the characteristics and attributes of the resources to be organized, designing a resource classification and taxonomy, specific for the envisioned IoT-fog-cloud scenario. The underlying objective of such a classification should be to describe a catalogue of resources, where resources are formally defined, thus easing the mapping of resources into services.

Categorization is usually handled considering hardware components (i.e. memory, storage, processor, etc.), software (i.e. APIs, OS, etc.), network aspects (i.e. protocol, standards, etc.), and sensors or actuators capacities (temperature, pressure, motion, etc.). Although many works may be found in the literature dealing with such aspects, a comprehensive survey showing the resource characteristics for distinct computing paradigms is presented in [33].

However, beyond the existing work, the large diversity and heterogeneity foreseen in scenarios putting together IoT edge/fog and cloud resources makes resources management in F2C a challenging effort. Broadly speaking, the near resources are to cloud, the higher the computing and storage capacities should be. Thus, we may assume that computation, processing, and storage capabilities are higher in cloud than in fog and higher in fog than in the edge. Interestingly, in the layered F2C envisioned scenario, this assessment is even more elaborated, leveraging the different layers fog is split into.

Moreover, when considering resources categorization, it is also worth noticing that the collaborative model envisioned for F2C fosters the concept of resources sharing, where devices may participate as either "consumer," "contributor," or "both" of them. A device acting as "consumer," is engaged into the F2C system to execute services, thus being a pure resources consumer. On the other hand, a device acting as "contributor" offers its resources to both itself and third users (in a future collaborative scenario) to run services. Finally, some resources can act as "both," hence not only accessing (i.e. consuming) some services but also contributing with their resources to support services execution.

2.3.1.3 Identification

A key aspect in fog computing refers to how devices are identified [34]. It looks evident that this aspect would also be inherited by F2C systems, thus making the management of the names used to identify the resources in the F2C topology pretty challenging. In short, the problem known as identity management (IDM) is the process of managing the resource identifiers. Certainly, many efforts have been already done in the IDM field, but the particular specifications of the envisioned F2C systems make these existing contributions not solid enough. Recognized the fact that some solutions have been proposed in cloud to manage resources, it is obvious that resource identifiers at the edge are not to be managed same way as resources do in datacenters, mainly rooted on both mobility and the huge amount of devices to be managed in fog. Indeed, devices mobility makes a big difference in both paradigms. A device in a geographical location at a certain tracking time may change its location very fast or just disappear from the network map, right after being tracked. In the same way, new devices may dynamically join the topology, thus continuously tweaking the network topology. Consequently, in an F2C scenario, it is not easy to set a static list of resources. On the other hand, the large envisioned set of devices to be identified at fog, may turn into a too excessively large string to grant identifiers uniqueness. This is an additional aspect that must surely be considered when designing the IDM strategy.

Therefore, since IDM plays a remarkable role in the design of the F2C control architecture, IDM becomes a key but challenging pillar in the definition of an F2C system.

Looking at the recent literature, most used IDM approaches are known as object-centric and user-centric, being the latter the most commonly used [35]. Related to the first set of approaches, [36] presents an object-centric IDM strategy for mobile cloud computing leveraging a two-step authentication process requiring zero knowledge proof and token verification for strengthening the authentication process and IDM. However, a too excessive overhead (around 30%) is added by the communication constraints. On the other hand, OpenID identifier [37] is an open source user-centric identity solution leveraging the openID concept widely applied in websites, endowing users with an identity URL, called OpenID identifier. However, this solution does not suitably meet F2C requirements since F2C systems require user devices (generally, more than one) to be identified. As a preliminary conclusion, we may argue that object-centric IDM solutions best suit the F2C systems requirements.

IoT devices identification is also handled in [38], leveraging the network traffic analysis and some machine learning–based classifiers. The main issue with this proposal is that devices are not individually identified but a classification of them, based on a list of known devices. Also the Cloud Security Alliance (CSA) presents a guideline for the identity and access management for the IoT [39]. However, there is no clue on the 23 points included in the guideline about a specific solution for managing the IoT device identities, rather only a set of steps that should guide the users in the process of choosing the IDM approach based on their use cases. A recent solution is based on partitioning the globally unique and commonly large IDs into smaller fragments [40]. Such fragmentation is aimed at allowing network resources to be identified using only a fraction of their name instead of the full identifier.

2.3.1.4 Clustering

The huge amount of devices at the edge must be managed according to certain rules as defined in the control and management strategy to be deployed. To that end, and for the sake of simplicity, it seems reasonable to cluster (setting the domains as shown in Figure 2.7) individual devices into larger infrastructures managed by specific devices, located a layer upper in the hierarchy. However, such a policy is not easy to be agreed mainly because of the large diversity and heterogeneity of resources and services to be executed. In fact, a potentially nice approach would be to start defining what the scope of the policy should be, for example, deciding if mobility, connectivity, energy savings, optimal performance or business models just to name a few should be considered.

The main concern however is that there is no easy way to converge on an optimal policy satisfying all potential requirements, what may turn into distinct policies in place, consequently requiring a preliminary prioritization of services and/or resources to consider what the approach suiting best a specific scenario should be. Certainly such diversity does not help enforce a concrete solution and may derive into difficulties in the clustering strategy.

Nevertheless, beyond selecting the policy for clustering resources, a policy must also be designed to decide what the node acting as leader for a specific cluster should be, and probably both policies will not be decoupled from each other. This is adding an additional concern, even more exacerbated when considering dynamic scenarios or scenarios defined on the fly (ad hoc) leveraging some sharing strategy in a resources collaboration model.

In short, the definition of the clustering policy is pretty complex due to both the many different aspects constraining the decision and the inherent links to other challenging policies somehow affecting the clustering policy itself.

2.3.1.5 Resources Discovery

Resource discovery is a hot topic in the recent literature, mainly enforced by the mobility attribute adopted as a key component in current communications systems. Consequently, several contributions can be found tailored to the demands specific scenarios have. Although no contributions exist for combined fog and cloud scenarios, OFRA [41] already emphasizes the need for both an autonomous and automated discovery strategy as well as a mechanism adopted by a fog element to broadcast information about its presence when added to a cluster deployment. Thus, before moving forward, two key bullets are mandatory. First, it is critical to commit on a clear definition of the problem to be solved, and second, to revisit current solutions in similar fields (fog or edge computing) to learn from past efforts.

The discovery problem may be stated as follows: design a solution to optimally find resources belonging to components (be it any element that may connect to an F2C system) willing to join an F2C area. Such a solution must consider the different characteristics inherent to F2C, such as for example mobility or collaborative models.

Regarding the existing literature, there are certainly many different approaches that deserve a detailed reading. For example, the cloudlet discovery strategy based on DNS-Service Discovery and Multicast DNS [42], although it may not be properly adopted by F2C since it assumes all elements must communicate through the same local network. The discovery of resources for tasks offloading in edge computing through a set of distributed home routers is proposed in [43], what looks too static to be adopted by an F2C scenario. Fog node discovery by mobile users is analyzed in [44] proposing a peer-to-peer approach, different from the hierarchical architecture envisioned for F2C systems. Foglets are proposed in [45] consisting in a discovery server responsible for maintaining an up-to-date list of the fog nodes available at different levels of the hierarchy for a given geographic area, what looks to be a promising approach although several key implementation details are not clearly defined yet. The recent solution in [46], improved in [8], particularly customized to F2C systems, focuses on using 802.11 beacon frames to embed additional information about the leader and also about the available F2C services in that cluster.

2.3.1.6 Resource Allocation

F2C systems are designed to benefit from distinct logical layers classifying resources from the edge up to the cloud according to several policies to be deployed (based on for example the underlying resources capacity, reliability, or volatility, among other parameters). Such a layering architecture is designed to both facilitate service decomposition and parallel execution in distinct layers as well as ease the control and management of resources and services. To that end, an F2C management system must find the set of resources best suiting the set of requirements for each subservice delivered by the service decomposition process. Although similar to parallel job execution in high-performing computing, the resources mobility inherent to F2C makes resource allocation a challenging problem. Actually, F2C exacerbates the challenges already identified in the fog computing field, by adding both the distinct requirements for each subservice and the distinct service types each F2C resource can allocate. Moreover, the specific constraints inherent to edge devices (in terms for example of volatility, scarcity of resources, etc.) must be also carefully analyzed.

Few contributions deal with the processing capacity of devices at the edge of the network, aimed at support services execution. Indeed, contributions in [47, 48] aim at achieving low delay on service execution, mainly focusing on the data acquisition delay, thus without considering fog resources for computing purposes.

On a different approach, [49] proposes a solution to split services into tasks that may be executed in parallel at edge devices. The results show that, as expected, the proposed distributed approach looks not so good for services with high processing requirements, instead a centralized approach would behave much better. In [50], a solution tailored to F2C systems is presented for the first time (although for a very limited scenario that simplifies the service allocation problem by assuming just one resource type) later extended in [51].

However, beyond the efforts done so far, resource allocation in F2C systems considering mobility, precise edge devices specifications, service decomposition, sharing models, and the heterogeneity of services and resources remains yet unsolved.

2.3.1.7 Reliability

It is with no doubt that the high dynamicity inherent to considering the edge claims for novel mechanisms dealing with failure recovery, specifically designed for distributed scenarios. Being yet an open challenge in fog computing, the protection of data and network resources inherited by F2C systems also demands low service allocation time and protection cost in terms of allocated resources and recovery delay.

Again, lessons learnt from past efforts in cloud and fog computing may contribute toward a proper solution for F2C scenarios. Proactive and reactive recovery techniques are commonly used in distinct scenarios, including those considered highly reliable systems (e.g. datacenters [52]) and those with unreliable resources (e.g. wireless sensor networks [53]). Backup paths are defined in [54] as a proactive strategy to recover from network failures. On a reactive side, service execution replication is applied in [55] in a fog computing scenario considering a remote server. Cloud resilience is analyzed in [56] by means of network virtualization. [57] proposes shared-path shared-computing (SPSC) protection for cloud. Multi-access Edge Computing (MEC) failures are discussed in [58] leveraging the offload of workload to neighboring resources (the availability of neighboring resources is not guaranteed though).

However, although not many works are devoted to address protection in F2C scenarios, a pioneering work in the area may be found in [59], where proactive and reactive strategies are confronted in order to illustrate pros and cons of each one. In that work, protection resources are always reserved in the same F2C layer of the resources actually allocated for service execution. An extension of that work may be found in [60] formalizing through linear programming, proactive and reactive protection strategies against failures (assuming only failures at fog infrastructure are possible, supported by the inherent vulnerability of fog nodes), and presenting an extensive evaluation related to the minimization of both service transmission delay and protection cost in protection scenarios, being both important factors to obtain low latency in reliable IoT service execution.

2.3.1.8 QoS

QoS delivery is one of the main (probably along with the one addressed in the next bullet) pillars a system must meet to guarantee the desired performance. As a concept, QoS has been a focus of research for long, with tones of proposals in different fields that have motivated an impressive amount of publications in many different areas, such as network routing. However, as services demands and user requirements evolve, existing QoS solutions become almost useless in F2C systems. As a kind of a never-ending story, QoS delivery will remain as a "yet an unaddressed problem."

Thus, F2C is not an exemption, and certainly novel solutions must be found to guarantee the expected QoS on the services executed through F2C systems. Unfortunately, since services execution in F2C encompasses many different aspects (from devices clustering or services decomposition to the resources allocation), an optimal solution for QoS delivery is pretty challenging as well. Also difficult to analyze what the characteristic should be to monitor the QoS in F2C systems. Shall we measure energy consumption, resources utilization, service performance, delay? This may be a tricky question, maybe requiring specific QoS service customization. It is also difficult to learn from past experiences (certainly traditional proposals in the routing area may be avoided), for example in the cloud arena (the scalability of SDN controllers in datacenters, cf. [61, 62]) or in fog (most of them data-plane oriented and thus not clearly useful for the envisioned F2C scenario, cf. [63]).

Indeed, we may argue that the specific F2C demands would require novel QoS delivery strategies that must be linked to other policies to be deployed in F2C, such as those mentioned previously regarding devices clustering, number of layers in the F2C hierarchy, or resources allocation. A preliminary attempt may be found in [10] intended to assess the distributed F2C control topology and its impact on QoS-aware end-to-end communication, by evaluating the effects on the latency produced by the amount of information to be processed in the controllers (i.e. the leaders), when executing services demanding high performance (the larger the information to be processed, the larger the delay).

2.3.1.9 Security

There is no doubt on the significance, and impact security has on any solution to be designed. Nowadays, certainly, security becomes a must on any technology to be moved to the market. Thus, with no need to justify the need for security, the challenging problem may be stated as how can we design a security architecture providing secure communication, confidentiality, integrity, availability, mutual fog, cloud and nodes authentication, and access control for F2C systems. Apparently, and recognized the vulnerabilities envisioned for an F2C scenario, the problem is not easy to solve.

As done before in this chapter, once the problem is identified, past experiences may help toward the design of a proper solution. However, although many recent works have assessed the design of security protocols and architectures to separately secure fog computing and cloud computing communications, unfortunately, a coordinated security solution, as demanded by F2C is not considered.

A solution providing a secure hierarchical architecture for fog and cloud communication is proposed in [64], although security for inter-fog communication is not granted. Contrarily, the solution proposed in [65] only provides a secure inter-fog communication without considering the security on the communication between fog and cloud. The works in [66, 67] propose a fog user/fog server mutual authentication (using a long-lived master secret key) and a gateway-based fog computing (master/slave) for wireless sensors and actuator networks, respectively, therefore only focusing on fog, so with no chances to be applied to F2C either.

Several solutions focus on the decoupling concept inherited from SDN. The architecture in [68] faces fog communications through gateways without contemplating cloud in a coordinated way. The work in [69] focuses on an SDN approach for securing IoT gateways using a centralized SDN controller with no capacity to handle secure mobility issues. Authors in [70] propose merging fog computing and SDN into the IoT architecture aimed at facilitating traffic control, resource management, scalability, mobility, and real-time data delivery.

Based on the work done so far, an incipient work in the area of security provisioning in F2C systems may be found in [71], assessing that the highly distributed F2C nature can be properly managed by using an SDN-based strategy, leveraging a set of distributed controller nodes, through a master–slave strategy. The proposed strategy has been extended in [72] by adding a key management and authentication strategy easing the management of the different elements considered in the F2C hierarchical architecture.

2.3.2 Industry Challenges

Despite the large set of challenges highlighted in the aforementioned research section, it looks an evidence that there is a strong industrial interest in the fog computing area (thus also in F2C), as shown by the industrial consortiums already discussing about fog/edge-related topics.

However, beyond the already-mentioned research challenges, some key aspects are also envisioned to make F2C to become a reality for the industrial sector. Most of these challenges are related to both the role industry will play in a potential deployment of an F2C solution (more relevant when considering the fact that while cloud providers are well established, fog providers are yet unclear) and the strategy to manage distinct providers interactions. In the coming subsections, some of these challenges are highlighted emphasizing on the aspects where some interaction or interoperability will be needed.

2.3.2.1   What an F2C Provider Should Be?

As previously said, since there is no any real deployment yet for an F2C system, nor for any system managing fog and cloud resources in a coordinated way, it is not clear what the providers' role may be. Aligned to that uncertainty, it is also worth mentioning that the specific mobility scenario envisioned at the edge as well as the coordinated management among fog and cloud resources is even exacerbating the referred uncertainty.

From a theoretical perspective, an F2C provider should provide resources management encompassing cloud and fog resources, aimed at an optimal services execution, through a proper resources allocation. This means that the F2C provider must be able to accommodate services into such a set of heterogeneous resources, thus requiring some access to infrastructure that will be probably not owned by the F2C provider itself as usually done in the cloud arena, what would require some sort of agreement between both. This would be the traditional scenario for example for smart cities that are endowed with existing infrastructure that may be managed by a global cloud provider. In this scenario, for example a traditional cloud provider may become an F2C provider setting a collaboration (i.e. agreement) with the city, to offer both cloud infrastructure and fog devices and deploying an F2C management system responsible for properly managing the resources.

Nevertheless, this scenario may be also enriched or extended with two key functionalities embedded in the F2C system. The first refers to the clustering capability the devices have to cluster themselves becoming some sort of resources provider. This scenario is envisioned to be opportunistic and very dynamic, thus creating dynamic resources areas deployed on the fly (does it also mean resources providers created on the fly?), that, when useful, should be coordinated by the overall F2C management solution, hence integrated into the F2C provider set of resources to be considered for services execution. The second refers to the collaborative model foreseen in F2C that assumes individual user's share their resources or part of them to be offered to others, thus also becoming some sort of resources providers. Thus, in this scenario, a user may become an F2C provider, since contributing to the resources to be globally managed, hence requiring some previous agreement in place to start the collaboration.

Hence, the F2C scenario, as envisioned when including all its foreseen capabilities, is quite disruptive when compared to traditional cloud providers, what undoubtedly add some key aspects that require some actions to identify what an F2C provider should be.

2.3.2.2   Shall Cloud/Fog Providers Communicate with Each Other

Apparently, providers should communicate with each other. This is undoubtedly the easy and fast answer. However, when digging into the question, some problems arise. For example, it is clear that in a scenario where a city becomes a sort of fog provider and gets in touch with a traditional cloud provider to deploy the F2C system, thus becoming an F2C provider, and both entities are well established, static, and clearly located. However, assuming the collaborative model envisioned in F2C, a user may decide to share a percentage of its resources to others, thus eventually serving as a resources provider that must get connected to an F2C provider. In this scenario, some agreement must be set between both to ease the brokering of the resources owned by the user. One potential option would be cloud or fog providers to deploy some "resources collectors" looking for resources. That could be a solution facilitating the integration of resources set on the fly coming from dynamic devices clustering or the proposed collaborative model. Another solution would go on the area of making whatever entity providing resource to become a provider, claiming for an agreement with existing resources providers (be it either at cloud or fog). However, this option would add some entropy into the system, in terms of the negotiation (pre-established) and possibly some issues on QoS delivery.

Another key issue refers to the information that should be made open among the different providers. This is a very important aspect in the traditional network providers domain, not willing to disclose information. In the envisioned F2C model, information disclosure should be mandatory to accommodate the high level of demands linked to service execution in F2C systems. For example, it does not make sense not to have updated information about a resource whether the resource is included in the list of available resources to run, for example dependable services.

Thus, as a summary, we see that the interactions among providers, assuming the wide definition of what a provider may be in F2C sytems, raises many challenging questions that require additional efforts in the industrial side.

2.3.2.3 How Multifog/Cloud Access Is Managed

The envisioned F2C scenario is multitenant, handling heterogeneous resources from many different sources to execute many different services. From a theoretical perspective, the access to different resources should not be a problem as long as resources are made available. However, on the real world, resources belong to providers that define some policies setting the rules for resources utilization. Thus, a key problem also observed in F2C systems is how many different providers multifog/cloud can be managed. Shall the F2C system get engaged to individual providers? Shall a cloud provider set an agreement with a fog provider thus setting some sort of joint venture as F2C provider? Shall users convince a cloud or fog provider to set a collaboration turning into an F2C area where some services will be deployed? These are just some questions that may be posted dealing with the management of too many resources providers at different levels.

Certainly, the F2C system, as a model, should be able to accommodate any solution, thus facilitating its deployment in the industrial sector.

2.3.3 Business Challenges

As largely introduced in this chapter, one of the key benefits envisioned for any new architecture leveraging IoT devices deployment and the capacities brought by cloud and fog computing resides in the collaborative framework designed to facilitate resources sharing. It is with no doubt, however, that such a collaborative framework will not come to be easily adopted, rather some issues are expected to come up, actually, as it is nowadays experienced in other domains also pushing for such a model (see for example the problems Uber, Cabify, or AirBnB are facing in many countries). It is worth noting the fact that collaborative models should usually compete with traditional providers, what sets the ground for a nonfair fight, commonly also involving different country regulations.

That said, it is also incontestable that one of the main benefits expected from the deployment of the envisioned sharing model will reside on the users. Indeed, users may theoretically benefit from lower fees while also theoretically keeping same delivered quality, and also act as "providers" when sharing their resources. Thus, two points become very relevant, to the extent of being critical to support the development of the F2C system, both related to the users. The first relates to the cost for the users to join an F2C system and the second to the benefits obtained when joining. These two key bullets, along with the interest traditional resource providers may have to get engaged into an F2C deployment and the facilities given to create new ones, will define the potential set of business models and thus the potential success of any F2C-like deployment.

A key aspect, mandatory for a successful F2C deployment deals with user engagement. To that end, users are expected to be willing to pay for either new services or a much better performance on existing ones. Both aspects are indeed, key features embedded within the envisioned F2C system, supported by a specific catalogue of services and for parallel execution in the distributed set of resources, respectively. Thus, users must not be reluctant to pay for such a set of benefits or whenever possible, users may have the opportunity to negotiate agreements with F2C providers to be part of the proposed collaborative model. This is indeed, one of the key benefits a potential F2C user may have. But, since the collaboration will need another provider to collaborate to, some agreement should be set to benefit them both. Although many different business models may come in this objective, this is particularly challenging in nowadays world. Moreover, a new market segment is up, targeting people have not been interested in such collaborative context ever.

Also from the point of view of traditional providers, although there is a clear trend pushing for using edge resources, they must see their niche to get some profit from. This is pretty challenging as well, since it will involve different providers and will require potential interactions.

Finally, novel business models can be created for service developers aimed to deploy novel services specifically tailored to the F2C architecture, and therefore taking the most out of the F2C system. Developers should not be only concerned on designing novel services but also on enriching existing ones with new features (for example, adding proximate computation for AR enriched services).

In short, there are big opportunities for novel business models that must be carefully exploited to go beyond existing cloud and fog deployments.

2.4 Ongoing Efforts

The high relevance envisioned for the proposed collaborative scenario built by putting together cloud and fog computing has been largely recognized in the recent years. Such relevance is nowadays driving many efforts at different levels, aimed at delivering solutions dealing with some of the challenges already emphasized. Although many research teams worldwide are devoting efforts on this scenario turning into a countless set of scientific contributions, next we summarize main global energies in the area, led by well-established initiatives we expect to stay alive for long. While some of the activities currently done in these initiatives have been already introduced in this chapter, the objective here is to summarize the main contributors as well as to provide pointers to them. In fact, these initiatives collect most of the results in the area, hence putting together a useful set of pointers, readers willing to get knowledge on related aspects may feed from. Activities are listed alphabetically, with no aim to establish any preference or higher significance among them.

2.4.1 ECC

ECC puts together an industrial consortium aiming to drive "the prosperity of the edge computing industrial sector," "pushing forward the sustainable development of the edge computing industry," as read in [73]. Recently, the ECC has also published a white paper with the Edge Computing Reference Architecture 2.0 [5], assessing the need for edge computing and fog computing to collaborate, introducing the proposed architecture, and also identifying some industry developments and business practices with standardization efforts and potential vertical sectors for applicability.

The proposed architecture identifies some key components, including the concept of an Intelligent Edge Computing Node (ECN), describing its main characteristics, functions, and expected product implementation options. The architecture deployment is envisioned as a layered model to be deployed as a three-layer or a four-layer model, depending on what the applicability context is and thus identifying specific services for each one of the models. Nevertheless, the document does not particularly emphasize the collaborative model discussed in this chapter, although certainly this does not preclude its adoption.

2.4.2 mF2C

mF2C is an EU research project, lasting three years till the end of 2019, funded within the H2020 EU research program, aimed at designing and developing the F2C concept. As previously rolled out in this chapter, the project pushes for a layered, open, secure, and hierarchical architecture to control the large set of distributed heterogeneous devices distributed from the edge up to the cloud.

Four key pillars hold up the mF2C architecture. The first refers to the procedure used to develop the mF2C system that leverages the installation of an app in the different devices willing to join the mF2C system. In fact, this is a key assumption in the way mF2C foresees the IoT-fog-cloud management, since no specific hardware is required to that end. The second refers to the proposed hierarchical architecture, where security is managed as a separate dimension, through specific elements (referred to as Control Area Units, CAUs) that could be implemented in terms of either a separate hardware or as a piece of software embedded in some elements in the architecture. The third refers to the collaborative model considered in the F2C architecture, supporting the active participation of user's devices into the whole system (sharing model), through the definition of different discovery and clustering policies easing an on-the-fly configuration of dynamic resources. Last but not the least, mF2C is designed to deploy parallel execution of services, distributed throughout the whole set of resources aimed at maximizing resources efficiency and service execution.

Section 2.1 already introduces main functional blocks of the mF2C architecture, as illustrated in Figure 2.3. More information about this initiative may be found at the project website [11].

2.4.3 MEC

MEC is an Industry Specification Group (ISG) initiative within ETSI, inferred from past efforts on Mobile Edge Computing. MEC's main focus falls into the networking field at the edge, particularly on using the RAN context to provide IT and cloud capacities, thus enabling services to benefit from proximate information as well as from running close to the user.

MEC has already deployed a large set of standards, describing its deployment along with different solutions (e.g. NFV) or application programming interface (APIs) (e.g. radio or location, etc.), just to name a few. A framework and main reference architecture have been also published in form of a standard [74], as shown in Figure 2.10. The main focus, as shown in the figure, is on mobile systems, highlighting the functional blocks building the two core components at host level, namely the mobile edge platform and the mobile edge management,

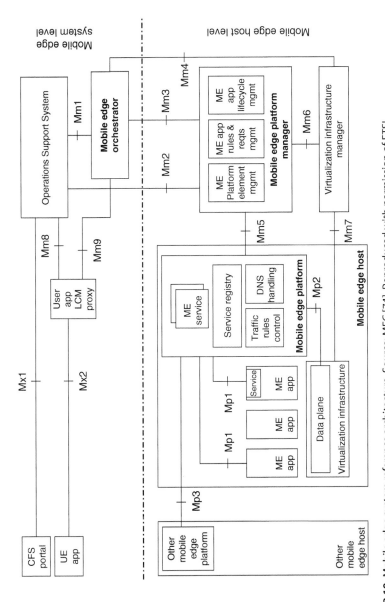

Figure 2.10 Mobile edge system reference architecture. Source: MEC [74]. Reproduced with permission of ETSI.

which along with the mobile edge orchestrator at the edge level set the whole framework.

The proposed architecture and hence, the overall framework envisioned by MEC also considers edge as a virtualized infrastructure, as widely analyzed in this chapter.

The work in MEC is pretty oriented to the edge and the technologies to be used to connect the edge, hence also dealing with CRAN (Centralized Radio Access Networks), security provisioning at the edge.

2.4.4 OEC

The Open Edge Computing (OEC) [75] is an industrial initiative, focusing on edge computing issues, especially motivated to drive new business opportunities and technologies around the edge computing concept. OEC's main rationale leverages "nearby" components offering their resources to support required computation, application agnostic, through standardized and open interfaces.

OEC defines edge computing as "a new network functionality that offers connected compute and storage resources right next to you," and in its vision, OEC considers that "Any edge node will offer compute and storage resources to any user in close proximity through an open and globally standardized mechanism," hence aligned with the collaborative model proposed in this chapter.

OEC has already delivered some technical documents describing main OEC insights and architectural concepts along with some white papers describing what the OEC is, but a remarkable outcome lies on the development of a Living Edge Lab, as a real-world edge computing testbed set in Pittsburg, putting together many key actors in the area, ready for experimentation and intended to demonstrate the benefits of bringing computation to the edge.

2.4.5 OFC

The OFC brings together a large set of companies and academic partners into a well-positioned forum in the fog arena. Notice that in January 2019 the OFC joined the Industrial Internet Consortium [76]. From the conceptual perspective, OFC defines fog computing as "a system-level horizontal architecture that distributes resources and services of computing, storage, control and networking anywhere along the continuum from Cloud to Things." Then, OFC tackles the whole resources scenario and extends the scope across multiple protocol layers, not only radio systems but also spanning across the final edge to the upper cloud.

OFC keeps a solid repository of research contributions and industrial white papers not only coming from the OFC consortium but also open to other teams developing activity in the area, to set a pointer to their work.

The OFC proposes a reference Architecture [41] describing the key components and functional blocks, the key pillars the architecture must meet, being security, scalability, openness, autonomy, programmability, RAS (reliability, availability, and serviceability), agility, and hierarchy as well as an end-to-end deployment use case.

Indeed, as of today, OFC is the most elaborated framework discussing about fog computing, addressing most of the identified challenges, collecting contributions from all over the world in the area, setting different real-world scenarios for solutions applicability and suitably reporting through both a large number of supported events and many relevant contributions in the area. In fact, OFC may be seen as the flagship initiative pushing for fog computing.

2.5 Handling Data in Coordinated Scenarios

The collaborative hybrid scenario presented so far consists of a vast set of resources, distributed from the fog layer up to the cloud, all deployed throughout a predefined area, for example a smart city, under a coordinated management. As mentioned before, such a scenario provides great variety and availability of resources for flexible and effective services execution, paving also the way to develop innovative services enriched with additional features, for example augmented reality or immediate data processing, referred to as smart services. Indeed, a proper data collection along with the capability to suitably process these data becomes key pillars in the development of smart services. However, this should not be done for granted. Indeed, aspects such as data volume, diversity, along with the fact that data processing systems might not know when data are expected to roll in, are all adding significant complexity in the way data should be processed. Browsing into this complexity issue, interestingly, smart services are fed mainly with fresh and rich data sets obtained from a variety of sources, from traditional data bases, to the current IoT, including dumb sensors, but also devices with some added functionalities, for example traffic lights, street lights, etc., which are deployed as part of the coordinated smart scenario.

A well-known and illustrative example matching the constraints to become a perfect smart services repository is a smart city. Indeed, there has been a tremendous push in the last years to deploy smart infrastructure in cities fueling the deployment of innovative services, aimed at empowering citizens with new "facilities" to make their daily activities perform better. Newer and more sophisticated sensors are being distributed along the cities, providing constantly and instantly detailed and comprehensive information about the city's vital signs. In fact, the more the sensors, the more the data, and consequently, the more the options for innovative and useful services, what is turning into many business opportunities and research avenues to handle the extremely large data sets, without substantially affecting the whole city performance. Thus, in this context, an effective data management is crucial for an efficient coordinated scenario. The key objective of a correct data management is to provide easy and safe access to data sources and repositories, to extract any form of value through complex computing and analytical processes over big data repositories. Therefore, an effective value generation will closely depend on how data are managed and organized. Recognized then the need for a proper data management in the proposed IoT-fog-cloud scenario, this section aims at first introducing the data nature (mainly in terms of the set of challenges inherent to considering heterogeneous data sources), then revisiting the insights of a complete data life cycle (DLC) model (describing challenges and policies for each one of the sequence of phases in the data life), and finally analyzing the impact the specific F2C collaborative scenario analyzed in this chapter may have on sorting out some of the identified data challenges (highlighting effects on each individual phase).

2.5.1 The New Data

A huge quantity of data is constantly being produced in the world, and this amount seems to keep increasing every day. Data are being generated from multiple sources, including scientific modeling, big data simulations, retail transactions, the IoT sensors network, surveillance cameras, but also from the users' social, professional, or daily activities, such as social media, third party applications, or participatory sensing (for instance, sensors integrated in citizens' smart phones), just to name a few. These daily fresh data can be used in real-time smart applications or preserved over historical repositories for future processing, setting up the complex universe of digital data.

Data sets can be private or public, or perhaps only partially shareable. Private data sets are managed with the own private repositories and used according to the particular private interests. Public data sets (or partially shareable data sets) used to be managed with a centralized repository from which accredited users can access these data. Furthermore, and for performance reasons, data sets can be partially replicated and geographically distributed throughout a set of sparse servers, providing high availability and achieving the following additional goals:

- Enhancing locality, by providing cached data copies closer to the end-users' locations;
- Increasing scalability, by providing different data access points to distribute the management load;
- Reducing the network backbone traffic and, therefore, the bandwidth costs; and,
- Improving fault tolerance, by maintaining several copies of the data set for backup purposes.

Aligned to this trend, content delivery networks (CDNs) are currently becoming quite popular, serving a large portion of today's Internet contents traffic and featuring such high availability and performance [77]. Note that the data movement pattern observed in these systems could be defined as a top-down model, where data have been initially created or collected in a centralized repository, and then spread downward along the geographically distributed servers network.

Notwithstanding, the scenario is getting worse when considering what data will be in the coming future. Indeed, extending the traditional data scenario, smart environments enriched with the IoT paradigm appear as an extraordinary source for a new data concept. No need to mention that sensed data that are continuously being generated from all corners of the city, providing all types of information about any city feature, all in all generating constantly and instantly detailed and comprehensive information about the city's vital signs adds a relevant complexity into the system. In fact the data scenario turns into vast and rich streams of heterogeneous data exposing three additional interesting and challenging features:

- *Multiple sources*: The same type of information is being generated from different sources, for instance, public sensors, private sensors, or even sensors from users' personal devices (such as a smartphone), each one using different data types, providing different accuracy, and probably obtained from different geographical locations.
- *Source positioning*: The physical location from where data have been obtained is, now, an important feature. In smart environments, it is not only enough to search for specific data content but also to know what their geographic origin is. For example, the temperature in a city may vary between different areas, especially in the case there is a fire nearby. A smart service should not settle for one single temperature information, but collect different temperature data from the appropriate city locations.
- *Data timing*: As sensed data are continuously being collected, the generation time is another important feature for these data sets. Now, data utility is different for fresh data, which feed interactive and real-time smart services, and old data, accumulated over historical repositories, which constitute the fuel for deep processing analytics.

In this context, the new data can be defined as a set of sensed data collections enriched with positioning and location attributes, plus additional metadata about the origin device, and including accuracy, trustiness, reliability, or any other attribute that may be of relevance.

Several research efforts are being devoted to designing frameworks aimed to properly manage data collected from the IoT, thus addressing partially the aforementioned features of the new data. These frameworks usually rely on a centralized repository, either a datacenter or cloud infrastructures, which constantly collects and organizes all sensed data, and an API to allow authorized smart services to access this information [78]. Note that unlike the top-down model of traditional systems, in this scenario, the data movement pattern observed could be defined as a bottom-up model, where data have been initially created from a massively spread sensors network, and then moved upward to the centralized data repository.

2.5.2 The Life Cycle of Data

Before going deep into the envisioned F2C scenario to analyze the potential benefits a potential F2C deployment may have on the data management, this subsection revisits the insights in the data life, analyzing the different phases over the whole data life, later used in next section to clearly show real improvements brought by F2C. Data management and organization during the entire data life, from creation to deletion, becomes a complex and challenging task [79]. The main objective of data management is to provide an efficient access to data repositories in order to facilitate extraction of value through more or less complex computing processes. DLC models have been proposed as an effective data management solution, defining the sequence of phases in the data life, specifying the management policies for each phase, and finally describing the relationship among phases [80]. In short, the main goal of a DLC model is to optimize data operation during all phases of the data life, providing high quality data through cleaning and waste removal, and offering end-users' data products best suiting the expected quality requirements.

A DLC model is specific for a particular field or scenario, addressing their particular requirements and challenges. In an IoT scenario, a basic DLC model could be defined as shown in Figure 2.11, which considers three main blocks: data collection, data storage, and data processing [81]. These blocks

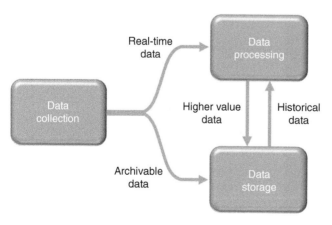

Figure 2.11 Basic DLC model for an IoT scenario.

are responsible for gathering, storing, and organizing data for processing purposes, while guaranteeing high data quality standards.

The data collection block is responsible for collecting IoT data into the system, gathering data from different sources, assessing data quality, and tagging data with any additional description required in the business model. Collected data can then be either stored, through the data storage block, or processed, through the data processing block. The data processing block is responsible for performing the main processing, extracting knowledge or generating additional value, through complex data analysis techniques. The outcome of the processed data can be either delivered to end users or stored for future additional data usage or reprocessing. The data storage block is responsible for data preservation, performing any eventual action related to data curation. These data are ready for publication or dissemination, or available for further processing.

The data flow in the DLC is as follows. Data are initially created and collected through the data collection block. If data are immediately processed, these are considered real-time data; otherwise, if data are stored, these are considered archivable data. Note that either all or part of the processed data can also be stored, and vice versa, i.e. these two data sets are not exclusive. When archived data in the data storage block are used for processing, these are considered historical data, as they have been stored for some time. So the data processing block is able to use both, real-time and historical data, for processing. Finally, the data processing outcome can be consumed by the end-user, which represents the end of the data life, or archived back through the data storage block – in this case, these data are considered higher value data.

2.5.3 F2C Data Management

As discussed in the previous subsections, managing data in such smart IoT scenarios becomes a complex challenge, which requires rethinking the traditional technologies and adapting them to the new context features and possibilities. Interestingly, the distributed hierarchical F2C resources management architecture provides an extraordinary framework for effective data management, addressing efficiently the particular features of the new IoT data concept. This is discussed in the following subsections for each stage in the DLC model described in the previous subsection.

2.5.3.1 Data Collection

The data collection process is responsible for collecting data from all sources and devices according to the business requirements. Data can be collected from sensors devices, surveillance cameras, users' smart devices or smart vehicles, or any other data source included in the system. As part of the data collection process, additional phases could be considered in the model, such as data filtering (which performs some valuable optimizations, such as data curation or data aggregation), data quality (aiming to appraise the quality level of the collected data), and data description (tagging data with some additional attributes).

In the IoT paradigm and in traditional data management systems, devices are typically connected to a computing device at the edge through some sort of low power communication technology, which, in turn is responsible for transferring the collected data to the centralized cloud data center. After moving the data to the cloud, these data can be filtered, if appropriate, appraised for quality, and organized and indexed in the corresponding data directories. Then, those data are ready to be used by the smart services.

Unlike traditional data management systems, in an F2C architecture, most data collection tasks can be made at the fog layer. In fact, the computing devices attached to the IoT devices that originally receive the data are part themselves of the fog layer. Hence, data are initially archived at the fog layer. If a real-time or critical data access is required, data will be immediately provided from such layer, thus reducing considerably the latency and, therefore, providing optimal performance.

Then, data can be filtered and appraised to meet the quality standards required. This process can be as complex and sophisticated as desired because fresh data are anytime available from this layer. As shown in [19], an appropriate data aggregation and compression process can drastically reduce some sensed data in more than 90% of their original size. After some time, configurable according to the city business model, the data are transferred to the following upper layer. This data transfer can be delayed as much as desired, so an additional optimization that can be made is sending the data when the network traffic is low. This process is repeated at each fog layer along the hierarchy until, finally, all data (aggregated and compressed, meeting high quality standards) reach the cloud layer. At this layer, all data can be permanently preserved and ready for future retrieval of the historical data.

As a summary, the advantages of an F2C data collection process are numerous and can be summarized as follows:

- Data can be effectively collected, in a distributed way along the whole city sensors network.
- Data are ready to be used immediately and may benefit from the advantages of locality, as opposed to a centralized platform where data must first be read to the cloud, and then, accessed from the cloud (two times a long distance transfer).
- The amount of data to be transferred to higher layers is less than 10% of the total data collected thanks to the filtering phase, which efficiently aggregates and compresses the data.
- The data filtering and data quality stages are not limited in time, as long as data are already accessible from the lower layers and, therefore, any sophisticated or complex ad-hoc process can be performed.

- Data transfer to higher layers can be done when the network load is low, therefore optimizing the global network traffic.

2.5.3.2 Data Storage

The data storage process is responsible for both archiving permanently the data and organizing them for any eventual future access. As part of the data storage process, additional phases could be considered in the model, such as data classification (aiming to organize and prepare data for efficient storage), data archive (storing data for short and long terms consumption), and data dissemination (publishing data for public or private accesses), providing efficient interfaces for the data organization and access.

In traditional IoT data management systems, all data are stored in the cloud datacenters before allowing end-users' accesses. The cloud provides unlimited storage capacity; however, this process adds unnecessary delays before being able to use these data. Alternatively, in an F2C architecture, data storage can be performed throughout the system hierarchy following a bottom-up pattern before storing the data permanently at the cloud layer.

In the F2C architecture, data are collected and initially stored from the lowest fog layer. Therefore, most fresh data are at the edge of the city, spread among numerous fog devices. If a smart service needs these fresh data (real-time), they can be accessed directly from some near fog device. Note that real-time services have an implicit feature of locality. Sensor data sources at the edge are continuously collecting data, so they can be continuously providing such data to any data consumer. However, the amount of stored data will be growing, and the storage capacity at the edge is somehow limited. Hence, a set of least recent data will be frequently transferred upward to higher layers of the hierarchy, i.e. to a fog area leader. The transfer frequencies can be flexibly adjusted, and they will be before the storage becomes full, but at any time, which could be convenient for the system performance – for instance, when the network load is low. In addition, as the data transfer is not urgent, the fog node can perform several data aggregation or data compression processes in order to drastically reduce the amount of data to transfer upward.

In a three-layer F2C scenario, the fog leader, in turn, will be receiving (aggregated and compressed) data from several lower layer fog devices within its fog area. The leader device will probably have greater storage and computing capacities, so this data accumulation should not be a problem. Now, if an application or service needs any least recent data, these can be obtained from the fog leader. Least recent data are not as critical as real-time data, so an extra delay for accessing the fog leader is not any inconvenience. Similarly, the fog leader's storage capacity is still limited. Hence, frequently, the oldest least recent data (i.e. historical data) will be transferred upward to the cloud. Again, if convenient, the data can be compressed before being transferred, and the exact transfer time can be adjusted according to the network preferences. Therefore, the cloud will be receiving historical data from the set of fog leaders along the hierarchy, and storing these data permanently for historical accesses.

The additional advantages of this model may be summarized as follows:

- Data can be cached at different layers of the architecture and accessed from the most appropriate layer (including real-time, least recent, and historical data).
- Data latencies and access times can be easily optimized by accessing data from the closest physical layer.

2.5.3.3 Data Processing

The data use block is in charge of providing an optimal data access interface for efficient data usage. This phase encompasses the set of processes to transform raw data into more sophisticated data or information.

In an F2C architecture, data processing can be performed at any layer of the hierarchy. The control system will decide at which fog or cloud devices the service should be executed, according to its performance cost model while also considering the SLA in place. With respect to the data management in the F2C framework during smart services execution, the F2C system is responsible for two main tasks:

- Providing detailed information about the geographical placement of the candidate data sources, in order to support the resources allocation process to select the most convenient data source for a specific service execution.
- Providing an appropriate API that allows selecting the best data source for that particular service requirement. Note that the IoT data should include at least additional attributes related to the physical location and the timing features of the data.

In order to fulfill these requirements, the F2C data management system should define an appropriate attribute-based naming system, and their corresponding directory structures to effectively organize these data sets. In this context, some CDNs technology should be enhanced to consider both, the performance information and the location and timing attributes of data.

2.6 The Coming Future

We all are witnessing an unstoppable evolution toward involving distributed infrastructure located at the edge to execute either innovative services or to enhance the performance of other already existing. This tendency, incipient in many different verticals (transport, housing, etc.), is strongly leveraging both the huge set of highly skilled devices deployed at the edge (close to the users) as well as the first step toward a collaborative model where devices owners are willing to share their resources. Although many different options come up, the most appealing scenario is the one when resources shared by final users (i.e. not owned by traditional providers) are used to deploy new services.

This complex scenario is additionally enriched by assuming connectivity as a commodity, whatever the technology may be, and also by recognizing the availability of large information collected either for analytics at backend systems or for real-time decision-making processes.

Certainly, looking at what the future will bring is not that easy and actually many companies, blogs, and reputed experts worldwide are spending significant efforts to deliver predictions on what is next. That said, it seems reasonable to widely adopt the idea that distributed computing at the edge will keep gaining ground in tight harmony with cloud computing. It is clearly evident that the unstoppable evolution of the hardware technology at the edge is undoubtedly easing such assessment. On the other hand, however, such a novel model will only succeed whether new business models are identified, what might issue some concerns considering what the scenario is. Indeed, engaging infrastructure not owned by traditional providers in the loop, raises some avenues for research but also for industrial and regulatory discussions. And apparently, the user will not act as the traditional services consumer, rather will extend that profile to adopt a novel, different one, actively contributing to the whole set of available resources. This assessment somehow includes the user (not the client) in the set of agreements to be considered what also gives room to claim for many questions. For example, would I be happy sharing my car or my smart phone resources? What would be the benefit I may get doing so? With whom shall I have to negotiate? Shall regulation be modified to consider such scenario?

Nevertheless, putting the focus on the user side should not hide the role new actors may have to empower this collaborative scenario. For example, local governments (Cities, regions, etc.), public services (police, hospital, schools, etc.), or even new actors, such as neighborhood communities or fleets may actively contribute to this model. As the interest of these new actors and the facilities (legal) to set them up may grow, the envisioned collaborative model will keep growing as well. Certainly, there are two key assessments we should not forget, the first "nothing must be ruled out when discussing about business models" and the second "we never know when innovative technology is going to roll in." Both assessments push toward keeping minds open, guessing where niche for innovation is.

In this context, beyond regulatory changes and business agreements, it is evident that some efforts in research are mandatory to accommodate the proper management of the highly distributed brought together by considering the IoT continuum toward cloud. Recognized the incipient efforts nowadays active in this area and the strong relevance such a context is acquiring not only in the academic sector but also in the industrial domain, it is obvious that the efforts currently done are only setting the first steps toward a more dynamic, heterogeneous, open, efficient (energy but also performance), participative, and probably cheaper communications and services delivery scenario. As of today, we do not see the end for this path to come.

Acknowledgments

The authors thank the members of the H2020 mF2C Project consortium for the work done toward the deployment of the F2C architecture, actively contributing to some of the concepts described in this chapter.

This work was partially supported by the H2020 EU mF2C Project ref. 730929 and by the Spanish Ministry of Economy and Competitiveness and by the European Regional Development Fund, under contract RTI2018-094532-B-I00 (MINECO/FEDER).

References

1 Heilig, L. and Voss, S. (2014). A scientometric analysis of cloud computing literature. *IEEE Transactions on Cloud Computing* 2 (3): 266–278.

2 Bonomi, F., Milito, R., Zhu, J., and Addepalli, S. (2012). Fog computing and its role in the Internet of Things. In: *Proceedings of the First Edition of the MCC Workshop on Mobile Cloud Computing, ser. MCC'12*, 13–16. New York, NY, USA: ACM.

3 Chiang, M. and Zhang, T. (2016). Fog and IoT: an overview of research opportunities. *IEEE Internet of Things Journal* 3: 854–864.

4 (2015). *Fog computing and Internet of Things: Extend the cloud to where the things are.* OpenFog Consortium.

5 ECC-RA (n.d.) Edge Computing Reference Architecture at http://en .ecconsortium.org/Uploads/file/20180328/1522232376480704.pdf (accessed 20 August 2019).

6 OpenFog Consortium (n.d.) Glossay of terms related to Fog Computing. https://www.openfogconsortium.org/wp-content/uploads/OpenFog-Consortium-Glossary-of-Terms-January-2018.pdf (accessed 21 August 2019)

7 Marín-Tordera, E., Masip-Bruin, X., Garcia, J. et al. (2017). Do we all really know what a Fog Node is? Current trends towards an open definition. *Computer Communications* 109: 117–130.

8 Z. Rejiba, X. Masip-Bruin, E. Marín-Tordera (2018) Analyzing the Deployment Challenges of Beacon Stuffing as a Discovery Enabler in Fog-to-Cloud Systems. *27th European Conference on Networks and Communications (EuCNC 2018)*, Ljubljana, Slovenia (June 2018).

9 Masip-Bruin, X., Marin-Tordera, E., Tashakor, G. et al. (2016). Foggy clouds and cloudy fogs: A real need for coordinated management of fog-to-cloud (F2C) computing systems. *IEEE Wireless Communications Magazine* 23 (5): 120–128.

10 Barbosa, V., Gómez, A., Masip-Bruin, X. et al. (2017) Towards a Fog-to-Cloud Control Topology for QoS-Aware End-to-End Communications. 2017 IEEE/ACM International Symposium on Quality of Service, (IWQoS'17), Vilanova i la Geltrú, Spain (June 2017).

11 mF2C (n.d.) The mF2C project. http://www.mf2c-project.eu (accessed 21 August 2019).

12 Marin-Tordera, E., Masip-Bruin, X., Sanchez-Lopez, S. et al. (2006). A hierarchical routing approach for optical transport networks. *Computer Networks Journal* 5 (2): 251–267.

13 Masip-Bruin, X., Marín-Tordera, E., Jukan, A., and Ren, G.J. (2018). Managing resources continuity from the edge to the cloud: architecture and performance. *Future Generation Computer Systems*: 37, 777–785.

14 O. Fratu, C. Pena, R. Craciunescu, S. Halunga (2015). Fog computing system for monitoring mild dementia and COPD patients, TELSIKS 2015, Nis, Serbia.

15 Rahmani, A.M., Gia, T.N., Negash, B. et al. (2017). Exploiting smart e-Health gateways at the edge of healthcare Internet-of-Things, a fog computing approach. *Future Generation Computer Systems* http://dx.doi.org/10.1016/j.future.2017.02.014.

16 Ferrer-Roca, O., Roca, D., Nemirovsky, M. et al. (2015). Fog computing in health 4.0 the role of small data. *International Journal of Embedded Systems*.

17 Perera, C., Qin, Y., Estrella, J.C. et al. (2017). Fog computing for sustainable smart cities. *ACM Computing Surveys* 50 (3): 32:1–32:43.

18 Sapienza, M., Guardo, E., Cavallo, M. et al. (2016). Solving critical events through mobile edge computing: an approach for Smart cities. *IEEE International Conference on Smart Computing*, SMARTCOMP, MO, St. Louis, USA 8–20 May 2016.

19 Sinaeepourfard, A., Garcia, J., Masip-Bruin, X. and Marín-Tordera, E. (2017). Fog-to-Cloud (F2C) Data Management for Smart Cities. Future Technologies Conference (FTC 2017), Vancouver, Canada (November 2017).

20 Stojmenovic, I. (2014) Fog computing: a cloud to the ground support for smart things and machine-to-machine networks. Australasian Telecommunication Networks and Applications Conference, Melbourne, Australia (November 2014).

21 Baccarelli, E., Cordeschi, N., Mei, A. et al. (2016). Energy efficient dynamic traffic offloading and reconfiguration of networked datacenters for big data stream mobile computing: review, challenges, and a case study. *IEEE Network* 30 (2): 54–61.

22 Al Faruque, M.A. and Vatanparvar, K. (2016). Energy
management-as-a-service over fog computing platform. *IEEE Internet of
Things Journal* 3 (2): 161–169.

23 Jalali, F., Hinton, K., and Ayre, R. (2016). Fog computing may help to save
energy in cloud computing. *IEEE Journal on Selected Areas in Communications* 34 (5): 1728–1739.

24 Baccarelli, E., Vinueza, P.G., and Scarpiniti, M. (2017). Fog of everything:
energy-efficient networked computing architectures, research challenges,
and a case study. *IEEE Access* 5: 9882–9910.

25 Bonomi, F., Milito, R., Natarajan, P., and Zhu, J. (2014). Fog computing: a
platform for Internet of Things and analytics. In: *Big Data and Internet of
Things: A Roadmap for Smart Environments*, vol. 546 (eds. N. Bessis and C.
Dobre), 169–186. Cham, Germany , Studies in Computational Intelligence
Book Series, SCI: Springer.

26 Dastjerdi, A.V. and Buyya, R. (2016). Fog computing: helping the Internet of
Things realize its potential. *Computer* 49 (8): 112–116.

27 Beauregard, G. (2015). *Fog Computing Virtualizing Industry*. LocalGrid
Technologies Inc.

28 Peng, M., Yan, S., Zhang, K., and Wang, C. (2016). Fog-computing-based
radio access networks: issues and challenges. *IEEE Network* 30 (4): 46–53.

29 Hou, X., Li, Y., Chen, M. et al. (2016). Vehicular fog computing: a viewpoint of vehicles as the infrastructures. *IEEE Transactions on Vehicular
Technology* 65 (6): 3860–3873.

30 Xiao, Y. and Zhu, C., Vehicular fog computing: vision and challenges. *IEEE
International Conference on Pervasive Computing and Communications
Workshops PerCom Workshops*, Kona, HI, USA, 13–17 March 2017.

31 Sookhak, M., Richard Yu, F., He, Y., and Kumar, N. (2017). Fog vehicular
computing: augmentation of fog computing using vehicular cloud
computing. *IEEE Vehicular Technology Magazine* 12 (3): 55–64.

32 Masip-Bruin, X., Marín-Tordera, E., Alonso, A. et al. (2016). Will it be
cloud or will it be fog? F2C, a novel flagship computing paradigm for highly
demanding services. In: *IEEE Future Technologies Conference (FTC) 2016,
San Francisco, USA (December 2016)*.

33 Sengupta, S., Garcia, J., Masip-Bruin, X. (n.d.) A Literature Survey on
Ontology of Different Computing Platforms in Smart Environments, https://
arxiv.org/pdf/1803.00087.pdf (accessed 21 August 2019).

34 Shi, W., Cao, J., and Zhang, Q. (2016). Edge computing: vision and challenges. *IEEE Internet of Things Journal* 3 (5): 637–646.

35 Kaffel-Ben Ayed, H., Boujezza, H., and Riabi, I. (2017). An IDMS approach
towards privacy and new requirements in IoT. *The Wireless Communications and Mobile Computing Conference (IWCMC), 13th International*,
Valencia, Spain (2017).

36 Suguna, M., Anusia, R., Mercy Shalinie, S., and Deepti, S. (2017). Secure identity management in mobile cloud computing. International Conference on Nextgen Electronic Technologies: Silicon to Software (ICNETS2), India, 23–25 March 2017.

37 Tapiador, A. and Mendo, A. (2011). A survey on OpenID identifiers. The 7th International Conference on Next Generation Web Services Practices, Salamanca, Spain, 2011.

38 Meidan, Y., Bohadana, M., Shabtai, A. et al. (2017) ProfilloT: A Machine Learning Approach for IoT Device Identification Based on Network Traffic Analysis. *The Proceedings of the Symposium on Applied Computing*, Marrakech, Morocco, 2017.

39 Cloud Security Alliance (2016). *Identity and Access Management for the Internet of Things – Summary Guidance*. IoT Working Group.

40 Gómez, A., Masip, X., Marín, E. et al. (2017). A Hash-Based Naming Strategy for the Fog-to-Cloud Computing Paradigm. *F2C-DP Workshop, Euro-Par 2017*, Santiago de Compostela, Spain (August 2017).

41 OFRA (2017) Open fog consortium working group: OpenFog Reference Architecture for Fog Computing White paper, Feb 2017.

42 Lewis, G., Echeverría, S., Simanta, S. et al. (2014). Tactical cloudlets: moving cloud computing to the edge. In: *Proceedings – IEEE Military Communications Conference MILCOM*, 1440–1446. IEEE.

43 Gedeon, J., Meurisch, C., Bhat, D. et al. (2017). Router-based brokering for surrogate discovery in edge computing. In: *Proceedings of IEEE 37th International Conference on Distributed Computing Systems Workshops, ICDCSW 2017*, 145–150. IEEE.

44 Soo, S., Chang, C., and Srirama, S.N. (2016). Proactive service discovery in fog computing using mobile ad hoc social network in proximity. In: *2016 IEEE International Conference on Internet of Things (iThings) and IEEE Green Computing and Communications (GreenCom) and IEEE Cyber, Physical and Social Computing (CPSCom) and IEEE Smart Data (SmartData)*, 561–566. IEEE.

45 Saurez, E., Hong, K., Lillethun, D. et al. (2016). Incremental deployment and migration of geo-distributed situation awareness applications in the fog. In: *Proceedings of the 10th ACM International Conference on Distributed and Event-based Systems-DEBS'16*, 258–269. ACM.

46 Rejiba, Z., Masip-Bruin, X., Marín-Tordera, E., Ren, G.J. (2018). F2C-Aware: Enabling discovery in Wi-Fi-powered Fog-to-Cloud (F2C) systems. *IEEE Mobile Cloud*, Bamberg, Germany (March 2018).

47 Simoens, P., Herzeele, L.V., Vandeputte, F., and Vermoesen, L. (2015). Challenges for orchestration and instance selection of composite services in distributed edge clouds. In: *IFIP/IEEE International Symposium on Integrated Network Management (IM)*, 1196–1201. Otawa, Canada: IEEE.

48 Tanganelli, G., Vallati, C., and Mingozzi, E. (2014). Energy-efficient QoS-aware service allocation for the cloud of things. In: *IEEE 6th International Conference on Cloud Computing Technology and Science (CloudCom), 2014*, 787–792. IEEE.

49 Mukherjee, A., Paul, H.S., Dey, S. et al. (2014). Angels for distributed analytics in iot. In: *IEEE World Forum On Internet of Things (WF-IoT)*, 565–570. IEEE.

50 Vitor Barbosa Souza, Wilson Ramírez, Xavi Masip-Bruin, et al. (2016). Handling Service Allocation in Combined Fog-Cloud Scenarios. In: *IEEE International Conference on Communications (ICC), 2016*. IEEE, 22–27 May 2016.

51 Souza, V.B., Masip-Bruin, X., Marín-Tordera, E., Ramírez, W. (2016). Towards Distributed Service Allocation in Fog-to-Cloud (F2C) Scenarios, *IEEE Global Communications Conference*, Globecom 4–8 December 2016.

52 Akyildiz, I.F., Lee, A., Wang, P. et al. (2016). Research challenges for traffic engineering in software defined networks. *IEEE Network* 30 (3): 52–58.

53 Younis, M., Senturk, I.F., Akkaya, K. et al. (2014). Topology management techniques for tolerating node failures in wireless sensor networks: a survey. *Computer Networks* 58: 254–283.

54 Sgambelluri, A., Giorgetti, A., Cugini, F. et al. (2013). OpenFlow-based segment protection in Ethernet networks. *Journal of Optical Communications and Networking* 5 (9): 1066–1075.

55 Kwon, Y.W. and Tilevich, E. (2012). Energy-efficient and fault-tolerant distributed mobile execution. In: *IEEE 32nd International Conference on Distributed Computing Systems (ICDCS)*, 586–595.

56 Harter, I.B.B., Schupke, D.A., Hoffmann, M., and Carle, G. (2014). Network virtualization for disaster resilience of cloud services. *IEEE Communications Magazine* 52 (12): 88–95.

57 Natalino, C., Monti, P., França, L. et al. (2015). Dimensioning optical clouds with shared-path shared-computing (SPSC) protection. In: *2015 IEEE 16th International Conference on High Performance Switching and Routing (HPSR)*, 1–6. Budapest: IEEE.

58 Satria, D., Park, D., and Jo, M. (2016). Recovery for overloaded mobile edge computing. In: *Future Generation Computer Systems*, vol. 70, 138–147. Elsevier.

59 Souza, V.B., Masip-Bruin, X., Marín-Tordera, E. et al. (2017). Proactive vs reactive failure recovery assessment in combined Fog-to-Cloud (F2C) systems. *22nd International Workshop on Computer Aided Modelling and Design of Communication Links and Networks (CAMAD)*, Lund, 19–21 June 2017.

60 Barbosa, V., Masip-Bruin, X., Marín-Tordera, E. et al. (2017) Towards Service Protection in Fog-to-Cloud (F2C) Computing Systems. *Future Technologies Conference (FTC)*, Vancouver, Canada (November 2017).

61 Hu, J., Lin, C., Li, X. et al. (2014). Scalability of control planes for software defined networks: modeling and evaluation. In: *2014 IEEE/ACM 22nd International Symposium of Quality of Service (IWQoS)*, 147–152.

62 Shalimov, A., Zuikov, D., Zimarina, D. et al. (2013). Advanced study of SDN/OpenFlow controllers. In: *Proceedings of the 9th ACM Central & Eastern European Software Engineering Conference in Russia*, 1.

63 Aazam, M., St-Hilaire, M., Lung, C.-H. et al. (2016). MeFoRE: QoE based resource estimation at fog to enhance QoS in IoT. In: *23rd IEEE International Conference on Telecommunications (ICT)*, 1–5. Thessaloniki, Greece: IEEE.

64 Ola Salman, Sarah Abdallah, Imad H. Elhajj, et al. (2016). Identity-based authentication scheme for the Internet of Things. *27–30 June 2016 IEEE Symposium on Computers and Communication*, Messina, Italy.

65 Moosavi, S.R., Gia, T.N., Nigussie, E. et al. (2016). End-to-end security scheme for mobility enabled healthcare IoT. *Future Generation Computer Systems* 64: 108–124.

66 Ibrahim, M.H. (2016). Octopus: an edge-fog mutual authentication scheme. *International Journal of Network Security* 18 (6): 1089–1101.

67 Lee, W., Nam, K., Roh, H.-G., and Kim, S.-H. (2016). A gateway based fog computing architecture for wireless sensors and actuator networks. In: *2016 18th International Conference on Advanced Communication Technology (ICACT)*. Pyeongchang, South Korea: IEEE.

68 Li, Y., Su, X., Riekki, J. et al. (2016). A SDN-based architecture for horizontal Internet of Things services, Communications (ICC). In: *2016 IEEE International Conference on Communications (ICC)*. Kuala Lumpur, Malaysia: IEEE.

69 Vilalta, R., Ciungu, R., Mayoral, A. et al. (2016). Improving security in Internet of Things with software defined networking. *2016 IEEE Global Communications Conference (GLOBECOM)*. IEEE, Washington, DC, USA (December 2016).

70 Tomovic, S., Yoshigoe, K., Maljevic, I., and Radusinovic, I. (2017). Software-defined fog network architecture for IoT. *Wireless Personal Communications* 92 (1): 181–196.

71 S. Kahvazadeh, V. Barbosa, X. Masip-Bruin et al. (2017). An SDN-based architecture for security provisioning in fog-to-cloud (F2C) computing systems. *Future Technologies Conference (FTC)*, Vancouver, Canada (November 2017).

72 Kahvazadeh, S., Masip-Bruin, X., Diaz, R. et al. (2018). Towards an efficient key management and authentication strategy for combined fog-to-cloud continuum systems. *3rd Cloudification of the Internet of Things (CIoT 2018)*, Paris, France (July 2018).

73 ECC (n.d.) The Edge Computing Consortium at http://en.ecconsortium.org/Content/index/cid/2.html (accessed 20 August 2019).

74 MEC (n.d.) MEC: Framework and Reference Architecture. http://www.etsi .org/deliver/etsi_gs/MEC/001_099/003/01.01.01_60/gs_MEC003v010101p .pdf (accessed 21 August 2019).

75 OEC (n.d.) Open Edge Computing, www.openedgecomputing.org.

76 Industrial Internet Consortium (n.d.) http://www.iiconsortium.org/index .htm (accessed 20 August 2019).

77 Hofmann, M. and Beaumont, L.R. (2004). *Content Networking: Architecture, Protocols, and Practice*. Morgan Kaufmann Publisher. ISBN: 978-1558608344.

78 Jin, J., Gubbi, J., Marusic, S., and Palaniswami, M. (2014). An information framework for creating a Smart City through Internet of Things. *IEEE Internet of Things Journal* 1: 112–121.

79 Grunzke, R., Aguilera, A., Nagel, W.E. et al. (2015). Managing complexity in distributed Data Life Cycles enhancing scientific discovery. In: *IEEE 11th International Conference on E-Science (e-Science)*, 371–380.

80 Levitin, A. and Redman, T. (1993). A model of the data (life) cycles with application to quality. *Journal of Information and Software Technology* 35: 217–223.

81 Sinaeepourfard, A., Garcia, J., Masip-Bruin, X., Marín-Tordera, E. (2016). A comprehensive scenario agnostic Data LifeCycle Model for an efficient data complexity management. *IEEE conference on eScience 2016*, Baltimore, USA (October 2016).

3

Computation Offloading Game for Fog-Cloud Scenario

Hamed Shah-Mansouri and Vincent W.S. Wong

Department of Electrical and Computer Engineering, The University of British Columbia, Vancouver, Canada

3.1 Internet of Things

The IoT is the network of devices, vehicles, buildings, and other items equipped with communicating and computing capabilities [1, 2]. The things in the IoT cover a wide range of physical devices including home appliances, surveillance cameras, actuators, displays, monitoring sensors (e.g. implanted heart monitor), vehicles, or any other objects that can be assigned a unique identifier and provided with the ability to transfer data over a network. The IoT is promising to build a connected smart world in foreseeable future and to provide variety of benefits by connecting everything to the Internet. Providing reliable services, enhancing the flexibility of services, safety, and security are the major benefits of IoT systems. Numerous applications are developing in various areas to take the advantages that IoT can provide. The applications of IoT include health-care [3], smart transportation [4], smart homes [5], smart cities [6], industrial automation, and so on. Patient vital sign data can be logged and monitored to better understand the cause and effect of certain health conditions. For example, an implanted heart monitor device can report the patients status in real-time for processing and taking appropriate treatments. Traffic lights connected to the Internet can be remotely controlled to avoid traffic congestion. Traffic information can also be collected from vehicles connected to the traffic controller. Such intelligent transportation system with connected vehicles can improve the quality of lives in smart cities. Buildings, bridges, railway tracks, and other urban and rural infrastructures can be monitored continuously for the maintenance purposes. Smart homes will enable their residents to remotely control the air conditioning systems and other appliances.

Billions of heterogeneous smart devices providing a variety of applications and services will generate huge volume of data in the IoT era. Cisco forecasts

Fog and Fogonomics: Challenges and Practices of Fog Computing, Communication, Networking, Strategy, and Economics, First Edition. Edited by Yang Yang, Jianwei Huang, Tao Zhang, and Joe Weinman.

that there will be 28.5 billion networked devices by 2022 [7]. Such growing number of connected devices introduces challenging issues to further develop IoT systems. A critical bottleneck for realizing the IoT systems is the explosive growth of IoT objects that generate enormous data to be transferred over the network and be processed. Novel technologies must be designed in order to meet stringent delay requirements of mission critical IoT applications, be robust and reliable, and be able to balance the trade-off between the capital/operational expenditure cost and service availability.

The IoT is promising to form the network of smart connected devices with powerful data analytics and intelligent decision-making capabilities. To realize the IoT systems and support their applications, powerful computing resources are required. Nonetheless, the limited processing capacity of IoT objects is a barrier to develop IoT applications as it cannot provide the required computation power. Furthermore, addressing the challenging issues of IoT such as reliability, scalability, security, and privacy requires powerful computing resources. The future IoT systems and their challenges are studied in [8].

The ubiquitous and pervasive computing services offered by cloud computing can relieve the computing limitations of IoT devices [9, 10]. Although cloud servers can provide the required computation power of IoT applications, the conventional cloud computing systems are not designed for the big data generated by IoT. Utilizing the remote cloud services introduces the following challenging issues in development of IoT systems: First, delay-sensitive IoT applications require low-latency and real-time computing services, while the relatively long delay of remote cloud computing cannot support their stringent delay requirements. Second, a huge amount of bandwidth is required to transfer the data from a large number of IoT objects to the cloud servers. An idea to alleviate these shortages is to analyze and process data at the edge of the network rather than dispatching to the cloud servers.

Fog computing, which is proposed by Cisco [11] and is now being promoted by the OpenFog Consortium [12], can mitigate the aforementioned drawbacks of conventional cloud computing frameworks. Fog computing offers cloud services at the edge of the network where data are generated [11]. The IoT systems can take the advantages of fog computing in several aspects. Processing data in the edge of network can significantly reduce the latency and meet the stringent requirements of delay-sensitive IoT applications. Fog computing also prevents the IoT users to send a vast amount of data to the cloud servers, which consequently conserves the network bandwidth. These result in enhanced IoT users' experiences. Although fog computing can provide low-latency computing service, it is not a substitute for cloud computing. The integration of fog and cloud computing can provide computation services required in IoT and enable IoT applications and businesses to further develop the IoT systems. The delay-tolerant IoT applications can be processed in the cloud servers to benefit from huge computation power provided by cloud computing. The cloud servers can also store the information for a relatively long time. This new computational paradigm, which is referred as fog-cloud computing, promises to alleviate the challenges that IoT systems face due to the tremendous growth of IoT devices and applications and improves the scalability of IoT applications [13, 14].

3.2 Fog Computing

In this section, we introduce the fog computing in IoT systems and discuss the computation task offloading to fog nodes, which is challenging due to the heterogeneity of IoT devices and applications and highly dynamic IoT environment. We further review the existing computation task offloading mechanisms in the literature.

3.2.1 Overview of Fog Computing

Fog computing is a novel framework in which data can be analyzed and processed by other IoT devices located in the edge of the network. This can be considered as the extension of cloud computing paradigm to the network edge. Figure 3.1 illustrates the coexistence of fog computing and cloud computing in a smart city. Those IoT devices that can run the applications on behalf of other devices are called fog nodes. The fog nodes with processing and storage capability can be deployed in close proximity of IoT devices and can provide computing services in a distributed manner. The delay-sensitive applications can be dispatched to the fog nodes located in close proximity, and the results can be retrieved in relatively short time. This significantly enhances the delay performance of real-time applications, while remote cloud computing may not meet their latency requirements. In addition to the fog nodes that are located close to the IoT object, fog aggregation nodes are the network edge devices (e.g. routers, smart gateways) with computing and storage capabilities. They usually have a higher processing power and can further provide computing services to those processing jobs with more relaxed latency requirements. Cloud computing complements this computation paradigm by providing computing services through remote cloud servers. The cloud computing can offer huge processing power to IoT systems. However, the relatively long delay to dispatch the processing jobs to the cloud servers and retrieve the results limits the advantages of cloud computing. Processing jobs of those applications which are less sensitive to latency can be sent to the cloud. Cloud computing also provides IoT applications long-term storage. Furthermore, fog nodes and fog aggregation nodes may send information periodically to the cloud for big data analysis.

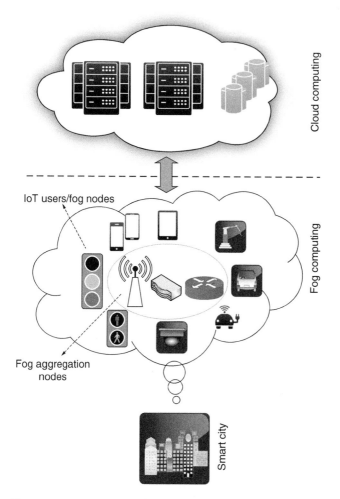

Figure 3.1 A smart city with fog-cloud computing. IoT users can use the computing services offered by fog nodes in their close proximity. The fog aggregation nodes (e.g. access points, routers, and gateways) provide more powerful services to IoT users and fog nodes. The processing jobs of delay-tolerant applications can be offloaded to the remote cloud computing servers.

3.2.2 Computation Offloading

Fog computing promises to provide low-latency computing services to IoT devices to enable emerging IoT systems. Computation-intensive tasks generated by heterogeneous IoT applications can be performed in either on-site or off-site computing resources on-behalf of IoT devices. This is referred as computation task offloading. Computation task offloading to either fog nodes or remote cloud servers can overcome the resource limitation of IoT devices, provide powerful processing power, and enhance the lifetime of battery-operated IoT devices. Moreover, task offloading to fog nodes which are located in close proximity of IoT devices can substantially reduce the computing services latency in order to meet the requirements of delay-sensitive IoT applications.

Computation task offloading to remote cloud servers has proven its benefits in enhancing the users' experiences [15]. Several cloud-assisted platforms have been proposed for smartphones to enable task offloading to cloud servers. These platforms include but are not limited to ThinkAir [16], MAUI [17], CloneCloud [18], and cloudlets [19]. Nonetheless, they cannot be applied directly to the computation tasks of IoT devices due to the highly dynamic environment of IoT systems. The following challenging issues make the aforementioned task offloading mechanisms inapplicable to IoT systems. First, IoT devices and fog nodes are highly dynamic. Second, they span a wide range of physical objects from different vendors equipped with different equipment and capabilities. Third, the computation and computing resources in IoT systems are also highly dynamic. The IoT devices not only suffer from the limited processing power but also face network connectivity issues due to limited communication resources. In order to fully exploit the advantages of fog computing, efficient mechanisms are required to allocate computation and communication resources and schedule the computation tasks.

3.2.2.1 Evaluation Criteria

We now introduce a set of criteria to evaluate the performance of computing services offered by the fog-cloud computing. In particular, these criteria are used to investigate the performance of computation task offloading mechanisms.

- *Quality of experience*: The QoE for each IoT users reflects its satisfaction of using computing services. The QoE indicates how computing services can address the requirements of IoT applications and services. The reduction in delay and energy consumption can contribute to the QoE that users perceive when utilizing the computing services.
- *Heterogeneity*: The IoT users and fog nodes are heterogeneous in terms of their computation requirements and processing power. Furthermore, a wide range of IoT applications are developing with heterogeneous requirements. The fog-cloud computing paradigm should be able to cope with this heterogeneity.
- *Scalability*: Due to the tremendous growth of IoT devices and applications, the scalability of IoT systems is a critical development criterion in order to make such system operational at large scale. This urges the need of proper architectural design for the fog-cloud federation.

It should be mentioned that there are some other evaluation criteria such as reliability, preserving security and privacy, and supporting mobility, which are beyond the scope of this chapter.

3.2.2.2 Literature Review

Computation task offloading in either fog or cloud computing has been widely studied in recent years due to the rapid development of IoT systems. Deng et al. [20] studied the cooperation between fog nodes and remote servers in the fog-cloud computing paradigm. They proposed an optimal allocation of the workload arrived at computing servers. The proposed allocation mechanism aims to minimize the power consumption of fog nodes and cloud servers under service delay constraints. To tackle the computational complexity of the optimal allocation mechanism, the authors further developed a suboptimal mechanism. The proposed mechanism demonstrates that fog computing can complement the cloud computing by offering low-latency computing services. Tan et al. [21] studied a system where a set of servers are deployed at the edge of network such that the users can offload their computation-intensive tasks and benefit from low-latency services. The authors proposed an online job dispatching and scheduling mechanism with the objective to minimize the total weighted response time over all the jobs. They assigned each processing job a weight based on its sensitivity to delay. The proposed mechanism is able to significantly reduce the total weighted delay of all jobs. However, this mechanism is only suitable for those IoT applications, for which the social performance of computing services is important. Xiao and Krunz [22] proposed a distributed mechanism that efficiently allocates the fog computing resources to processing jobs. The objective is to minimize the response time of fog nodes under a given power efficiency constraint for the battery-operated fog nodes. Chang et al. [23] proposed an energy efficient computation offloading scheme in a multi-user fog computing system. The mechanism aims to minimize the energy consumption, while satisfying strict delay constraints. The authors studied both communication and computation delay by utilizing a queuing theory approach. The proposed mechanism significantly reduces the total energy consumption of all users and demonstrates the benefits of fog-cloud computing paradigm. In Ref. [24], Du et al. studied computation task offloading from smartphone users in fog-cloud computing systems. They jointly optimized the offloading decisions and the allocation of computing resource, transmit power, and radio bandwidth while guaranteeing user fairness and maximum tolerable delay. The objective of the proposed mechanism is to minimize the maximum weighted cost of all users, while the users' cost consists of energy consumption and service delay.

Fog networks have also received much attention in recent years. The cooperation of interconnected fog nodes in providing computing services forms the fog networks. Lee et al. [25] proposed an online framework that enables the fog nodes to form a pool of computing resources. The proposed framework aims to minimize the maximum delay of all computation tasks generated by all users within the network. Elbamby et al. [26] also studied the fog networks. They considered a fog network that provides computing services and is able to cache the popular computation tasks and proposed a latency constrained resource allocation mechanism. The objective is to minimize the aggregate delay of all tasks.

The aforementioned computation offloading mechanisms satisfy a subset of the criteria presented in the previous subsection. They mainly focused on optimizing the social performance of the system (e.g. minimizing the total cost of the IoT system) and do not consider the strategic behavior of IoT users. Selfish IoT users only follow the strategies that enhance their own QoE. As the fog computing resources are limited, the IoT users compete against each other with the goal of maximizing their QoE. To guarantee the success of IoT systems, it is important to consider this competition when designing offloading mechanisms.

3.3 A Computation Task Offloading Game for Hybrid Fog-Cloud Computing

We now propose a computation task offloading mechanism for hybrid fog-cloud computing. The proposed mechanism makes an offloading decision for each task arriving at the IoT users such that their QoE is maximized. We consider that each user is connected to multiple fog nodes and the remote cloud servers using multi-radio access technology. In this section, we first introduce the system model. To determine the offloading decision in each IoT user, we then formulate a QoE maximization problem. We use a game theoretic approach to model the competition among IoT users and propose the computation offloading game. We further analyze the properties of this game and show the existence of a pure NE. We finally present the numerical results in this section to evaluate the performance of the formulated computation offloading game.

3.3.1 System Model

Figure 3.2 illustrates an instant of an IoT system with hybrid fog-cloud computing and a set of users. The computing servers include a set of fog nodes located in the close proximity of IoT users and the remote cloud servers. Each IoT user chooses an offloading strategy to either perform its tasks locally or offload them to the computing servers.

Remote cloud servers

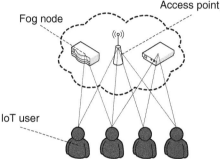

Access point

Fog node

IoT user

Figure 3.2 An instance of IoT systems with IoT users, fog nodes, and the remote cloud servers. The fog nodes provide computing services to their nearby IoT users, while the users may also use the remote cloud computing services.

3.3.1.1 Hybrid Fog-Cloud Computing

We assume an IoT system with S fog nodes and the remote cloud servers. We denote the set of all computing servers as $S = \{0, 1, \dots, S\}$, where 0 represents all remote cloud servers that IoT users can use for their computation task offloading, and $\{1, \dots, S\}$ is the set of fog nodes. Each cloud server is modeled as a virtual machine with dedicated processing power f_0, while we assume that the cloud service provider owns sufficient servers.

We further model the computing resource of each fog node $s = 1, \dots, S$ as a virtual machine with processing power f_s. We consider that the processing power of each fog node is equally shared among those IoT users which are offloading their tasks to that node. We further consider that the fog nodes do not prioritize applications of an IoT user since all users are selfish.

3.3.1.2 Computation Task Models

We denote the set of IoT users as $\mathcal{N} = \{1, \dots, N\}$, where N is the total number of users in the system. We denote the computation task of IoT user $n \in \mathcal{N}$ as $\mathcal{T}_n = (l_n, p_n)$, where l_n is the size of task in bits and p_n is the number of centralized processing unit (CPU) cycles required to process the task. The size of each task and its required processing cycles depend on the application type and are known to the user upon arrival of the task. Based on these parameters, each IoT user decides to either perform its task locally or offload it to the computing servers. The user also chooses a computing server, which can either be one of the fog nodes in close proximity or belong to the remote cloud. We assume that each computation task should either be served entirely by the IoT device of the user or be offloaded to one computing server. To model the offloading decision, we define the offloading indicator $a_{n,s} \in \{0, 1\}$ for each IoT user $n \in \mathcal{N}$ and computing server $s \in S$. Let $a_{n,s} = 1$ indicate that IoT user n offloads its task to the server s. Moreover, when $a_{n,s} = 0$ for all $s \in S$, then task $\mathcal{T}_n$ is performed locally by the user. We also define the offloading vector of user $n \in \mathcal{N}$ as $\mathbf{a}_n = (a_{n,0}, a_{n,1}, \ldots, a_{n,s})$. The following constraint indicates that each task n can be offloaded to at most one server.

$$\sum_{s \in S} a_{n,s} \le 1$$

Each user is able to offload its tasks to a computing server if they are connected through a wireless interface. We introduce constant $b_{n,s}$ to show the connectivity of user n to server s, where $b_{n,s} = 1$ indicates that user n is able to offload its task to the server s. For each IoT user n, we have $a_{n,s} \le b_{n,s}$ for all $s \in S$.

We now consider the two different cases of local and cloud computing and calculate the cost associated with each case. The cost of performing a task consists of the energy consumed to either perform the task locally or offload it to the cloud as well as the delay that the user experiences.

Case 1. Local Computing In this case, we assume that the task is locally performed in the IoT device CPU. We denote the processing power of IoT device n as f_n. The computation energy for performing task $\mathcal{T}_n$ is $\alpha_n p_n / f_n$, where α_n is a user-dependent constant. Furthermore, the computation time to process task $\mathcal{T}_n$ is p_n / f_n as it requires p_n CPU cycles. We define the cost of performing the task $\mathcal{T}_n$ locally as the weighted sum of computation energy and time. Let λ_n^E and λ_n^T denote the constant weights of computation energy and computation time, respectively, which depend on the IoT users and the type of application. The cost of local computing for task $\mathcal{T}_n$ is

$$c_n^L \triangleq \lambda_n^E \frac{\alpha_n p_n}{f_n} + \lambda_n^T \frac{p_n}{f_n}$$

We assume that $\lambda_n^E \in [0, 1]$, while $\lambda_n^T \in (0, 1]$. Notice that in the case of $\lambda_n^T = 0$, the computation offloading may result in a huge delay.

Case 2. Fog-Cloud Computing We now study the case that the IoT user offloads its computation task to either a fog node in close proximity or the remote cloud server. The energy required to offload the task to a computing server is the transmission energy consumed in the corresponding wireless interface. We denote the data rate of the wireless interface of user n connected to the computing server s as $r_{n,s}$. The transmission energy is $\beta_{n,s} l_n / r_{n,s}$, where constant $\beta_{n,s}$ depends on the type of the wireless interface. We assume that the transmission rate can be obtained by measurement and is known to the user. The same assumption is used in [21, 27]. We now determine the delay each user experiences when offloading its tasks. The delay consists of two terms: the time required to dispatch the task to the allocated computing server and the computation time spent on the server. The former includes the transmission time of the wireless interface and the roundtrip delay between the IoT device and the server as denoted by $d_{n,s}^{rt}$. The time required to transmit the task $\mathcal{T}_n$ over the wireless interface is $l_n/r_{n,s}$ for each $s \in S$.

We further determine the computation time required in the computing servers. To do so, we denote the offloading strategies of all IoT users excluding user n as matrix $\mathbf{A}_{-n} = (a_{n,s})_{n \in \mathcal{N}\setminus\{n\}, s \in S}$. We also define $d_{n,s}^c(\mathbf{A}_{-n})$ as the computation time required to process task $\mathcal{T}_n$ in server s. We focus on a particular time instant and assume that the processing power of fog nodes is equally shared among the workload that is arrived from IoT users in this time instance. Let us assume that task $\mathcal{T}_n$ is offloaded to fog node $s \in S\setminus\{0\}$. The processing power of the fog node is entirely assigned to this task if it is the only task in the server. Notice that in this case, $a_{m,s} = 0$ for all $m \in \mathcal{N}\setminus\{n\}$ and $d_{n,s}^c = p_n/f_s$. If other IoT users are also allocated to this server, the computation time may increase as the processing power of the fog node is shared among them. The number of the tasks including task $\mathcal{T}_n$ that are offloaded to this server is $1 + \sum_{m \in \mathcal{N}\setminus\{n\}} a_{m,s}$. However, those tasks which require less processing power than task $\mathcal{T}_n$ (i.e. $p_m < p_n$) depart the server sooner. We reassign the processing power to other tasks which are still under processing once a task leaves the server. Thus, the computation time task $\mathcal{T}_n$ spends in fog node $s \in S\setminus\{0\}$ is

$$d_{n,s}^c(\mathbf{A}_{-n}) = \frac{p_n}{f_s}\left(1 + \sum_{m \in \mathcal{N}\setminus\{n\}} \min\left\{\frac{p_m}{p_n}, 1\right\} a_{m,s}\right)$$

The second term of $d_{n,s}^c(\mathbf{A}_{-n})$ determines how the computation time of each task is affected by the other tasks offloaded to the same fog node. We have so far obtained the computation time if the task is offloaded to a fog node. The computation time spent in the remote cloud servers can also be determined using a similar approach. However, unlike fog nodes, the cloud servers have sufficient computing resources. Thus,

$$d_{n,0}^c(\mathbf{A}_{-n}) = \frac{p_n}{f_0}$$

Using the offloading energy and delay calculated in this subsection, we now determine the cost imposed to IoT user n when offloading the task $\mathcal{T}_n$ as follows.

$$c_n^C(\mathbf{a}_n, \mathbf{A}_{-n}) \triangleq \lambda_n^E \sum_{s \in S} a_{n,s} \frac{\beta_{n,s} l_{n,s}}{r_{n,s}} + \lambda_n^T \sum_{s \in S} a_{n,s}\left(\frac{l_{n,s}}{r_{n,s}} + d_{n,s}^{rt} + d_{n,s}^c(\mathbf{A}_{-n})\right)$$

3.3.1.3 Quality of Experience

We define the QoE as the cost reduction achieved from using computing services. In particular, the QoE of a user is the amount of its cost reduction when offloading the tasks. The QoE reflects the benefits the IoT user achieves by offloading its tasks to the computing servers. We denote the user n's QoE as $q_n(\mathbf{a}_n, \mathbf{A}_{-n})$.

$$q_n(\mathbf{a}_n, \mathbf{A}_{-n}) \triangleq \begin{cases} c_n^L - c_n^C(\mathbf{a}_n, \mathbf{A}_{-n}), & \text{if } \sum_{s \in S} a_{n,s} = 1 \\ 0, & \text{otherwise} \end{cases}$$

The task is offloaded when $\sum_{s \in S} a_{n,s} = 1$, and the QoE is the cost of local computing minus the cost imposed by offloading the task. Otherwise, when $\sum_{s \in S} a_{n,s} = 0$, the user performs the task locally, and the perceived QoE from computing services is zero.

3.3.2 Computation Offloading Game

In this section, we use a game theoretic approach and model the competition between the IoT users as a strategic game. We utilize a potential game approach as an effective tool to model the interactions between IoT users. We then study the existence of an NE for this game and analyze the price of anarchy (PoA).

3.3.2.1 Game Formulation

We formally define the computation offloading game $\mathcal{G} \triangleq (\mathcal{N}, \prod_{n \in \mathcal{N}} \mathcal{A}_n, \{q_n\}_{n \in \mathcal{N}})$ as follows:

- *Players*: IoT users, as denoted by the set $\mathcal{N}$, are the players.
- *Strategy*: The strategy of each user n is $\mathbf{a}_n$ that belongs to the feasible strategy space $\mathcal{A}_n$.
- *Payoff*: The payoff of each user n achieved from using computing services is its QoE as denoted by q_n.

A strategic IoT user always follows its best response strategy, which is introduced as follows.

Definition 3.1 (Best Response Strategy [28]). Player n's best response strategy in response to the strategy of other players, $\mathbf{A}_{-n}$, is

$$\mathbf{a}_n^* = \arg \max_{\mathbf{a}_n} q_n(\mathbf{a}_n, \mathbf{A}_{-n})$$

$$\text{subject to } \sum_{s \in S} a_{n,s} \leq 1$$

$$a_{n,s} \leq b_{n,s}, \forall s \in S$$

$$a_{n,s} \in \{0, 1\}, \forall s \in S,$$

which is the choice of $\mathbf{a}_n$ that results in the highest possible QoE.

We refer the aforementioned problem as the QoE maximization problem. The best response strategy of each user is obtained by solving the QoE maximization problem. In this problem, the first constraint ensures that each task is either performed locally or offloaded to one computing server. By introducing the second constraint, we guarantee that user n can offload its task to computing server s if there is a wireless link connecting the user to this server. The feasible region of the QoE maximization problem represents the feasible strategy space of player n, as denoted by $\mathcal{A}_n$.

We now introduce the concept of NE with the help of Definition 3.1.

Definition 3.2 (Nash Equilibrium [28]). A strategy profile $\{\mathbf{a}_n^*\}_{n \in \mathcal{N}}$ is an NE of Game $\mathcal{G}$ if it is a fixed point of best response strategies. In other words, for all $\mathbf{a}_n' \in A_n, n \in \mathcal{N}$,

$$q_n(\mathbf{a}_n', \mathbf{A}_{-n}^*) \leq q_n(\mathbf{a}_n^*, \mathbf{A}_{-n}^*)$$

Definition 3.2 states that when an equilibrium is achieved, none of the players has incentive to change its strategy as it cannot obtain a higher QoE by following any other strategy. We now study the Game $\mathcal{G}$ and investigate the existence of an NE. To do so, we utilize a potential game approach [29, 30], which is a powerful tool to analyze the properties of NE. In particular, we model the competition between IoT users as a weighted potential game, which is defined as follows.

Definition 3.3 (Weighted Potential Game). A weighted potential game is a game with a potential function P such that for every player $n \in \mathcal{N}$ and strategies $a_n, a_n' \in A_n$, there exists a vector $w = (w_n)_{n \in \mathcal{N}}$ that satisfies the following equality.

$$q_n(\mathbf{a}_n, \mathbf{A}_{-n}) - q_n(\mathbf{a}_n', \mathbf{A}_{-n}) = w_n(P(\mathbf{a}_n, \mathbf{A}_{-n}) - P(\mathbf{a}_n', \mathbf{A}_{-n}))$$

Such function is called a w-potential function. In order to show that Game $\mathcal{G}$ is a weighted potential game, we define the following two functions. Let $\mathbf{A} = (\mathbf{a}_n, \mathbf{A}_{-n})$. We denote the weighted aggregate QoE of all users as $Q(\mathbf{A})$.

$$Q(\mathbf{A}) \triangleq \sum_{n \in \mathcal{N}} \frac{1}{\lambda_n^T} q_n(a_n, \mathbf{A}_{-n})$$

We further assume that each user is alone in the game and define function $\overline{Q}(\mathbf{A})$ as the aggregate QoE of all users in this case.

$$\overline{Q}(\mathbf{A}) \triangleq \sum_{n \in \mathcal{N}} \frac{1}{\lambda_n^T} q_n(a_n, 0)$$

where 0 is an all zero $(N-1) \times (S+1)$ matrix. In order to show that Game $\mathcal{G}$ is a weighted potential game, we need to introduce a w-potential function that meets the condition of Definition 3.3. Through the following theorem, we propose a w-potential function for Game $\mathcal{G}$.

Theorem 3.1 *With the choice of*

$$w = (w_n)_{n \in \mathcal{N}} = (\lambda_n^T)_{n \in \mathcal{N}}$$

the following function is a w-potential function for Game $\mathcal{G}$.

$$P(\mathbf{A}) = \frac{Q(\mathbf{A}) + \overline{Q}(\mathbf{A})}{2}$$

Proof: Please refer to [31] for the proof of Theorem 3.1.

Utilizing Definition 3.3, we can conclude that Game $\mathcal{G}$ is a weighted potential game. To show the existence of a pure NE in Game $\mathcal{G}$, we use the following lemma, which is proven in [29].

Lemma 3.1 *Every finite potential game possesses a pure strategy NE.*

According to Lemma 3.1, we conclude that there exists an NE for Game $\mathcal{G}$. The NE can be obtained by determining the users' best response strategies.

Algorithm: Best Response Adaptation for an IoT user n.

1 initialization: $t \leftarrow 0$, and $a_n^{*(0)}$

2 do

3 $t \leftarrow t + 1$

4 User n collects $d_{n,s}(\mathbf{A}_{-n})$ from fog nodes.

5 User n updates its best response strategy by solving the QoE maximization problem.

$$a_n^{*(t)} \neq \underset{a_n}{\mathrm{argmax}}\, q_n(a_n, \mathbf{A}_{-n})$$

$$\text{subject to } \sum_{s \in S} a_{n,s} \leq 1,$$

$$a_{n,s} \leq b_{n,s}, \quad \forall s \in S,$$

$$a_{n,s} \in \{0,1\}, \quad \forall s \in S,$$

6   User n submits its strategy to the neighboring fog nodes.

7 while $a_n^{*(t)} \neq a_n^{*(t-1)}$

8 output: $a_n^{NE} \leftarrow a_n^{*(t)}$

Figure 3.3 Best response strategy algorithm.

3.3.2.2 Algorithm Development

We now develop a best response strategy algorithm to obtain the NE. Figure 3.3 illustrates the proposed best response algorithm. Each IoT user iteratively updates its best response strategy according to the strategies of other users. Each user collects the information from the nearby fog nodes and solve the QoE optimization problem in order to determine its best response. Each user n only needs to know the value of computation delay $d_{n,s}^c(\mathbf{A}_{-n})$ as well as the roundtrip delay for available computing servers. By utilizing the proposed best response algorithm in all users, we can determine their best response strategies, while the fixed point of these strategies corresponds to the NE of Game $\mathcal{G}$ as stated in Definition 3.2.

The convergence of the best response algorithm is guaranteed using the following lemma, which states that each potential game has the finite improvement property.

Lemma 3.2 *Every finite potential game has the finite improvement property.*

The proof of Lemma 3.2 can be found in [29].

The finite improvement property implies that every improvement path is finite, and since the best response strategies always increase the potential function (i.e. provide an improvement), the algorithm terminates in finite time. Moreover, according to Lemma 3.1, the algorithm reaches a pure strategy NE. For each IoT user $n \in \mathcal{N}$, we denote the NE as $\mathbf{a}_n^{\text{NE}}$.

3.3.2.3 Price of Anarchy

In the previous subsections, we showed that Game $\mathcal{G}$ possesses at least an NE, and proposed an algorithm to obtain the equilibrium. To investigate that how good is the obtained NE in terms of overall system performance, we use the PoA. The PoA is a concept in game theory that measures how the efficiency of a system degrades due to strategic behavior of its players. This reveals the efficiency of the obtained equilibrium in the presence of selfish players. To evaluate the PoA, we focus on the cost of IoT users and update their best response strategies in order to minimize their cost. Notice that this will result in the same equilibrium as the QoE maximization framework. The cost of user $n \in \mathcal{N}$ is defined as

$$c_n(\boldsymbol{a}_n, \mathbf{A}_{-n}) \triangleq \begin{cases} c_n^C(\boldsymbol{a}_n, \mathbf{A}_{-n}), & \text{if } \sum_{s \in S} a_{n,s} = 1, \\ c_n^L, & \text{otherwise.} \end{cases}$$

We further define the total cost of all users as

$$c(\mathbf{A}) \triangleq \sum_{n \in \mathcal{N}} c_n(\boldsymbol{a}_n, \mathbf{A}_{-n})$$

The PoA is the ratio between the worst social cost obtained in an NE and the minimum social cost. The minimum total cost of all users, as denoted by $c^{\min}$, can be obtained by solving the following optimization problem.

$$c^{\min} = \text{minimize}_{\mathbf{A} \in \mathcal{A}} c(\mathbf{A})$$

where $\mathcal{A} = \prod_{n \in \mathcal{N}} \mathcal{A}_n$. Let $\mathcal{A}^{\text{NE}}$ denote the set of all equilibria. The PoA can be obtained as

$$\text{PoA} = \frac{\max_{\mathbf{A} \in \mathcal{A}^{\text{NE}}} c(\mathbf{A})}{c^{\min}}$$

Through the following theorem, we show that the PoA, hence the degradation of the social cost due to the strategic behavior of players, is no worse than a constant.

Theorem 3.2 *The PoA of the computation offloading game $\mathcal{G}$ is bounded as follows.*

$$\text{PoA} \le \min \left\{ N_s, \frac{\sum_{n \in \mathcal{N}} c_n^L}{\sum_{n \in \mathcal{N}} \min_{\mathbf{a}_n \in \mathcal{A}_n} c_n(\mathbf{a}_n, 0)} \right\}$$

where N_s is the maximum number of IoT users that are using the computing services of a fog node.

Proof: Please refer to [31] for the proof of Theorem 3.2.

3.3.2.4 Performance Evaluation

In this section, we investigate the computation offloading game by evaluating the performance of the proposed best response strategy algorithm. In particular, we study the IoT users' QoE at the equilibrium and that of the socially optimum mechanism as well as the delay that IoT users experience. We use the following simulation parameters. The CPU clock speed of IoT users is randomly and uniformly chosen from [100 MHz, 1 GHz]. According to the measurement results reported in [32], the energy consumption constant $\alpha_n = 0.33 f_n^3 + 1$ for each IoT device $n \in \mathcal{N}$. We further assume that each IoT device has Long Term Evolution (LTE), Wi-Fi, and Bluetooth interfaces, while the LTE interface is used to communicate with the remote cloud servers. The IoT users can also use the Wi-Fi and Bluetooth interfaces to connect to at most two nearby fog nodes. We further assume that for each user n, $\beta_{n,0} = 2605$ mJ/s, while $\beta_{n,s} = 1224.78$ mJ/s [33] if the user is connected to fog node s via Wi-Fi interface and $\beta_{n,s} = 84$ mJ/s if the Bluetooth interface is used. The other simulation parameters are as follows. The transmission rate of LTE, Wi-Fi, and Bluetooth interfaces are uniformly and randomly distributed over [4.85, 6.85]Mbps, [2.01,

4.01]Mbps, and [0.7, 2.1]Mbps, respectively. The processing power of fog nodes is uniformly distributed in [2, 3] GHz, and the processing power of each cloud server is 4 GHz. The task size is uniformly distributed in [100 B, 0.5 MB], while the processing density of each task is uniformly distributed over [100, 600] cycles per bit. Moreover, λ_n^E and λ_n^T are randomly chosen from [0, 1] and [0.5, 1], respectively.

We first investigate the average QoE of IoT users perceived at the NE as obtained by the best response algorithm. Figure 3.4 illustrates the average QoE of IoT users when offloading their computation tasks to the computing servers. As can be observed, in an IoT system with a large number of fog nodes, a higher average QoE is obtained than that of a system with fewer fog nodes. This is because more computing resources are available to users when there are more fog nodes in their close proximity. However, due to the limited resources of fog nodes, the average QoE of each user decreases when we increase the number of users.

We next evaluate the delay each task experiences to be processed. To do so, we consider delay-sensitive applications and set $\lambda_n^E = 0$ and $\lambda_n^T = 1$ for all $n \in \mathcal{N}$. We also assume that the roundtrip delay between the IoT users and the remote cloud servers, as denoted by $d_{n,0}^{\mathrm{rt}}$, is the same for all $n \in \mathcal{N}$. Let d^{rt} denote this roundtrip delay. In Figure 3.5, we show the average delay when different number of fog nodes exists in the system. We vary the roundtrip delay between IoT devices and the remote cloud servers (i.e. d^{rt}) to investigate how it affects the total computation delay. As can be observed from Figure 3.5, in the case that only cloud computing services are available and there are no fog nodes, each

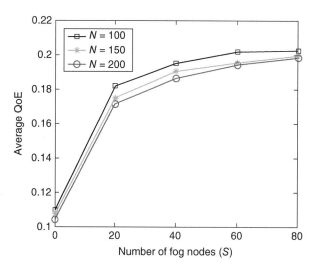

Figure 3.4 The average perceived QoE of IoT users at the NE. The IoT users perceive a higher QoE when a larger number of fog nodes offer computing services.

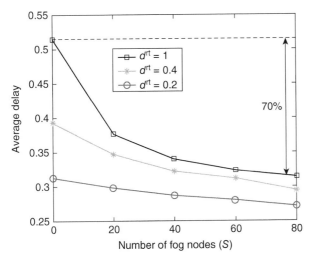

Figure 3.5 The average delay each task experiences for different roundtrip delay of remote cloud servers when $N = 200$. Fog computing can significantly reduce the delay compared to the case that users can only offload their computation tasks to the cloud servers (i.e. $S = 0$).

task experiences a huge delay. However, when fog computing services are also available, the delay significantly reduces as the users can offload their computation tasks to their nearby fog nodes with low latency. When there are $S = 80$ fog nodes in the system, the delay is reduced by 70% compared to the case that $S = 0$. This demonstrates that delay-sensitive IoT applications can benefit from low-latency computing services provided by the fog nodes.

To further clarify how fog nodes can provide low-latency services, we investigate the different terms that contribute to the delay. To do so, we study the computation time, communication time, and the roundtrip delay that each task experiences. The average of these metrics for different number of fog nodes is shown in Figure 3.6 when there are $N = 200$ IoT users in the system. Figure 3.6 demonstrates that the computation time is reduced when there are more fog nodes in the system. This is because more IoT users offload their task rather than local execution, which consequently reduces the computation time. The average roundtrip delay each task experiences also reduces in this case. Unlike these two metrics, the communication time increases when more tasks are offloaded. This is because in the local computing case, the communication time is zero as the tasks do not leave the IoT devices.

We now evaluate the computation offloading game in terms of the number of users that can benefit from computing services. Note that these users offload their tasks as they can achieve a QoE greater than zero. Figure 3.7 illustrates the number of beneficial users when we vary the number of fog nodes. In this

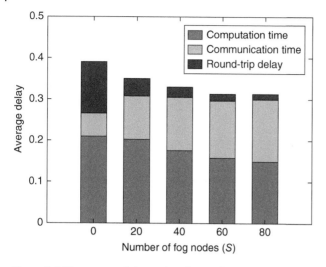

Figure 3.6 The average delay each task experiences versus different number of fog nodes when $N = 200$. We categorize the average delay into three terms: the computation time spent in either users' CPU or computing servers, communication time to dispatch the tasks to computing servers, and the roundtrip delay.

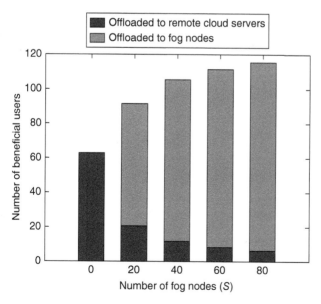

Figure 3.7 The number of beneficial users versus the number of fog nodes when $N = 200$. We categorize the users based on the computing services they choose for task offloading.

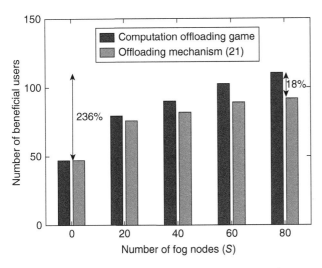

Figure 3.8 The number of beneficial users in the proposed computation offloading game in comparison to that of [21] when $N = 200$. As can be observed, our proposed mechanism outperforms [21] by 18% when there are 80 fog nodes in the system. In both mechanisms, the same number of users offload their tasks to the remote cloud when there is no fog node.

figure, the beneficial users are categorized into two sets based on the computing services they use. As can be observed from Figure 3.7, the number of users that offload their computation tasks to the remote cloud servers significantly reduces when there are more fog nodes offering computing services. In addition, more IoT users benefit from the computing services when there are more fog nodes in the system. This demonstrates that deploying more fog nodes can alleviate the challenges imposed by tremendous growth of IoT systems.

We finally evaluate the performance of our proposed computation offloading game in comparison to an offloading mechanism proposed in [21]. Two offline and online mechanisms are proposed in [21]. In order to compare in a fair manner, we choose the offline mechanism as it always outperforms the online mechanism. The offline mechanism proposed in [21] aims to minimize the total weighted delay of all tasks when the tasks are prioritized based on their sensitivity to delay. We choose λ_n^T randomly and uniformly from [0.01, 1], while $\lambda_n^E = 0$ for all users. The weight λ_n^T indicates the user n's priority when investigating the performance of the offloading mechanism [21]. Figure 3.8 shows the number of beneficial users in our proposed computation offloading mechanism in comparison to [21]. As shown, fog computing services significantly increase the number of beneficial users. Furthermore, we can observe that our proposed computation offloading mechanism outperforms [21] by up to 20% in terms of the number of beneficial users.

3.4 Conclusion

In this chapter, we studied fog computing in IoT systems. We first provided an overview of fog computing in a hybrid fog-cloud computing paradigm and discussed the challenges that future IoT system face. We then proposed a computation offloading game utilizing a potential game approach. We further analyzed the properties of the game and proposed a best response strategy algorithm to determine the equilibria. We also evaluated the performance of the proposed algorithm through numerical experiments. Our results demonstrate that fog computing is promising in providing low-latency computing services for emerging IoT systems. More details of the proposed computation task offloading game as well as additional simulation results can be found in [31].

References

1 Gubbi, J., Buyya, R., Marusic, S., and Palaniswami, M. (2013). Internet of Things (IoT): a vision, architectural elements, and future directions. *Future Generation Computer Systems* 29 (7): 1645–1660.

2 Al-Fuqaha, A., Guizani, M., Mohammadi, M. et al. (2015). Internet of Things: a survey on enabling technologies, protocols, and applications. *IEEE Communication Surveys and Tutorials* 17 (4): 2347–2376.

3 Yeh, L.-Y., Chiang, P.-Y., Tsai, Y.-L., and Huang, J.-L. (2018). Cloud-based fine-grained health information access control framework for lightweight IoT devices with dynamic auditing and attribute revocation. *IEEE Transactions on Cloud Computing* 6 (2): 532–544.

4 Guerrero-Ibanez, J., Zeadally, S., and Contreras Castillo, J. (2015). Integration challenges of intelligent transportation systems with connected vehicle, cloud computing, and Internet of Things technologies. *IEEE Wireless Communications* 22 (6): 122–128.

5 Zafari, F., Papapanagiotou, I., and Christidis, K. (2016). Microlocation for Internet- of-Things-equipped smart buildings. *IEEE Internet of Things Journal* 3 (1): 96–112.

6 Pham, T.N., Tsai, M.F., Nguyen, D.B. et al. (2015). A cloud-based smart-parking system based on Internet-of-Things technologies. *IEEE Access* 3: 1581–1591.

7 Cisco (2019). Cisco Visual Networking Index: Forecast and Trends, 2017–2022 White Paper. https://www.cisco.com/c/en/us/solutions/collateral/service-provider/visual-networking-index-vni/white-paper-c11-741490.html (accessed 16 August 2019).

8 Wong, V.W.S., Schober, R., Ng, D.W.K., and Wang, L.-C. (2017). *Key Technologies for 5G Wireless Systems*. Cambridge University Press.

9 Botta, A., De Donato, W., Persico, V., and Pescape, A. (2014). On the integration of cloud computing and Internet of Things. *Proceedings of International Conference on Future Internet of Things and Cloud (FiCloud)*, Barcelona, Spain (August 2014).

10 Razzaque, M., Milojevic Jevric, M., Palade, A., and Clarke, S. (2016). Middleware for Internet of Things: a survey. *IEEE Internet of Things Journal* 3 (1): 70–95.

11 Cisco (2015). Fog computing and the Internet of Things: extend the cloud to where the things are,. http://www.cisco.com/c/dam/enus/solutions/trends/iot/docs/computingoverview.pdf (accessed 16 August 2019).

12 OpenFog Consortium. https://www.openfogconsortium.org/ (accessed 16 August 2019).

13 Yi, S., Li, C., and Li, Q. (2015). A survey of fog computing: concepts, applications and issues. *Proceedings of ACM Workshop on Mobile Big Data*, Hangzhou, China (June 2015).

14 Mouradian, C., Naboulsi, D., Yangui, S. et al. (2018). A comprehensive survey on fog computing: state-of-the-art and research challenges. *IEEE Communication Surveys and Tutorials* 20 (1): 416–464.

15 Shah-Mansouri, H., Wong, V.W.S., and Schober, R. (Aug. 2017). Joint optimal pricing and task scheduling in mobile cloud computing systems. *IEEE Transactions on Wireless Communications* 16 (8): 5218–5232.

16 Kosta, S., Aucinas, A., Hui, P. et al. (2012) ThinkAir: dynamic resource allocation and parallel execution in the cloud for mobile code offloading. *Proceedings of IEEE INFOCOM*, Orlando, FL (March 2012).

17 Cuervo, E., Balasubramanian, A., Cho, D.-K. et al. (2010). MAUI: making smartphones last longer with code offload. *Proceedings of ACM Int'l Conference on Mobile Systems, Applications, and Services (MobiSys)*, San Francisco, CA (June 2010).

18 Chun, B.-G., Ihm, S., Maniatis, P. et al. (2011). CloneCloud: elastic execution between mobile device and cloud. *Proceedings of ACM Conference on Computer Systems (EuroSys)*, Salzburg, Austria (April 2011).

19 Satyanarayanan, M., Bahl, P., Caceres, R., and Davies, N. (2009). The case for VM-based cloudlets in mobile computing. *IEEE Pervasive Computing* 8 (4): 14–23.

20 Deng, R., Lu, R., Lai, C. et al. (2016). Optimal workload allocation in fog-cloud computing toward balanced delay and power consumption. *IEEE Internet of Things Journal* 3 (6): 1171–1181.

21 Tan, H., Han, Z., Li, X.-Y., and Lau, F. C. (2017). Online job dispatching and scheduling in edge-clouds. *Proceedings of IEEE INFOCOM*, Atlanta, GA (May 2017).

22 Xiao, Y. and Krunz, M. (2017). QoE and power efficiency tradeoff for fog computing networks with fog node cooperation. *Proceedings of IEEE INFOCOM*, Atlanta, GA (May 2017).

23 Chang, Z., Zhou, Z., Ristaniemi, T., and Niu, Z. (2017). Energy efficient optimization for computation offloading in fog computing system. *Proceedings of IEEE GLOBECOM*, Singapore, Singapore (December 2017).

24 Du, J., Zhao, L., Feng, J., and Chu, X. (2018). Computation offloading and resource allocation in mixed fog/cloud computing systems with min–max fairness guarantee. *IEEE Transactions on Communications* 66 (4): 1594–1608.

25 Lee, G., Saad, W., and Bennis, M. (2017). An online secretary framework for fog network formation with minimal latency. *Proceedings of IEEE ICC*, Paris, France (May 2017).

26 Elbamby, M. S., Bennis, M., and Saad, W. (2017). Proactive edge computing in latency-constrained fog networks. *Proceedings of IEEE European Conference on Networks and Communications (EuCNC)*, Oulu, Finland (July 2017).

27 Chen, X., Jiao, L., Li, W., and Fu, X. (2016). Efficient multi-user computation offloading for mobile-edge cloud computing. *IEEE/ACM Transactions on Networking* 24 (5): 2795–2808.

28 Shoham, Y. and Leyton-Brown, K. (2008). *Multiagent Systems: Algorithmic, Game-Theoretic, and Logical Foundations*. Cambridge University Press.

29 Monderer, D. and Shapley, L.S. (1996). Potential games. *Games and Economic Behavior* 14 (1): 124–143.

30 Bodine-Baron, E., Lee, C., Chong, A. et al. (2011). Peer effects and stability in matching markets. *Proceeding of Int'l Conference on Algorithmic Game Theory*, Amalfi, Italy (October 2011).

31 Shah-Mansouri, H. and Wong, V.W.S. (Aug. 2018). Hierarchical fog-cloud computing for IoT systems: a computation offloading game. *IEEE Internet of Things Journal* 5 (4): 3246–3257.

32 Kim, Y., Kwak, J., and Chong, S. (2015). Dual-side dynamic controls for cost minimization in mobile cloud computing systems. *Proceeding of IEEE WiOpt*, Mumbai, India (May 2015).

33 Kwak, J., Kim, Y., Lee, J., and Chong, S. (2015). DREAM: dynamic resource and task allocation for energy minimization in mobile cloud systems. *IEEE Journal on Selected Areas in Communications* 33 (12): 2510–2523.

4

Pricing Tradeoffs for Data Analytics in Fog–Cloud Scenarios

Yichen Ruan[1], Liang Zheng[2], Maria Gorlatova[2], Mung Chiang[3], and Carlee Joe-Wong[1]

[1] *Department of Electrical and Computer Engineering, Carnegie Mellon University, Moffett Field, CA, USA*
[2] *Department of Electrical Engineering, Princeton University, Princeton, NJ, USA*
[3] *Department of Electrical and Computer Engineering, Purdue University, West Lafayette, IN, USA*

4.1 Introduction: Economics and Fog Computing

The proliferation of devices (sometimes called "things") and enriched applications has created an expectation of near-ubiquitous connectivity among users around the world. Until recently, this activity has largely been enabled by cloud computing, which allows users to virtually share computing resources on a large scale. Applications can then easily access these resources without spending significant amounts of money and effort on setting up and maintaining their own host servers. Cloud resources can also be easily scaled up, allowing applications to rapidly grow their user base. For instance, applications can use cloud servers to collect, store, and analyze data about their users; or to host large files like videos to be streamed. Applications and devices ranging from remote-access home surveillance cameras to health monitoring smartwatches rely on the cloud to serve their users today.

The traditional cloud paradigm, in which applications utilize servers in a remote data center, may not meet new performance requirements in the Internet of Things (IoT), 5G, artificial intelligence (AI), and other emerging technologies. For instance, homeowners may use surveillance cameras to remotely track movement in their backyards in real time. Constantly streaming camera footage to the cloud, from which it can be viewed on homeowners' apps, requires significant amounts of network bandwidth. In another example, running data analytics algorithms on the cloud incurs latency due to the need to transfer data from local devices to cloud servers.

Real-time analytics, e.g. analyzing heartbeat patterns collected by a smart-watch to detect medical emergencies, may not be able to tolerate this latency.

Fog computing aims to meet these new performance requirements by distributing some application functionality to edge devices, instead of running them solely in remote datacenters. Thus, instead of applications only utilizing resources at a client and server, they can leverage computing devices that lie across the cloud-to-things (C2T) continuum. For instance, instead of sending data to a cloud server to be analyzed, a smartwatch might send data to a nearby smartphone. Devices' placement on the C2T continuum can be roughly determined by their computational capabilities, e.g. devices like low-power sensors would lie at one extreme, with powerful yet battery-constrained smart-phones, connected vehicles, laptops, local servers, and finally remote servers lying close to the edge of the network [1]. Figure 4.1 illustrates this hierarchy of devices. Fog computing is predicted to be the next multi-billion-dollar business [2, 3].

In this work, we define *fog computing* [4–7] as a computing architecture that locates functionalities (e.g. storage, computation, communication, and management) at two or more types of devices along the C2T continuum. Fog

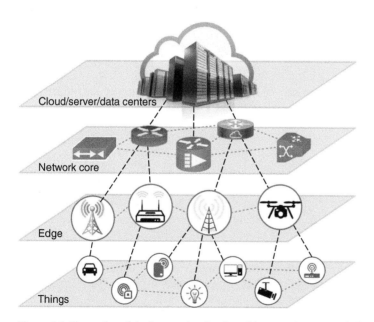

Figure 4.1 Illustration of devices on the cloud-to-things continuum. Each device may contribute computing, storage, and communication capabilities; and the devices are generally connected to each other over the Internet. Cloudlets and points-of-presence may be located at the edge of the network, e.g. colocated with mobile base stations.

computing thus represents a generalization of traditional cloud computing, in which application functionality resides at a local device and a remote cloud server. While this architecture shows promise for addressing applications' bandwidth and latency challenges, it also raises new research challenges, e.g. security and handling device heterogeneity. The past two years have seen progress on these technological challenges, but little progress has been made on understanding the economics of fog networks (i.e. fogonomics). Yet the economics of fog computing will be crucial to its success: just as cloud computing's popularity was built on its promise of cost saving, the long-term future of fog computing will rely on its ability to deliver economic as well as technological benefits. Little existing work, however, has explored the fundamental trade-offs that fog-based applications face between economic and technological performance measures. We use a case study of distributed data processing to illustrate these trade-offs and inform a broader survey of the research challenges in fogonomics.

4.1.1 Fog Application Pricing

Applications utilizing fog architectures will require access to both computing devices at different levels of the C2T continuum and network bandwidth to connect these devices together. The cost of deploying fog services must then account for both of these types of resources. Yet the heterogeneity inherent in fog architectures leads to significant pricing challenges:

Coexistence of heterogeneous networks: Fog devices may have different network interfaces (LTE, Wi-Fi, Bluetooth, femtocells, etc.). Thus, application developers or users will need to choose between the interfaces used, e.g. according to their costs or provided quality-of-service (QoS). Integrating connectivity between different interfaces, however, may be challenging, as different types of network infrastructure will be operated by different network providers. The providers themselves may negotiate agreements to supplement each other's connectivity, further complicating network selection decisions.

Serverless cloud computing: Cloud computing providers have traditionally used an infrastructure-as-a-service (IaaS) model, in which applications specify the amount of time compute jobs need to run on various types of available servers. Recent offerings like Amazon Lambda, which charges per function call, are more flexible but still may not meet fog applications' needs. In a fog architecture, application requirements may include rapidly scaling up or down their computing needs, combining resources at multiple devices along the C2T continuum, or choosing where to place compute functionalities depending on device costs. Simple usage-based pricing, as generally offered by cloud providers like Amazon, is likely inadequate for meeting these needs.

4.1.2 Incentivizing Fog Resources

Devices that wish to provide fog resources to users can range from edge servers and cellular base stations to smartphones, sensors, laptops, vehicles, and cloud servers. Notably, many of these devices will want to use their resources for their own applications, as well as applications that originate from other devices (e.g. smartphones aiding the transmission of sensor data from smart buildings). The heterogeneity of fog devices can make these resource exchanges especially useful, e.g. devices with restricted network access, like sensors with Bluetooth low-energy, may incentivize smartphone gateways who can connect to IP networks to forward their data. Yet these devices will not provide resources for free: they must be incentivized to help applications.

Monetary incentive mechanisms are not new in network economics research [8–10] and many practical incentive mechanisms in mobile data pricing [11–14] and cloud pricing [15–17] can be adaptively applied to fog networks. In fog scenarios, however, incentives may not be monetary. Fog devices could collaborate to compensate for each other's inadequate provisioning of certain types of resources. For instance, smartphones might act as network gateways for low-power sensors, in exchange for access to the information gathered by these sensors. The "right" form of this collaboration and exchange, however, is an open research question.

4.1.3 A Fogonomics Research Agenda

The above trends in the pricing schemes and incentive mechanisms enabled by fog computing will inevitably impact several stakeholders in the network ecosystem, including not only network operators and cloud providers but also device manufacturers, equipment vendors, system integrators, and chip suppliers. These stakeholders' decisions are moreover not made independently: the way cloud providers charge fog devices, for instance, would determine the services deployed of device manufacturers, who would in turn request certain features from their chip suppliers. While several organizations like the [5] have attempted to bridge this dependency technologically by standardizing fog architectures, it is unclear how the economic dependencies will play out. Regulatory concerns may also shape the development of fog computing. A comprehensive theory of fogonomics must account for all these stakeholder incentives and concerns.

The heterogeneity of fog-based architectures may give rise to varied market structures. Some architectures may involve loosely coupled ecosystems with simple charging and payments, such as subscriptions to free Wi-Fi at all Starbucks stores or to proprietary architectures provided by a single fog service provider. Others, however, may require complex and dynamic pricing transactions, e.g. an augmented reality application mounted in a vehicle that continually requests compute offloading from nearby devices as the vehicle moves down a freeway. Such market structures may give rise not only to application and resource providers, but also to middlemen, brokers, and speculators who facilitate monetary transactions between users and the compute and network resources that their applications need.

Unlike physical network performance metrics, many economic quantities, such as user utility, market competitiveness, and social welfare, are highly non-linear, difficult to directly measure in practice, and may vary greatly from user to user. This lack of physical measurements complicates the development of mathematical models and introduces the need for inference methodologies that can detect the economic factors at play in fog computing's deployment and incorporate them into fogonomics research.

This book chapter presents an initial survey of fogonomics, using a case study of distributed data processing to illustrate its research challenges. In particular, we demonstrate the trade-off between balancing QoS and service cost when distributing application tasks between cloud and edge devices, given a set of service prices. In the next sections, we first discuss the economics faced by fog applications today. We then give an overview of typical fog application architectures on which economic markets will be constructed. We illustrate how current price schemes impact the design of fog applications through our case study. We finally conclude by posing several research questions for the research community.

4.2 Fog Pricing Today

As outlined in the introduction, fog applications today must pay for both access to network bandwidth and access to computing resources on a variety of devices. In this section, we outline current pricing schemes for both of these types of resources and their implications for fog computing.

4.2.1 Pricing Network Resources

For the past few decades, Internet service providers in the United States have offered simple flat-rate unlimited data plans for accessing wireline and wireless networks. Yet the proliferation of mobile applications has dramatically increased demand for network bandwidth. Coupled with limited network capacity, these developments have caused increasing congestion, particularly on wireless networks [10]. To relieve this congestion, mobile users today are discouraged from consuming too much data with a mix of capped and usage-based data plans that impose an expensive base payment per month for a limited data quota and steep overage fees or severe QoS degradation above this quota. Other variations of static data pricing include shared data plans, rollover data plans, and sponsored data, all bearing the same economic principle behind it: the more data that is sent through the network, the more users should pay.

By dynamically pooling local resources, fog computing can relieve increasing network congestion: running computations on local devices will reduce the amount of data that need to be transferred between devices. Today's data pricing schemes, however, largely do not address different devices' varying bandwidth requirements for their data flows. They also do not address the bandwidth and latency requirements of different applications: for instance, users of a real-time analytics application may pay more in exchange for guaranteed low-latency connectivity. Yet matching different requirements to the right prices is itself a nontrivial research problem. The design of such pricing schemes should also be sensitive to network neutrality concerns: while differentiated pricing may be needed to ensure acceptable QoS for all applications, it naturally raises neutrality and fairness concerns.

In fog networks, the coexistence of heterogeneous network interfaces gives devices more choices for network connectivity, further complicating network pricing questions. The US Federal Communications Commission (FCC) has recently opened a broader range of high-frequency licensed bands, such as TV white space, as long as the interference caused to licensed users is below a given threshold. Users select the best radio access technology (RAT) after observing their local conditions, either in a game-theoretic way [18, 19] or by joint optimization [20, 21]. Though these networks may not always be available, they are often free to use, offering users a way to reduce their network communication costs and creating new business opportunities for Internet service providers (ISPs) [22].

Another way to gain access to more networks is to rely not on individual devices' opportunistic access to free network interfaces but on infrastructure sharing through ISPs. Many ISPs have proposed to lower their users' costs by supplementing cellular connectivity with access to Wi-Fi networks [23, 24]. Other ISPs have begun to go even further and pool access to their own cellular networks, in addition to access to Wi-Fi hotspots. Users subscribing to these "virtual ISPs" can then choose the best network among all partner ISPs' networks. For example, Google's Project Fi [25] runs on T-Mobile's, Sprint's, and US Cellular's network infrastructures as well as offloading its users' traffic to free open Wi-Fi networks whenever available. It is unclear whether this type of cross-carrier data plan can make a profit in long term, but it can lead to better QoS for users [26, 27]. Fog applications may find virtual ISPs particularly attractive, as they guarantee access to multiple network interfaces, which may be necessary for the heterogeneous devices that work together to enable fog architectures.

4.2.2 Pricing Computing Resources

Most payments for computing resources today involve variations of cloud computing, which allows users to access a variety of IT resources (e.g. CPU, memory, and storage). Cloud service providers (CSPs) offer a variety of different pricing schemes, such as simple usage-based pricing, more advanced auction-based pricing [15], and volume-discount pricing [16].

Most major CSPs have also begun to offer cloud solutions that aim to meet the needs of IoT devices [28, 29]. For example, Microsoft Azure's IoT Hub [28] charges based on the total number of messages transmitted to the cloud per day, while data analytics using HDInsight or machine learning is charged separately. Such a pricing model would make sense for sensors sending data to a CSP in a fog context. Likewise, Amazon's AWS Lambda pricing [29] consists of the request price, which is the unit price charged for the total number of requests sent to the cloud, and the compute price, which is the unit price per second charged for executing preinstalled on the cloud given an amount of memory: devices or cloud-based services can trigger this code whenever their applications need it. Thus, cloud computing becomes serverless; its use may be controlled by edge devices as well as cloud servers.

These types of cloud pricing, however, do not fully account for fog computing's use of resources at multiple types of devices along the C2T continuum. It is reasonable to suppose that fog applications will pay different monetary and perhaps nonmonetary prices for the use of resources at different devices, which in turn will influence their application design. Indeed, one could imagine cost optimization services that determine the application configuration that optimizes a combination of cost and performance.

4.2.3 Pricing and Architecture Trade-offs

In a heterogeneous fog computing scenario, individual fog devices can be selected to play the role of gateways that carry data to other devices, or as computing nodes that carry out some preliminary computation tasks. These computational tasks can be performed on preconfigured dedicated instances or via executing serverless functions instantiated on request. The design of a fog application, i.e. decisions of what resources to utilize at available fog devices, should account for these factors and the trade-offs between relevant performance metrics [30]. For example, both accuracy and latency are relevant to most fog applications.

Accuracy: Too much local computation on gateways may lead to errors in the decision taken by local devices, since these local decisions are made based on partial information of the network and without the knowledge of other data collected by the application on other devices. A centralized computation that accounts for all collected data may be more accurate but may also be more expensive.

Latency: Different functional decompositions of a given computation can lead to longer or shorter communication and computation latencies. Local devices, for instance, may have slower processors but may ultimately be able to process data faster as computations can be parallelized across several devices. Moving significant amounts of data to centrally located devices, on the other hand, may lead to significant communication latencies. Moreover, the latency will be different for different types of computations.

In Section 4.4, we outline an example model of the trade-offs between accuracy, latency, and cost in our case study of distributed data processing.

4.3 Typical Fog Architectures

As discussed in the introduction, the economics of fog computing will depend on the architectures of different fog applications: some applications will need dynamic access to a wide range of resources, while others may have more stable resource requirements and need only limited computing or network resources. Thus, in this section, we outline some representative examples of fog applications and resources on the cloud-to-things continuum.

4.3.1 Fog Applications

Fog computing aids local deployments of IoT nodes and mobile devices, by providing additional intelligence that is close to the end users. Fog computing also aids the performance of wide-area networks, as processing data close to where it is originated reduces the load on the wide-area network.

Fog computing architectures serve as the necessary computing foundation for the broad vision of smart cities and smart environments, including smart transportation [31–33]. Among other things, fog is used for distributed analytics [34], including video and image processing [35–37], and as an aid for mobile devices [38], including unmanned autonomous vehicles (drones) [39]. It is particularly useful for devices with heterogeneous QoS requirements that can be met by different devices, e.g. virtual reality's low latency processing can be done at edge servers, with the bulk of high-bandwidth streaming done from cloud servers.

4.3.2 The Cloud-to-Things Continuum

The cloud-to-things continuum can include the following computing options, each corresponding to different trade-offs in performance and costs.

- *Cloud services*: Modern cloud services offer a wide range of hardware configuration options, from "pay-as-you-go" serverless computing [29, 40] to dedicated core collections with hardware accelerators [41]. Cloud services are based in a set of centralized locations. These services are accessible via traditional large-scale wired network infrastructures, architectured for resilience and uptime guarantees.

- *Points of Presence* (Pops): Content delivery network (CDN) and telecommunication providers' geographically distributed *Pops*, traditionally used for networking and for content placement, are recently starting to be viewed as possible locations for computing operations [42]. The Pops are located closer to the users than cloud services: for example Amazon currently augments its 18 regional datacenters with Pops placed in 56 cities across 24 countries [43]. The Pops contain high-performance server-grade hardware and are well connected and well managed.

- *On-premise server-grade computing points ("cloudlets")*: Despite the rising popularity of cloud computing services, on-premise computing remains an essential feature of many organizations. Ranging from stand-alone server racks to on-site datacenters, local server computing infrastructures offer high-performance computing options in relative proximity to the users, for example, in-building, or on the same campus [36]. The on-premise computing points are connected via wired networks but may not be as reliable and as well-managed as cloud or PoP computing points.

- *Wired gateways*: Gateway-based architectures are becoming a de-facto standard for pervasive IoT applications, including smart homes and smart factories [4, 44, 45]. The plugged-in stationary gateways can perform substantial amounts of processing for the IoT devices. The gateways are, however, usually *consumer-grade* devices and are restricted in computing capabilities compared to the previously listed server-grade computing options. They are usually connected via traditional Ethernet networks.

- *Mobile gateways*: Phones, tablets, and other mobile consumer electronic devices, connected to the wider networks via cellular or Wi-Fi connections, serve as gateways for many types of IoT devices, including wearables and toys [4, 46, 47]. For mobile gateways, energy consumption is a priority, and thus they may seek to offload processing to nearby nonmobile devices or to the cloud [1, 48]. Energy efficiency- and form factor-related performance and cost trade-offs are usually built into these devices at design time: for example they are likely to include computing cores that are optimized for energy efficiency, rather than performance (the so-called heterogeneous multicore designs) [49]. They are also usually relatively expensive in proportion to their computing performance characteristics, as they are optimized for user-friendly features and form factor, rather than raw computing capabilities. Additionally, as such mobile gateways are wirelessly connected to the wider area networks, and their connections may be interrupted and unreliable.

- *IoT nodes ("things")*: The IoT endpoints of the cloud-to-things continuum, usually connected via Bluetooth or Wi-Fi, include embedded microprocessors. These microprocessors are usually computationally restricted [50]; the operation of these nodes is usually geared toward achieving ultra-low energy consumption. Individual IoT devices tend to be inexpensive, as many individual devices need to be deployed for functional large-scale IoT applications. The devices' connectivity may be unreliable, and, in order to save energy, they may maintain only intermittent contact with the wired or mobile gateways.

Additional computing options may be available in emerging distributed fog computing deployments and in special circumstances and specific applications. In emerging smart city architectures, for example street corner servers could be available to support user needs [6, 31, 33] – the equipment in these cases is likely to be similar to on-premise server-grade computing points. In emergency response scenarios and in military applications, vehicles equipped with dedicated server-grade hardware can also be available.

4.4 A Case Study: Distributed Data Processing

4.4.1 A Temperature Sensor Testbed

To experiment with different fog computing scenarios, we designed and developed a custom fog computing testbed, where data analytics applications can be partially computed on local nodes, and partially on commercial cloud services. The photo of the developed testbed, which we presented in a demonstration session of ACM SenSys'18 [51] is shown in Figure 4.2. An annotated video of an operational testbed is available online [52].

The testbed uses three nodes from the [53] as local computing devices. The Raspberry Pis emulate both the sensor nodes and the computing gateways. For capturing environmental parameters, we outfitted the Raspberry Pis with Sense HAT add-on boards that measure temperature, humidity, and air pressure.

Cloud computing elements are implemented via Amazon Lambda [29], a computing service, and DynamoDB, a data storage service [54]. All computing operations are implemented in Python. The sensor nodes communicate with the Raspberry Pi computing gateways over Bluetooth. The computing gateways reach cloud services over Wi-Fi via standard HTTP request/response mechanisms.

We examine scenarios where local IoT devices $S = \{s_1, s_2, \dots\}$ and remote cloud services carry out data sensing, collection, and analytics. We select a subset of the IoT devices to act as computing gateways. Each computing gateway j performs a computing operation h_j over the data Ψ_i received from each sensor node i connected to the gateway and obtains the result $\Psi'_j = h_j(\cup_{s_i \in S_j} \Psi_i)$. Here, S_j denotes the set of sensor nodes that send their sensed data to gateway j. The

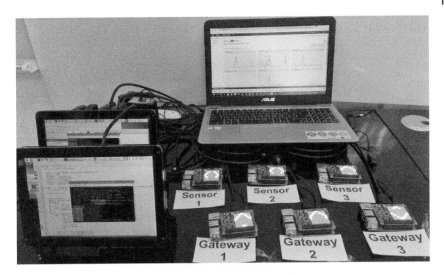

Figure 4.2 A fog computing testbed with local sensors and local computing gateways processing data in collaboration with Amazon cloud services. Source: Chang et al. 2017 [51]. Reproduced with permission of ACM publication.

functional operations h_j can be different for the different gateways j. We focus on the trade-offs that arise between latency (as both computing and communications can be time-consuming), quality (as different decompositions lead to different results), and costs (as cloud services are charged for data transmission, storage, and computing). We instrumented our testbed to measure all related parameters.

We chose a linear regression as a demonstrative case study due to its widespread use and well-understood mathematical properties. In the decomposition method, we adapt from [55] for fog computing settings, the linear regression is solved jointly by the computing gateways and the cloud services, with the cloud service operations calculating the final result based on partial results calculated by the computing gateways.

We consider both fog- and cloud-based linear regression; in the next section, we quantify conditions under which one would prefer to use either of these two methods. We calculate the linear regression using the data collected by the sensor nodes' Sense HAT add-on boards. As shown in Figure 4.3, all six Raspberry Pis collect a timeseries of temperature data, represented by the vector $\mathbf{y}$, as well as a corresponding timeseries of the time of the day, humidity, and air pressure, denoted by the matrix $\mathbf{X}$. Thus, $\Phi = \{\mathbf{X}, \mathbf{y}\}$. The goal of each regression algorithm is to infer a linear predictor β that minimizes the squared error $|\mathbf{y} - \beta\mathbf{X}|^2$.

In the cloud-based regression algorithm, we simply concatenate the data collected by each sensor at the cloud. The cloud can then solve for the optimal

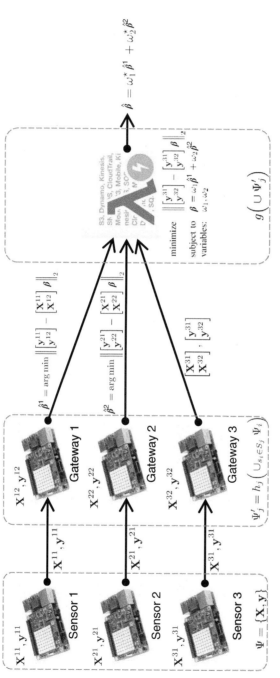

Figure 4.3 Decomposing data analytics between the computing gateways and the cloud services in fog networks: network diagram for the linear regression case study. Source: Chang et al. 2017 [51]. Reproduced with permission of ACM publication.

β by running standard gradient descent algorithms to minimize the squared error. In the fog-based regression algorithm, Ψ_1 and Ψ_2 on Gateways 1 and 2 run the gradient decent algorithm to calculate features $\hat{\beta}^1$ and $\hat{\beta}^2$, while Ψ_3 on Gateway 3 aggregates its data without processing it. The cloud services combine the features received from Gateways 1 and 2 that minimize the least-squared error on the data received from Gateway 3. The combined $\hat{\beta} = \omega_1^\star \hat{\beta}^1 + \omega_2^\star \hat{\beta}^2$ is the final data analytics result.

4.4.2 Latency, Cost, and Risk

Depending on its preferences, a business or user may adapt either the cloud or fog regression setting; formally, either setting may maximize its utility. In this and the following Sections 1.4.2 and 1.4.3, we quantify the performance of both settings and how they compare with each other under different circumstances.

We consider three types of performance: latency, cost, and prediction risk (error). Intuitively, given a set of data to be analyzed, the fog setting reduces both latency and cost by finishing part of the regression locally. On the other hand, the cloud setting not only requires more computing and processing time but also incurs less prediction risk as it can access all the data when doing the regression. Particularly fast cloud servers may also incur less computing latency than the fog setting, as they are more powerful than fog devices. If a user is more sensitive to latency and cost, and cares less about accuracy, the fog setting will thus likely be a good choice. Otherwise, the cloud setting may be preferred.

Users' preferences for the fog or cloud setting also depend on the amount of data collected. In contrast to latency and cost, which have a linear relationship with the size of the data collected, the prediction risk of either type of linear regression generally decreases sublinearly with respect to the number of observations made. As more data are collected, both settings realize similar prediction errors, and the performance trade-off is therefore dominated by latency and cost. In that case, the fog setting usually outperforms the cloud setting with lower latency and cost. We quantify these relationships for our temperature sensor testbed.

In our analysis, we assume that the number of samples collected is a hyper-parameter determined by the user. For instance, the user may determine the size of a fixed time window for collecting samples, or the frequency of collecting samples within a given time window of interest. We suppose that the devices start processing or uploading data only after all the data has been collected, which is generally done in practice since more frequent transmissions require more energy.

For both settings, we assume that each of the six devices collects n data entries. We model latency, cost, and risk as follows:

- *Latency*: The latency is defined as the length of the time interval that starts when all gateways finish collecting data and ends at the moment the final regression results from the cloud are delivered. We omit the time to deliver the intermediate (in the fog setting) and final values of β as well as packet headers. For the cloud setting, the latency is simply the Wi-Fi transmission time plus the computing processing time of Amazon Lambda. We write it in an equation as follows:

$$T_{\text{cloud}}(n) = T_{\text{wifi}}^{(\text{cloud})}(n) + T_{\text{lambda}}^{(\text{cloud})}(n) \tag{4.1}$$

Note that both the computing and transmission latencies are functions of n, the number of observations. For the fog setting, the Wi-Fi transmission from the third node and local data processing at the first two nodes can be done in parallel, but Amazon Lambda can start running on the cloud only after both are done. The latency is therefore given by:

$$T_{\text{fog}}(n) = \max\{T_{\text{wifi}}^{(\text{fog})}(n) + T_{\text{pi}}^{(\text{fog})}(n)\} + T_{\text{lambda}}^{(\text{fog})}(n) \tag{4.2}$$

Suppose that the link between the fog nodes and the cloud is fully multiplexed, i.e. any single node can take up the full link capacity. Then, the Wi-Fi latency is an affine function of the total number of samples, from all nodes, sent over the network: $2n$ for the fog setting, and $6n$ for the cloud setting. We also model the PI runtime and Lambda runtime with affine functions since both the local and remote gradient descent algorithms have $O(n)$ complexity.

- *Cost*: We assume all local resources (e.g. PIs, Wi-Fi) are free, i.e. we can use them with zero marginal cost. We also omit the cost of storing data on DynamoDB, as this storage is not strictly necessary to run the regression algorithms. We therefore only consider the cloud computing bill charged by Amazon Lambda. In practice, Lambda uses a pay-as-you-go pricing scheme [29], which implies that its cost is proportional to the computing time. Given some marginal price p_c, we can model the cost for each setting as

$$C_{\text{lambda}}^{(\text{cloud})}(n) = p_c T_{\text{lambda}}^{(\text{cloud})}(n) \tag{4.3}$$

$$C_{\text{lambda}}^{(\text{fog})}(n) = p_c T_{\text{lambda}}^{(\text{fog})}(n) \tag{4.4}$$

AWS sets different p_c for different types of cloud instances. Instances with better configurations (e.g. more powerful CPUs) are charged higher prices. Thus, p_c may also impact T_{lambda} since better machines usually, but not always, require less computing time. For example, increasing the clock rate does not help much for an IO bound task. In this analysis, we will take p_c as a hyper-parameter predetermined by the user based on a given AWS configuration. We will see later that our model can abstract away the exact value of p_c.

- *Risk*: We quantify the prediction risk by viewing $X \in \mathbb{R}^d$, $Y \in \mathbb{R}$ as random variables. Our linear regression problem is equivalent to estimating the conditional expectation $m(x) = \mathbb{E}[Y|X = x]$. We assume that $\text{Var}(Y|X = x)$

is bounded for all x and that all random variables are bounded (note that in our case study, they represent bounded physical quantities). We then use Theorem 11.3 in [56] to characterize the least-square regression solution $\hat{\beta}$:

$$\mathbb{E} \int |\hat{\beta}^T x - m(x)|^2 dP(x) \leq 8 \inf_\beta \int |\beta^T x - m(x)|^2 dP(x) + \frac{\gamma(\log(N) + 1)}{N} \tag{4.5}$$

where γ is some constant, and N is the total number of observations x. We assume the real regression function $m(x)$ is a linear function of x, and thus that the infimum term equals zero. This bound provides a metric for the prediction risk of the cloud setting. For the fog setting, however, the risk analysis is more difficult. Thus, for simplification, we use the same form of the bound (4.5). Then for some linear function $R_{ols}^{(cloud)}(\cdot)$ and $R_{ols}^{(fog)}(\cdot)$, we have

$$R_{cloud}(n) = R_{ols}^{(cloud)} \left(\frac{\log(6n) + 1}{6n} \right) \tag{4.6}$$

$$R_{fog}(n) = R_{ols}^{(fog)} \left(\frac{\log(6n) + 1}{6n} \right) \tag{4.7}$$

We have now introduced five affine functions for the cloud and fog latency and two linear functions for the prediction risk. Table 4.1 shows the values of the $5 \times 2 + 2 \times 1 = 12$ coefficients estimated for these functions from 8 days and 206 measurement runs of experiment data. The intercept terms of the latency functions should be interpreted as the booting up time, which must be nonnegative. We note that, as we would expect, the computing latency of the Raspberry Pis has both a higher intercept and slope compared to the computing latency of Amazon Lambda.

Figure 4.4 shows our fitted affine function compared to the empirical data of the Wi-Fi transmission time versus the total number of data observations sent

Table 4.1 Fitted values.

Function	Intercept	Slope
$T_{wifi}^{(fog)}$	1.062 13	0.000 60
$T_{wifi}^{(cloud)}$	1.062 13	0.001 80
$T_{pi}^{(fog)}$	4.036 22	0.010 61
$T_{lambda}^{(fog)}$	0.214 25	0.000 43
$T_{lambda}^{(cloud)}$	0.563 83	0.013 85
$R_{ols}^{(fog)}$	0.000 00	2640.17
$R_{ols}^{(cloud)}$	0.000 00	1978.99

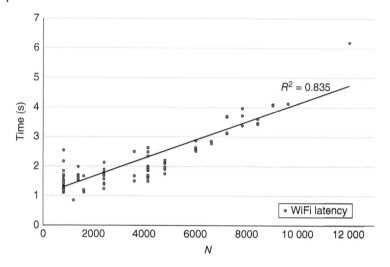

Figure 4.4 Wi-Fi transmission latency versus the total number of entries sent through the gateways. $N = 2n$ for the fog setting, $N = 6n$ for the cloud setting.

through the gateways. Here, we take N to be $2n$ and $6n$, respectively for the fog and cloud settings.

4.4.3   System Trade-off: Fog or Cloud

In order to determine when the fog and cloud settings yield higher user utility, we first formulate users' disutility function for running the linear regression. We model the disutility of users as a weighted linear function of the realized latency, cost, and risk shown in Eq. (4.8); we use a linear model as they are frequently used in human behavior models, but our framework extends to general disutility formulations. The linear model offers the additional advantage of being easily interpretable, as the ratio of the coefficients weighting each term can be interpreted as the marginal rate of substitution between these performance factors.

$$F(n) = C(n) + c_T T(n) + c_R R(n) \tag{4.8}$$

where $C(n)$ denotes the cost, $T(n)$ the total latency, and $R(n)$ the prediction risk; of either the fog or cloud setting. Here, c_T measures users' value of time and c_R measures users' aversion for prediction risks. We set $c_T = \tau p_c$, $c_R = \rho p_c$. Note that since $C(n) = p_c T_{\text{lambda}}(n)$, the effect of p_c can be eliminated, and we may equivalently analyze the normalized disutility:

$$G(n) = T_{\text{lambda}}(n) + \tau T(n) + \rho R(n) \tag{4.9}$$

We then substitute Eqs. (4.1)–(4.7), as well as the estimated coefficient values in Table 4.1, to obtain users' disutilities for the fog and cloud settings, as functions of n:

$$G_{\text{cloud}}(n) = (0.5638 + 0.0138n) + \tau(1.6259 + 0.0156n) + \rho\left(1978.9\frac{\log(6n) + 1}{6n}\right)$$

$$(4.10)$$

$$G_{\text{fog}}(n) = (0.2142 + 0.0004n) + \tau(4.2504 + 0.0110n) + \rho\left(2640.1\frac{\log(6n) + 1}{6n}\right)$$

$$(4.11)$$

Figure 4.5 plots $G(n)$ for both settings with $\tau = 1$ and $\rho = 100$. The curves intersect at around $n_* = 2600$, and the fog system has lower disutility when $n > n_*$. Thus, as we collect more and more measurements, users would prefer the fog over the cloud setting. In such scenarios, the prediction risk is lower for both settings due to the number of measurements, so users' disutility is dominated by the latency and cost, which are both lower in the fog setting.

To further analyze the implications of our results, we define the *equilibrium point* as the number of observations n for which the fog and cloud settings

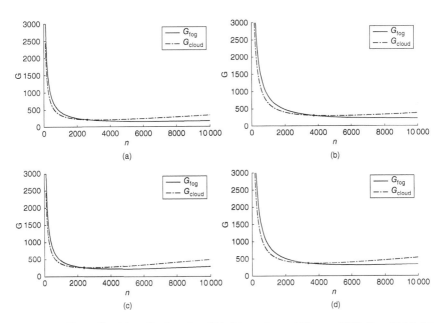

Figure 4.5 Normalized disutility function $G(n)$. (a) $\tau = 1, \rho = 100, n_* = 2617, G_* = 213.7$; (b) $\tau = 1, \rho = 200, n_* = 3733, G_* = 307.0$; (c) $\tau = 2, \rho = 100, n_* = 2379, G_* = 257.7$; (d) $\tau = 2, \rho = 200, n_* = 3370, G_* = 369.5$. n_* is rounded to the nearest integer.

have the same levels of disutility. In other words, it is the point where the two curves in Figure 4.5 intersect. We can calculate the equilibrium by solving $n_* = \arg(G_{fog}(n) = G_{cloud}(n))$, which upon substituting Eqs. (4.10) and (4.11) allows us to find the equivalent condition

$$\rho = \frac{(0.00461\tau + 0.01342)n_*^2 - (2.62451\tau - 0.34958)n_*}{110.197(\log(6n_*) + 1)} \tag{4.12}$$

For the coefficients in Table 4.1, $G_{fog}(n) - G_{cloud}(n)$ is monotonically decreasing when $n \geq 1$. Thus, Eq. (4.12) has at most one root, n_*. The fog setting is then preferred when $n > n_*$. In the case when n_* does not exist, the user will always choose the cloud setting.

Figure 4.6 plots the sensitivity of n_* with respect to different user disutility parameters ρ and τ. Each curve represents a set of (n, ρ) pairs where the fog and cloud settings yield the same disutility for the user. For values of n above the curve $(n > n_*)$, the fog setting is preferred. We note that as the parameter τ increases, the curve shifts gradually downward, creating more space for the fog setting to win the trade-off. This result is not surprising, as τ parameterizes the importance of latency: as the user's disutility becomes more latency-sensitive, the user would tend to prefer the fog setting, which has lower latency. In contrast, as ρ increases, n_* also increases and users would tend to prefer the cloud setting, since that yields lower prediction risk.

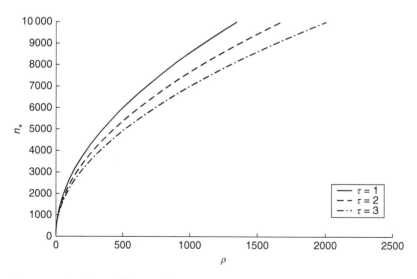

Figure 4.6 Sensitivity of the equilibrium with respect to ρ and τ.

4.5   Future Research Directions

Economic aspects of fog computing can include anything from designing cost-aware fog applications, to understanding their cost trade-offs, as illustrated in our case study previously, to proposing new pricing plans for fog resources. While these research questions are not directly related to each other, the common thread through these topics is that they involve both economic and technological questions related to fog computing. Building a comprehensive theoretical framework for fogonomics will thus require collaboration from researchers in many fields, including distributed computing, network economics, wireless networks, human–computer interaction, machine learning, and multimedia applications. In this section, we outline some possible directions for the research community to explore.

Architecting cost-aware applications: Fog-based applications require both computing and resources provided by multiple devices with heterogeneous capabilities. The distribution of application functionalities among these devices, however, can influence the realized cost and performance of these applications. Putting too much functionality at low-power edge devices, for instance, would lead to significant energy drain and possibly computation latency. Our case study of distributed data processing takes some initial steps in this direction by proposing a model to evaluate the latency, error, and cost trade-offs when data processing is distributed between edge and cloud devices.

Compute and network resource pricing: As discussed in prior sections, current pricing plans for computing and network resources are likely suboptimal for fog applications. Many fog devices may also be owned by individuals, and thus not subject to the economies of scale that can occur in the cloud. Dedicated fog providers, such as an ISP fog data service or a cloud provider's fog-specific compute service, would likely wish to introduce new types of pricing that account for factors like the heterogeneity of the networking or computing resources required and the stringency of the application's performance requirements. Sophisticated pricing plans for fog applications, which are often automated, may also be more widely accepted than sophisticated pricing for traditional applications, which generally assume direct human involvement and thus require relative simplicity. Real-time auction ideas, for instance, have been widely studied for wireless spectrum allocation but have never been widely deployed, in large part due to their complexity for users. Determining the optimal pricing plans must take into account the cost-performance trade-offs experienced by the application, as well as the cost of providing data and computing resources on the part of the provider.

Mechanisms for device collaboration: The defining feature of fog computing is its integration of resources from multiple heterogeneous devices. These devices must then collaborate in order to provide fog applications with the resources that they need. While fog applications could enforce collaboration between their desired resources, negotiations between the application and each individual device would likely be burdensome and result in suboptimal solutions. Involving all devices in deciding how the application or user would pay for the resources required might lead to a pricing solution that better reflects the dependencies between these resources. Indeed, devices might even agree to exchange resources without explicit monetary pricing: for instance, applications running on one user's smartphone could temporarily make use of another user's machine, with the understanding that the trade would be reversed later.

New revenue models for fog computing: Fog computing is projected to generate hundreds of billions of dollars in revenue for fog providers [57]. However, it is unclear how this revenue will be generated. Businesses looking to deploy fog computing applications will need to find new revenue models, aided by research into the value provided by fog architectures and the new technologies that they will require. A software platform that automatically decomposes data processing algorithms to run across multiple devices, for instance, might represent a new source of revenue [58]. Research on fog business models and gaps in existing fog technologies will be needed to identify these opportunities.

4.6 Conclusion

In this work, we take the first steps toward outlining the research challenges of fogonomics. We first contextualize the problem by describing current fog architectures and pricing practices, as well as outlining some of the trade-offs that arise between performance and pricing under these current practices. We then present a case study in which we build a testbed of distributed temperature sensors and empirically evaluate the trade-offs between the latency of data processing, computation time, and monetary cost that are realized when the temperature data are processed at local or at cloud devices. In doing so, we create a preliminary theoretical framework for evaluating these trade-offs in more general data processing systems. We finally outline four promising research directions for fogonomics, which build on our case study and go beyond it. Our hope is that this work can help create a community of researchers who will fill the gap between a technological and economic understanding of fog computing.

Acknowledgments

This work was partially supported by NSF CNS-1751075.

References

1 Satyanarayanan, M. (2015). A brief history of cloud offload: a personal journey from odyssey through cyber foraging to cloudlets. *GetMobile: Mobile Computing and Communications* 18 (4): 1923.

2 Bort, J. (2016). The next multibillion-dollar tech market was quietly born this year, says A-list VC Peter Levine. Business Insider. http://www .businessinsider.com/edge-computing-is-the-next-multi-billion-tech-market-2016-12.

3 Levine, P. (2016). Return to the edge and the end of cloud computing. a16z .com/2016/12/16/the-end-of-cloud-computing/.

4 Chiang, M. and Zhang, T. (2016). Fog and IoT: an overview of research opportunities. *IEEE Internet of Things Journal* 3 (6): 854864. https://doi.org/10.1109/JIOT.2016.2584538.

5 Open Fog Consortium (2017). OpenFog. https://www.openfogconsortium .org/.

6 OpenFog Consortium (2017). OpenFog reference architecture. https://www .openfogconsortium.org/ra/.

7 Cisco (2015). Fog computing and the Internet of Things: extend the cloud to where the things are. http://www.cisco.com/c/dam/en_us/solutions/ trends/iot/docs/computing-overview.pdf.

8 Sen, S., Joe-Wong, C., Ha, S., and Chiang, M. (2012). Incentivizing time-shifting of data: a survey of time-dependent pricing for internet access. *IEEE Communications Magazine* 50 (11): 91–99.

9 Sen, S., Joe-Wong, C., Ha, S., and Chiang, M. (2013). Smart data pricing (SDP): economic solutions to network congestion. Recent Advances in Networking, ACM SIGCOMM, pp. 221274.

10 Sen, S., Joe-Wong, C., Ha, S., and Chiang, M. (2013). A survey of smart data pricing: past proposals, current plans, and future trends. *ACM Computing Surveys* 46 (2): 15.

11 Zheng, L., Joe-Wong, C., Tan, C.W. et al. (2017). Customized data plans for mobile users: feasibility and benefits of data trading. *IEEE Journal on Selected Areas in Communications* 35 (4): 949–963.

12 Joe-Wong, C., Ha, S., and Chiang, M. (2015). Sponsoring mobile data: an economic analysis of the impact on users and content providers. Proceedings of IEEE INFOCOM.

13 Andrews, M., Jin, Y., and Reiman, M.I. (2016). A truthful pricing mechanism for sponsored content in wireless networks. Proceedings of IEEE INFOCOM.

14 Harishankar, M., Srinivasan, N., Joe-Wong, C., and Tague, P (2018). To accept or not to accept: the question of supplemental discount offers in mobile data plans. IEEE INFOCOM 2018-IEEE Conference on Computer Communications, pp. 2609–2617. IEEE.

15 Zheng, L., Joe-Wong, C., Tan, C.W. et al. (2014). How to bid the cloud. Proceedings of ACM SIGCOMM'14, August 2014.

16 Zheng, L., Joe-Wong, C., Brinton, C.G. et al. (2016). On the viability of a cloud virtual service provider. Proceedings of ACM SIGMETRICS'16, July 2016.

17 Huang, Z., Weinberg, M., Zheng, L. et al. (2017). Discovering valuations and enforcing truthfulness in a deadline-aware scheduler. Proceedings of IEEE INFOCOM'17, May 2017.

18 Aryafar, E., Keshavarz-Haddad, A., Wang, M., and Chiang, M. (2013). RAT selection games in HetNets. Proceedings of IEEE INFOCOM.

19 Cheung, M.H., Hou, F., Huang, J., and Southwell, R. (2017). Congestion-aware distributed network selection for integrated cellular and Wi-Fi networks. *IEEE Journal on Selected Areas in Communications* 35 (6): 1269–1281.

20 Madhavan, M., Ganapathy, H., Chetlur, M., and Kalyanaraman, S. (2015). Adapting cellular networks to whitespaces spectrum. *IEEE/ACM Transactions on Networking* 23 (2): 383–397.

21 Bayhan, S., Zheng, L., Chen, J. et al. (2017). Improving cellular capacity with white space offloading. Proceedings of IEEE WiOpt.

22 Luo, Y., Gao, L., and Huang, J. (2016). *Economics of Database-Assisted Spectrum Sharing*. Springer.

23 Joe-Wong, C., Sen, S., and Ha, S. (2015). Offering supplementary network technologies: adoption behavior and offloading benefits. *IEEE/ACM Transactions on Networking* 23 (2): 355–368.

24 Shah-Mansouri, H., Wong, V.W.S., and Huang, J. (2017). An incentive framework for mobile data offloading market under price competition. *IEEE Transactions on Mobile Computing* 16 (11): 2983–2999.

25 Google (2015). Project Fi. https://fi.google.com/.

26 Zheng, L., Chen, J., Joe-Wong, C. et al. (2017). An economic analysis of wireless network infrastructure sharing. Proceedings of IEEE WiOpt.

27 Zheng, L., Joe-Wong, C., Chen, J. et al. (2017). Economic viability of a virtual ISP. Proceedings of IEEE INFOCOM.

28 Microsoft Azure (2017). IoT Hub. https://azure.microsoft.com/en-us/services/iot-hub/ (accessed 29 August 2019).

29 Amazon (2018). AWS Lambda. aws.amazon.com/lambda/ (accessed February 2018).

30 Weinman, J. (2017). The 10 laws of fogonomics. *IEEE Cloud Computing* 4 (6): 8–14.

31 Yannuzzi, M., van Lingen, F., Jain, A. et al. (2017). A new era for cities with fog computing. *IEEE Internet Computing* 21 (2): 54–67. https://doi.org/10.1109/MIC.2017.25.

32 Bonomi, F., Milito, R., Zhu, J., and Addepalli, S. (2012). Fog computing and its role in the Internet of Things. Proceedings of ACM MCC'12, August 2012.

33 Moustafa, H., Gorlatova, M., Byers, C. et al. (2017). OpenFog consortium fog use case scenarios: autonomous driving. www.openfogconsortium.org/new-use-cases/.

34 McMahan, H.B., Moore, E., Ramage, D. et al. (2017). Communication-efficient learning of deep networks from decentralized data. Proceedings of the 20th International Conference on Artificial Intelligence and Statistics (AISTATS). http://arxiv.org/abs/1602.05629.

35 Hu, W., Amos, B., Chen, Z. et al. (2015). The case for offload shaping. ACM HotMobile'15, February 2015.

36 Streiffer, C., Srivastava, A., Orlikowski, V. et al. (2017). ePrivateEye: To the edge and beyond! IEEE/ACM SEC'17, October 2017.

37 Satyanarayanan, M., Simoens, P., Xiao, Y. et al. (2015). Edge analytics in the internet of things. *IEEE Pervasive Computing* 14 (2): 24–31.

38 Gao, Y., Hu, W., Ha, K. et al. (2015). Are Cloudlets Necessary? Technical Report CMU-CS-15-139. CMU School of Computer Science.

39 Wang, X., Chowdhery, A., and Chiang, M. (2016). SkyEyes: Adaptive video streaming from UAVs. Proceedings of ACM HotWireless'16, September 2016.

40 Microsoft (2018). Azure functions.

41 Di Tucci, L., Rabozzi, M., Stornaiuolo, L., and Santambrogio, M.D. (2017). The role of CAD frameworks in heterogeneous FPGA-based cloud systems. 2017 IEEE International Conference on Computer Design (ICCD), November 2017, pp. 423–426. https://doi.org/10.1109/ICCD.2017.74.

42 Amazon (2018). Lambda at edge. docs.aws.amazon.com/lambda/latest/dg/lambda-edge.html (accessed February 2018).

43 Amazon (2018). Amazon cloudfront edge infrastructure. https://aws.amazon.com/cloudfront/details/ (accessed 29 August 2019).

44 Samsung (2018). Smart things. www.smartthings.com/ (Accessed February 2018).

45 Chen, C.H., Lin, M.Y., and Liu, C.C. (2018). Edge computing gateway of the industrial internet of things using multiple collaborative microcontrollers. *IEEE Network* 32 (1): 24–32. https://doi.org/10.1109/MNET.2018.1700146.

46 Zachariah, T., Klugman, N., Campbell, B. et al. (2015). The Internet of Things has a gateway problem. Proceedings of ACM HotMobile'15, February 2015.

47 Islam, N. and Want, R. (2014). Smartphones: past, present, and future. *IEEE Pervasive Computing* 13 (4): 89–92. https://doi.org/10.1109/MPRV.2014.74.

48 Gordon, M.S., Anoushe Jamshidi, D., Mahlke, S. et al. (2012). COMET: Code offload by migrating execution transparently. Proceedings of USENIX OSDI'12, October 2012.

49 Muthukaruppan, T.S., Pathania, A., and Mitra, T. (2014). Price theory based power management for heterogeneous multi-cores. *SIGARCH Computer Architecture News* 42 (1): 161–176. https://doi.org/10.1145/2654822.2541974.

50 Adegbija, T., Rogacs, A., Patel, C., and Gordon-Ross, A. (2018). Microprocessor optimizations for the internet of things: a survey. *IEEE Transactions on Computer-Aided Design of Integrated Circuits and Systems* 37 (1): 7–20. https://doi.org/10.1109/TCAD.2017.2717782.

51 Chang, T., Zheng, L., Gorlatova, M. et al. (2017). Demo: decomposing data analytics in fog networks. Proceedings of ACM Conference on Embedded Networked Sensor Systems (ACM SenSys'17), November 2017.

52 Chang, T.-C., Zheng, L., Gorlatova, M. et al. (2017). Demo video: decomposing data analytics in fog networks. https://youtu.be/CBpCVyFvGXI.

53 Raspberry Pi Foundation (2017). Raspberry Pi. https://www.raspberrypi.org/.

54 Amazon AWS (2017). DynamoDB. https://aws.amazon.com/dynamodb/ (accessed 29 August 2019).

55 Yang, Y. (2003). Regression with multiple candidate models: selecting or mixing? *Statistica Sinica* 13: 783–809.

56 Györfi, L., Krzyżak, A., Kohler, M., and Walk, H. (2002). *A Distribution-Free Theory of Nonparametric Regression*. Springer.

57 451 Research (2017). Fog computing global market will exceed $18 billion by 2022. OpenFog Consortium.

58 Jeong, T., Chung, J., Hong, J.W.-K., and Ha, S. (2017). Towards a distributed computing model for fog. Proceedings of the Fog World Congress.

5

Quantitative and Qualitative Economic Benefits of Fog

Joe Weinman

XFORMA LLC, Flanders, NJ, USA

Fog computing is a rapidly emerging computing paradigm, which represents an evolution of the last wave of computing – cloud, and its benefits [1] – from dozens or hundreds of highly centralized facilities to tens of billions of endpoint devices, edge processing and storage, intermediate layers, and the networks to tie them all together.

To pigeonhole fog computing as just a technology architecture, however, would be a mistake. The reason for its rapid emergence is that it delivers a broad variety of quantitative and qualitative benefits to businesses, consumers, governments, and societies. Some may be exactly characterized, such as cost optimization or backhaul traffic reduction. Others are more situation-dependent, such as revenue generation through user experience enhancement through latency reduction via geographic dispersion.

It would be a mistake to characterize fog computing as an improvement to prior existing solutions. Instead, it expands the toolkit available to enterprise architects and solution builders including product and service developers. With this expanded toolkit come numerous choices, and with those choices, come numerous trade-offs [2]. As one example, placing applications and data close to the edge helps reduce latency and improve reliability for edge and endpoint applications that leverage edge and endpoint capabilities such as processing, sensing, storing, and actuating. On the other hand, placing them in the cloud reduces latency and improves reliability for cloud-resident applications and data.

Consider a video surveillance device used to trigger a home alarm. If the triggering workflow required the cloud, burglars could be long gone if, say, the Internet connection was down. On the other hand, consider a mesh of cooperating micro-services conducting a complex computation on a massive data set. Keeping the ensemble cloud-resident is clearly preferred to dispersing microservices and data globally.

Fog and Fogonomics: Challenges and Practices of Fog Computing, Communication, Networking, Strategy, and Economics, First Edition. Edited by Yang Yang, Jianwei Huang, Tao Zhang, and Joe Weinman.
© 2020 John Wiley & Sons, Inc. Published 2020 by John Wiley & Sons, Inc.

To these trade-offs, one must add real-world constraints, such as the fact that there are relatively few greenfield solutions, and those greenfield solutions will soon become legacy ones. For legacy solutions, it is often uneconomical to perform a "forklift upgrade," i.e. to rip out the old approach and completely replace it. Instead, most real-world solutions are a complex mix of old and new elements, such as legacy systems, high performance computing, custom architectures, virtualized environments, private clouds, colocation, and public clouds, leading to the "hybrid multi-cloud fog," [3] comprising hybrids of private and public, multiple public clouds at the IaaS, PaaS, and/or SaaS layers, and a hybrid of cloud and fog.

In fact, it is even more complex than this because processing can occur in devices and things, the edge, intermediate fog layers, or consolidated cloud facilities. Thus, for example, compute resource utilization can be lower in the fog than in the cloud, but higher in the fog than at edge devices. To understand this, consider one network of sensors that processes only during daylight hours, and another one that only performs processing during nighttime. Roughly speaking, each set of sensors only has 50% utilization. If that processing was brought up one layer, theoretically, a utilization of 100% could be achieved.

5.1 Characteristics of Fog Computing Solutions

To understand benefits and trade-offs, we must look at a variety of characteristics in turn. These include the following:

Strategic Value

- *Information excellence*: How can better processes, asset utilization, and organization structures be achieved?
- *Solution leadership*: How can better smart, digital, connected products be built and services offered?
- *Collective intimacy*: How can we improve customer relationships?
- *Accelerated innovation*: How can we achieve faster, cheaper, better innovation?

Bandwidth, Latency, and Response Time

- *Network latency*: How much time is required for physical propagation of data across a network?
- *Server latency*: How much time is required at the server, based on processing capacity and/or congestion?
- *Total latency*: How long does it take to perform a task, including network and server latency?
- *Data traffic volume*: What is the total data volume cached, transported, bandwidth required, and related data transport costs?

- *Nodes and interconnections*: How complex do network topology and architecture become with a plethora of connected nodes?

Capacity, Utilization, Cost, and Resource Allocation

- *Capacity requirements*: What quantity of resources is required to meet demand?
- *Utilization*: How well used are those resources?
- *Resource cost*: How expensive is it to deliver a used resource?
- *Resource allocation, sharing, and scheduling*: How to best allocate tasks to resources?

Information Value and Service Quality

- *Precision and accuracy*: Does it generate the best possible information and insights?
- *Survivability, availability, and reliability*: How much of the time is the service working, and does it produce the desired results?

Sovereignty, Privacy, Security, Interoperability, and Management

- *Data sovereignty*: Does it comply with regulations concerning cross-border data transport?
- *Privacy and security*: How secure and invulnerable is the architecture?
- *Heterogeneity and interoperability*: How can multiple elements from different vendors and providers work together?
- *Monitoring, orchestration, and management*: How can fog architectures be managed?

5.2 Strategic Value

Before we delve into fine degrees of cost and performance optimization, we first need to examine the strategic value of the fog.

A recent McKinsey analysis shows that the economic impact of the Internet of Things could be over US$10 trillion annually by 2025. This includes in factories, through better operations management and predictive maintenance; in smart cities, through things like traffic control; in human health; in retail, through self-service and better store layouts; in autonomous vehicles and better navigation, such as for delivery companies; in vehicles such as cars and trucks, in homes such as with energy management, and in offices, thanks to better training and worker monitoring [4].

In my book Digital Disciplines [5], I have defined four generic strategies that exploit digital technologies: (i) information excellence, i.e. better processes and asset utilization; (ii) solution leadership, i.e. better products and services; (iii) collective intimacy, i.e. better customer relationships; and (iv) accelerated innovation, i.e. faster, cheaper, and better innovation.

These four generic strategies, or *"digital disciplines,"* create unique customer value, and thereby strengthen competitive advantage, ultimately driving accelerated profitable revenue growth. Fog computing is a key enabler for such strategic growth.

5.2.1 Information Excellence

Information excellence represents the evolution of operational excellence in a digital world. Operational excellence includes processes which are faster, cheaper, higher quality, more repeatable, more standardized or conversely, more flexible, and so forth. It also involves better asset utilization, or more flexible organizations.

In a digital world, operational excellence evolves to information excellence, where digital technologies can be used to replace, fuse, mirror, manage, optimize, or monetize processes; where assets can become better utilized or more flexible; and where new organizational structures such as dynamic networked virtual organizations can be enabled through capabilities such as online markets and matching engines.

Fog technologies have a key role to play in enabling information excellence. An excellent example is the use of irrigation sensors tied to optimization software in agricultural applications such as vineyards to reduce water requirements, while increasing yield and quality. Another is factory or warehouse operations, with robotic cells or materials handling systems tied to ordering systems.

5.2.2 Solution Leadership

Solution leadership represents the evolution of product leadership in a connected digital world. Product leadership in a legacy context involves the very "best" products and services, defined by excellence in design, usability, materials, aesthetics, reliability, and so forth.

In a digital world, product leadership evolves to solution leadership, where standalone products become smart, digital, and connected. Good examples are the thermostat, which used to be a device only connected to electricity and the HVAC (heating, ventilating, and air conditioning); the smoke detector, which used to be only connected to electricity (and a battery), or the watch, which used to be connected to nothing. These products have all evolved to connect across the Internet to an ecosystem of other products and services, such as the fire department, remote control smartphone apps, insurance companies, text messaging services and the like.

5.2.3 Collective Intimacy

Collective intimacy represents the evolution of customer intimacy in a digital world: customer relationships evolve from anonymous transactions to ongoing relationships to customer intimacy, where deep knowledge is acquired regarding customer needs to custom create or co-create solutions.

Finally, customer intimacy evolves to collective intimacy, where data regarding customer attitudes, behaviors, contexts, etc. are processed collectively across all customers to better develop recommendations and solutions for each customer. As one example, Netflix can deduce attitudes and intents from search queries; can monitor specific behaviors such as entertainment menu navigation; and can understand contexts such as the time of day a movie was watched, whether it was at the home or while traveling, and which device it was watched on. This may seem obvious now, but consider the dramatic change from Nielsen ratings, or even the days when DVDs were shipped in the mail so that Netflix knew intent but not behavior. Netflix could not tell for sure whether the movie was even watched, whereas now, through connected devices, it can determine the exact scene where a movie was paused, or other behaviors such as repeated watching, rewinding, and so forth.

5.2.4 Accelerated Innovation

Accelerated innovation is the latest wave of innovation in the innovation process. From the early days of the solitary inventor, to corporate research, to open innovation, digital technologies create new possibilities in innovation, such as innovation networks, idea markets, contests, and hackathons.

Traditionally, innovation has involved understanding current customer use of products and services, and concomitant issues, "pain points," or opportunities. In a disconnected world, such understanding to feed design thinking requires live events such as focus groups, observations of use in practice, or written feedback. However, in a world of smart, connected, digitalized products and service delivery infrastructure, actual use of such products and services can be exactly characterized.

5.3 Bandwidth, Latency, and Response Time

Data generated at the edge may be used for closed-loop local control at the edge, such as the way that a standalone thermostat determines the temperature and adjusts the heating and air conditioning. It also may be sent over a LAN or WAN to a control node or data collection and storage facility. Edge or fog processing can be used to reduce the amount of data sent to a central location such as a data center or the cloud, or even eliminate the need to send data to that location. As a result, the total amount of network data traffic can be reduced; the product of the amount of traffic and distance it is sent can be reduced, thus leading to reduced capital expenditures for network builds and/or reduced data transport costs for pay-per-use services; and the volume of data stored can be reduced.

In addition, many types of applications are time-sensitive. Sometimes these are human-oriented, such as delivering a web page or search query result, or autocompleting a query or autocorrecting a typo. Increasingly, they are machine-oriented, such as coordinating robots in a flexible manufacturing cell or autonomous communicating vehicles on a highway or soon, in airspace. Humans generally operate in ranges of tens to hundreds of milliseconds. While some machines operate in those ranges, such as the time needed to brake a car, some digital entities operate in the microsecond range, as with algorithmic trading programs. As the world becomes more interconnected, the end-to-end response time performance can smoothly support human or machine requirements, or can cause customer or user experience to suffer or create safety issues, as with manufacturing cell robot coordination or autonomous vehicle braking.

In general, we can partition total response time into (i) endpoint latency, i.e. the speed at which the user device or thing conducts necessary processing, such as rendering graphics, processing mousedowns or keystrokes, compressing or decompressing a video stream; (ii) network latency, i.e. the time it takes for a complete message to get from source to destination or a transaction query to traverse the network and the response to be sent back; and (iii) server latency, i.e. processing time on the remote end.

Endpoint latency can be reduced by using higher powered processing, caching frequently used objects, preprocessing images such as prerendering for a given size, or using hardware acceleration, and is not the focus of this chapter. Network latency can be partly reduced through route optimization between the endpoint and the service node, lighter-weight protocols, optimal selection of a service node, reductions in congestion or packet loss, or even reduction in physical propagation delays. For example, the speed of propagation in fiber is actually slower than in copper because it is typically 2/3 C, the speed of light, whereas electrical signal propagation is essentially C, but free-space optics can match that. However, the main strategy for reducing network latency is to reduce the distance between the endpoint and the service node. Finally, the server-side latency can be reduced through strategies such as application tuning and parallelism [6, 7].

All these have implications regarding fog architecture and economics. Briefly, centralized, consolidated cloud servers create long distance round trips for transactions from highly distributed endpoints. In contrast, a dispersed fog is more likely to have a service node "nearby" a requesting endpoint. However, it is important to remember that dispersed fog nodes reduce latency to dispersed endpoints but *increase* latency to and from cloud applications, services, and data. If the data are at the edge and so is the application driving it, say, autonomous vehicle control, then the fog may well be the best place to host the application. *But* if the data are in the cloud and component microservices are running in the cloud and the processing requires a massive number of parallel servers, then the cloud is the best place to run the application.

We will look at the economics of data traffic volume, network latency, server latency, and total response time.

5.3.1 Network Latency

There are numerous approaches to reducing network latency N, i.e. the time delay induced by the network between, say, an endpoint and a cloud server. One is to select the shortest route. Another is to select a path with little congestion or packet loss. Another is to select the fastest transmission medium and relevant protocol.

Assuming all these and related techniques have been utilized, the best thing to do is to reduce the distance between the endpoint and the server. However, endpoints can be anywhere, so it is not possible to just move a single server into an optimal position, but rather the only feasible approach is to use a strategy of dispersing the service nodes comprising fog infrastructure to reduce average (or worst-case) latency.

The trade-off, then, is between improving customer experience by reducing latency and increasing the investment in service nodes. Unfortunately, the economics of this trade-off become increasingly unfavorable. The reason is that to reduce latency by 50% requires four times as many service nodes on a plane and "roughly" that on the surface of a sphere.

The calculations on a plane are straightforward. The area a covered by a circle is proportional to the square of the radius, since $a = \pi r^2$. With n such circles, the total area covered is $A = na = n\pi r^2$. With a more in-depth analysis, we would also consider η (eta), the packing density, i.e. degree of overlap of the circles. Since latency is proportional to the distance to the center (i.e. the service node), worst-case latency is represented by the radius of the circle. Therefore, for a given area A, there is an inverse relationship between the latency and service nodes, namely, since A is fixed and π is a constant, we have $r \propto 1/\sqrt{n}$. This means that if a single node latency is N, as we increase fog service nodes n, the latency essentially follows $N/\sqrt{n}$. This means that, say, if n increases from 1 to 4, r goes from 1 to 1/2. Since worst-case latency on the surface of the earth from one point to an antipolar point is roughly 160 ms, the first few nodes, or even hundreds of nodes can provide major reductions in latency, perhaps down to a few or tens of milliseconds. However, at some point, even millions or billions of additional nodes will only reduce latency by a few microseconds.

One final point: Because the earth is not a plane, the aforementioned analysis is not exactly correct. Abstractly, we can consider the earth to be a sphere. The first node only requires one additional node to reduce latency by 1/2. When there are a large number of nodes, four times as many nodes are required, as per the planar formula. There is a trigonometric adjustment required of

$(1 - \cos(2\beta))/(1 - \cos(\beta))$, where β is the angular radius defining the separation between nodes, which depends on the radius of the sphere and the number of nodes. Even laying out the nodes to be equidistant and maximally packed is a challenge, known as the Tammes problem. It can be approximated by dividing the total surface area of the sphere by the number of nodes to get the surface area of a spherical cap and working back to its angular equivalent radius. It will be appreciated that as $\beta \to 0$, no adjustment is necessary from the base inverse root formula.

5.3.2 Server Latency

Server latency is a complex function of parameters such as I/O bandwidth, database connections, CPU congestion, virtual machine (VM) overhead, and the like. One very simple analysis, however, and benefit of the cloud is availability of elastic resources. If the total time for a computation is divided into a serial portion S and a parallel portion P, then the total time on p processors for the computation is $S + P/p$. If the computation is "embarrassingly parallel," $S = 0$, so the time for the computation can be reduced to the extent desired. In a private environment, this would require the acquisition of many processors and the concomitant capital expenditure. In the case of an infinitely elastic pay-per-use cloud with infinitely fine-grained billing increments, the marginal cost to accelerate a computation to run infinitely quickly is zero. This is an advantage of the public cloud.

5.3.3 Balancing Consolidation and Dispersion to Minimize Total Latency

If we consolidate fixed endpoint time F, the latency due to the serial portion of the computation S, the network latency $N/\sqrt{n}$, and the parallelized portion P/p of server latency, we can consider the total latency $T = F + S + N/\sqrt{n} + P/p$. In the real world, there are other subtleties, such as the number of threads and objects transmitted, etc., but this is a good general equation. Now suppose that we have a given number of processors, Q. We could put one processor in each of Q maximally distributed fog locations, or Q processors in only one consolidated location, or Q/n processors in each of n locations. What is the best way to reduce response time? It turns out that $T = F + S + N/\sqrt{n} + P/(Q/n)$ is minimized when $n = \sqrt[3]{[QN/(2P)]^2}$ [7]. Again, this is an approximation which is increasingly correct as the sphere increasingly behaves like a plane.

Of course, more complex solutions are likely to be found in the real world: elastic cloud resources, parallel and unitary processing in intermediate nodes in the fog, and a variety of endpoints. However, this kind of analysis shows that optimal performance requires a balance of dispersion for latency reduction and consolidation for parallelism, leading to a balance of cloud and fog.

5.3.4 Data Traffic Volume

Most people who move to another city first use up items such as food in the refrigerator, rather than transporting that food. They houseclean by throwing away, recycling, or donating unnecessary items. Finally, items such as furniture are disassembled if possible to take up less space in the van.

These same three strategies – use, discard, or compress – are also relevant to fog computing. Use of data at the edge means that there is no need to transport it to higher layers and reduces response time for real-time tasks. Even if it does need to be transported to a higher layer, discarding unnecessary data, i.e. cleansing or deduplication reduces data traffic volume.

Such techniques can be used to reduce backhaul traffic volume when moving data up to the cloud or higher layers. Other techniques can be used when bringing data from origin servers or the cloud out to the edge. Caching can exist at multiple layers, involving wireless and wireline networks. For example, Netflix is well known for using its Open Connect appliance to store entire copies of the Netflix streaming library close to the edge, in colocation and interconnection facilities [8]. This reduces backbone traffic and reduces latency to the end user device. Wireless base stations can also double as caching or task execution resources, bringing such content only one (potentially extremely low latency) hop away from the user.

One step further is to cache content in user devices. This content can be cached in the user's own device, as when a song or movie is downloaded into the device. Interestingly, it can also be cached in another user's nearby device. Such peer-to-peer or device-to-device strategies and architectures can be effective and will be even more effective when ultra-reliable, low-latency, high-bandwidth wireless networks such as are enabled by 5G are widely deployed.

Benefits from compression, deduplication, and caching can be substantial. For example, the Nyquist–Shannon sampling theorem essentially says that many analog signals can be reliably converted to digital by sampling at twice the maximum frequency of the signal. But often samples are taken much more frequently. For example, an IoT sensor on a jet engine, say, for oil pressure, may take scores or thousands of samples each second, even if that oil pressure does not change meaningfully over several hours of flight time. Discarding unneeded, duplicate, or spurious data can dramatically reduce the amount of data transported. In addition to all of these, the data that do need to be transported can be compressed at the edge first, further reducing transport costs.

A broader generic strategy can also be applied, namely collecting the data but transporting only the information or insight. At its extreme, one can think of the discovery of the Higgs boson by the ATLAS and CMS detectors at CERN's Large Hadron Collider. Although these experiments generate exabytes of data, the fundamental conclusion as to whether the Higgs boson exists is only 1 bit – yes or no. Less extreme examples include store retail data, which might be used to conclude that brown wool sweaters are not selling and should be discontinued. Such an approach can be pursued by running a distributed query on the partitioned data, i.e. running a query in each store and only aggregating the results, rather than bringing all the retail sales data to a single central data warehouse.

This also leads to the insight that such processing can be performed at multiple layers: endpoint devices, the edge layer, and higher layers in the fog, before reaching a central cloud. The degree of volume compression depends on various components: volume of spurious data, volume of duplicate data, compressibility, and so forth. In a multi-tier network [5, 9], the degree of data volume-route miles depends also on the distances between nodes, and thus their location, and ultimately the total cost depends on the costs of various links between this multi-tier hierarchy. Suppose that we can cover an area with either c more consolidated cloud nodes or f more dispersed fog nodes, i.e. $f > c$. As we saw earlier, the distance to be traveled is proportional to the inverse square root of the number of nodes. As a result, all other things being equal, the reduction in the product of transport capacity and route miles would decrease by $1/\sqrt{c} - 1/\sqrt{f}$. Moving from one node to 16 nodes, for example, would mean 3/4 less physical network infrastructure, conceptually.

Moving to actual hard dollar costs can be challenging, for a variety of reasons. One is that the cost to transport a packet can arguably be zero (the marginal cost for carriage in an uncongested network) but can also be nonzero (in a congested network). Pricing for bandwidth increases sublinearly, so quadrupling data volume does not necessarily quadruple network costs. Also, the direction can matter: some cloud providers do not charge to import data to their cloud, but do charge for egress.

5.3.5 Nodes and Interconnections

The decrease in latency and traffic volume comes about at a potential increase in the number of deployed nodes, and the interconnections between those nodes. For example, suppose we replace a single cloud data center connected to 1000 endpoints with a fog of 100 nodes, each connected to 10 endpoints.

The first obvious point is that we have gone from one centralized facility to 100 dispersed ones. Some costs will have a similar multiplier, say, building permits. Others will remain constant, perhaps compute resources that are merely being dispersed among facilities. Others will be somewhere between the two, such as storage costs, which may reflect some data that are replicated and other that is partitioned.

Suppose further that the fog nodes are all completely connected, i.e. there is a dedicated physical connection between every pair of fog nodes. In the first case, we had 1000 links. In the second case, each of the 100 fog nodes has 10 links to endpoints and 99 links to other nodes, divided by two to eliminate double counting of each link. In other words, there are now the same 1000 links to endpoints and $(n)(n-1)/2$ or 4950 additional links. Of course, it would be hard to find a real-world architecture with node counts at scale that was completely connected (in the graph theory sense, not just network reachability) at the physical layer, but this illustrates that network topology can grow more complex and costly through dispersion.

These costs can go away if rather than being net new costs, the fog computing capability is deployed on existing network, compute, and storage resources. As an example, an app that can be deployed on an existing smartphone and wireless Internet capability may have essentially zero marginal cost to deploy.

5.4 Capacity, Utilization, Cost, and Resource Allocation

5.4.1 Capacity Requirements

Let us assume that the resource demands of n different workloads vary over time and can thus be characterized as $D_1(t), D_2(t), \dots D_n(t)$. For example, the resource might be the number of web servers needed by a bank to support its online banking application. Let us further assume that they can be characterized as independent, identically distributed random variables, with mean μ and variance σ^2. For this illustrative analysis, let us further assume that the demands are normally distributed. Because they vary, for any i, $1 \leq i \leq n$, we know that the expected value of the demand function will be less than its maximum, i.e. $E(D_i(t)) < \text{Max}(D_i(t))$. In practice this means that a dedicated environment serving only this application that is sized to the maximum demand will always be underutilized. In contrast, one that is sized below this level will have insufficient resources, and thus experience degraded performance or be unavailable for some users, which may be external users (customers or partners) or internal ones (employees, contractors) and thus lead to either revenue loss or labor productivity losses.

In reality, there is typically no definite bound $\text{Max}(D_i(t))$. Consequently, IT managers build in a degree of headroom. More resources increase the likelihood that resources are available to meet demand but are also more costly. Let us assume that the capacity is defined as some number of standard deviations above the mean. If capacity is set to the mean μ, there will be sufficient capacity 50% of the time. If the capacity is set to $\mu + \sigma$, this rises to 84.3% (68.27% within one standard deviation plus 15.87% in the left tail). If we set the capacity to $\mu + 2\sigma$, this rises further to 97.7%, and so on.

Let us assume that a given level of capacity $\mu + k\sigma$ is considered to offer an appropriate balance between sufficient capacity and reasonable cost. If the n demands are all served out of siloed resources, the total capacity required for unaggregated resources is $C_U = n(\mu + k\sigma)$.

Conversely, let us consider what happens when those demands are aggregated and served out of pooled resources with dynamic allocation. We know that the mean of the sum is the sum of the means, so if $D(t) = D_1(t) + D_2(t) + \cdots + D_n(t)$, then the mean of the sum $\mu_D = n\mu$, and because the variance of the sum is the sum of the variances, $\sigma_D^2 = n\sigma^2$. But this means that the standard deviation of the sum is only $\sigma_D = \sqrt{n}\sigma$. So, to achieve the same "headroom," the capacity required for the aggregated resources is $C_A = n\mu + k\sqrt{n}\sigma$. The savings can be expressed as $C_U - C_A$, i.e. $n(\mu + k\sigma) - (n\mu + k\sqrt{n}\sigma) = kn\sigma - k\sqrt{n}\sigma = k\sigma(n - \sqrt{n})$. To put it another way, for any positive headroom k, and any nonzero (and thus positive) variation in demand levels, the more workloads we aggregate, the bigger the savings, or to put it differently, the less total capacity required [10].

These results can be interpreted as follows. For a given set of computing assets, it is better to pool them to reduce required capacity and thus investment. This includes pooling siloed servers into a private cloud, multiple private clouds into a public cloud, or multiple public clouds into a federated or intercloud [11]. It also means that device capabilities are best aggregated into a fog layer, say, the intelligence to run multiple machine tools, but also that, at least when viewed solely from a capacity requirements perspective, fog resources are better deployed in a connected and shared environment.

This analysis is for the case of independent, identically distributed demands; the degree to which pooling generates a reduction in capacity requirements obviously will differ under other circumstances based on the exact distributions.

The aforementioned discussion relates to processing resources; storage resource requirements ultimately depend on the degree to which data are partitioned or replicated. For example, a 10 Gigabyte reference data set might be replicated across 100 nodes, in which case roughly a Terabyte of total storage capacity is required. On the other hand, it might be partitioned across those nodes, in which case the *total* storage capacity needed does not change (subject to bin-packing effects based on available storage sizes).

5.4.2 Capacity Utilization

To continue the prior analysis, if a mean level of demand is μ and capacity is set to $\mu + k\sigma$ (with k, $\sigma > 0$), there is no way for all the capacity to be used. Complicating things further, when $D(t) < \mu + k\sigma$, the capacity utilization will obviously be less than 100%. However, because utilization cannot be more than 100%, when $D(t) \geq \mu + k\sigma$, utilization will only be 100%. Consequently, while the mean demand is μ, the mean capacity in use is not actually μ, but a new mean μ', which is the mean of the function $D'(t) = \min(D(t), \mu + k\sigma)$. However, to a first-order approximation, for "reasonable" k, we can (and will) just use μ.

It should be clear that if, for a given silo, the average demand is μ, but the fixed capacity is $\mu + k\sigma$, the utilization for that silo's resources will be $\mu/(\mu + k\sigma)$. This will be the case for each silo, so this is the average utilization. We of course will find the same result if we considered the n siloes in total because we just would have $n\mu/(n(\mu + k\sigma))$.

However, if we aggregate demand, again as a first order approximation, we get a better utilization because the same amount of demand is being served out of fewer pooled resources. The new utilization has the same aggregate total demand of $n\mu$, but the denominator is now $n\mu + k\sqrt{n}\sigma$. The utilization is therefore approximated by $n\mu/(n\mu + k\sqrt{n}\sigma)$. Multiplying this by unity, expressed as $\left(\frac{1}{\sqrt{n}}\right)/\left(\frac{1}{\sqrt{n}}\right)$, gives us a utilization of $\sqrt{n}\mu/(\sqrt{n}\mu + k\sigma)$[10]. Since $k\sigma$ is a constant, it should be clear that as an increasing number n of workloads is aggregated, the utilization approaches 100%. This contrasts extremely favorably with silos, where the average utilization remains the same.

As previously mentioned, this suggests that aggregation of fog resources at a higher layer in the fog is beneficial in the presence of variable demand. However, it also suggests that aggregating device resources at a higher layer such as the fog is beneficial.

5.4.3 Unit Cost of Delivered Resources

Another way to think of these phenomena is from a resource cost perspective. For the sake of making a point, let us ignore all the details around cores, and power costs and power usage effectiveness (PUE), and volume discounts, and Open Compute designs, and taxation, and the intensity of industry competitive rivalry and keep things simple. Let us also ignore the complications of considering conversion between capital expenditures and daily rates over a useful service life less depreciation.

In this simplistic world, if a server costs $1 an hour to an IT organization or cloud provider, they could charge $1.10 and turn a profit. But here is where utilization levels impact cost. If utilization is running at 50% (which is 5–10 times higher than the reality in many organizations), what that means is that, on average, for every revenue-generating server, there is also an idle server. Profitability needs to be examined not only in terms of the server that is revenue-generating but also the costs associated with the idle server.

In the real world, an increasingly significant cost component of computing is power costs; therefore, if power can be reduced or eliminated to idle compute infrastructure, the actual cost difference will be less than the utilization numbers would suggest. However, again assuming "reasonable" k, a good approximation of the cost penalty for hardware costs is [10]:

$$\frac{n(\mu + k\sigma)}{n\mu + k\sqrt{n}\sigma}$$

5.4.4 Resource Allocation, Sharing, and Scheduling

Regardless of the capacity, utilization, and cost of resources, there are benefits and challenges associated with allocating tasks along the cloud to thing continuum. Bounded compute and storage resources exist in smartphones, smart TVs, thermostats, and the like, fog nodes ranging from the edge to the center, and essentially limitless (although technically "bounded") scalable resources in the cloud.

These tasks come in a variety of flavors: ranging from embarrassingly parallel to highly serial; from small microservices to large monoliths; standalone or highly interconnected. They can be safety critical and mandatory, or optional and deferrable. They can be extremely low latency or batch. They can interact with a variety of objects which may be small or large, message-oriented or file-oriented, and so on and so forth. At one extreme, consider a smart meter, sending a few bytes every 15 minutes or even less frequently; at the other, a live 4K video surveillance stream.

Fog computing offers many more options as to where to run these components. This flexibility is both a blessing and a curse because the combinatorics explode to make resource allocation computationally complex, akin to the NP-complete knapsack problem and its variants. If we further try to optimally match CPU bandwidth and storage requirements to resources of various types, possibly under latency or network security constraints, we can end up with computationally complex problems akin to vertex cover or three dimensional matching problems [12].

In general, demand aggregation and resource pooling create utilization benefits, as we have discussed previously. Whether resources should be considered pooled or siloed can sometimes be application dependent. For example, a loosely coupled set of microservices can execute, communicate, and cooperate across a loosely coupled set of heterogeneous, ubiquitous, and decentralized fog nodes, which then behave as if they are pooled.

5.5 Information Value and Service Quality

5.5.1 Precision and Accuracy

Generally speaking, information has value. This value can be expressed in multiple ways, such as the marginal benefit attributed to actions taken based on decisions made based on that value. For example, knowing that the price of a stock is going to go up tomorrow could be worth millions if the stock itself or call options are purchased today. This value may also be achieved through cost avoidance. For example, knowing that a house being considered for purchase has termites can avoid the costs of treatment and repair. This value may be expressed probabilistically, for example, by avoiding an extended warranty based on the likelihood that a repair will be needed and the price of the warranty.

The edge is the first point at which data collection devices connect to global networks, and thus is the primary, initial source for real-time, geo-dispersed data. In general, there are five generic regimes for information value based on edge data. Their actual value will depend on the exact application. If there are n data collection nodes and the information value of those nodes is $I(n)$, then the typical value is either

$I(n) = c$, i.e. the value is a constant not dependent on n. An example might be the time, where the time is synchronized across all nodes. Once an application knows the time from a given node, there is no additional value in querying additional ones. Another example may be constant, such as the point release of the software running in each node.

$I(n) < kn$, i.e. the value is sublinear in the number of nodes queried, e.g. an S-curve that has an asymptote of a maximum value. An example might be understanding consumer buying habits. A few stores worth of buying behavior might be enough to deduce that customers typically buy peanut butter when they buy jelly; thousands of stores worth of additional data points would not necessarily improve on that insight. Another might be weather prediction. After sufficient weather station data to predict the weather over the next week, multiplying the number of stations will not multiply the length of the forecast or the precision of the prediction equally.

$I(n) = kn$, i.e. a linear increase in information value. If each node has data on potential revenue-generating prospects (without overlap), then the revenue generated is proportional to the number of nodes queried.

$I(n) = 0$, $n < k$; $I(n) > 0$, $n \geq k$. This can particularly happen if there is a critical state transition point. For example, suppose that a 512-bit cryptographic key is divided into eight 64-bit segments. Objects cannot be decrypted with only seven segments, only with all eight. Such a "secret sharing" system is employed in some distributed storage systems, such as Publius [13]. Some databases or filestores such as Cassandra [14] replicate data multiple times, and the degree of replication is particularly relevant for node-failure scenarios.

$I(n) > kn$, i.e. a superlinear value. Purported "laws" such as Metcalfe's (the value of a network is proportional to n^2) and Reed's (2^n) also lead to a superlinear value, although whether they accurately reflect the real world is debatable [15].

The extent to which one of these regimes applies to a given application scenario helps define the relationship between the value generated by and the costs of data acquisition, processing, and aggregation.

5.5.2 Survivability, Availability, and Reliability

The survivability, availability, and reliability of a fog solution are functions of its components and their location, the architecture of those components, its possible failure modes, and its applications and their requirements.

For example, for any given architecture, the less reliable the components are, the less reliable the overall architecture will be. On the other hand, extremely unreliable components can be architected into a reliable system through redundant components and links.

Fog solutions have advantages in terms of survivability. A hyperscale data center can be brought down by a power outage, a natural disaster such as a hurricane, tornado, flood or earthquake, or a terrorist attack. A fog solution is inherently more resilient and survivable because a globally dispersed architecture would require a planet-scale disaster.

An architecture can be resilient against a given failure mode, but defenseless against a different mode. For example, today's cloud architectures have multiple geographic regions, redundant connections between those regions, and multiple availability zones within a region. An earthquake or tornado or hurricane will not cause a service failure if the application has been correctly architected. However, systemic software failures can do so, and have done so, such as the Netflix Christmas Eve outage, which was caused by elastic load balancer control plane problems that rapidly spiraled out of control [16].

Conversely, physical problems can impact availability. For example, even when companies use multiple network service providers for vendor diversity to protect against a systemic provider problem, an issue such as a fiber cut or digging issue along, say, a railway right-of-way can take down "diverse" links from the different providers who actually all have routes along the same right-of-way or even the same fiber cable.

Finally, outages or planned downtime may or may not meaningfully impact availability. For example, a complete loss of airplane control systems would be disastrous in flight, but be inconsequential while the plane is in the hangar. A local fog node that is processing inter-vehicle coordination could cause an accident by failing just as two cars were about to collide, but its downtime would not be noticed if no cars were near the intersection.

Given all these, the system may lie at one of two extremes, or somewhere in the middle. Suppose that every one of n nodes has an independent nonzero probability of failure p. The probability that it has not failed is then $(1 - p)$. The probability that all nodes have not failed is somewhat dismal as the fog scales because the probability that everything is running perfectly smoothly is

$(1 - p)^n$, which approaches 0 as $n \to \infty$. On the other hand, if we do not need *all* nodes to be working, only, say one, probability works in our favor because the probability that all nodes are down is p^n, which also approaches 0 as $n \to \infty$.

Autonomy is an additional consideration in favor of fog. When network connections are uncertain – as they often seem to be – having local processing resources sufficient for a given need is essential. A mine or factory must continuously operate regardless of connectivity issues; as must traffic lights, autonomous vehicles, user productivity tools such as laptops and smartphones, and so forth.

The reliability of the results generated ultimately depends on which information value regime applies. If, as previously described, $I(n) = c$, i.e. the information value is constant, or $I(n) < kn$, i.e. information value is sublinear, say, i.e. has decreasing marginal returns to scale, then having "most" nodes functional or perhaps even "some" or a "few" nodes will give us just about as good an answer as we would get with "all" nodes, then the reliability of the insight or decision is pretty high even in the presence of node failures.

5.6 Sovereignty, Privacy, Security, Interoperability, and Management

5.6.1 Data Sovereignty

Put simply, if countries require their citizens' data to be maintained within the country and to never traverse its borders, a centralized cloud-type of solution will not work; only a dispersed one will.

5.6.2 Privacy and Security

In the same way that it is easier to break into a car than into Fort Knox, a proliferation of endpoint devices and heterogeneous fog nodes can potentially increase the potential attack surface and make vulnerabilities harder to lock down and apply patches. Also, as fog and devices essentially democratize the production of digitally enabled endpoints and multilayer nodes to numerous global manufacturers who may be more concerned with features and packaging than with security, various vulnerabilities have been multiplying. Thus have hackers been able to gain access to airliner and automobile control systems through their entertainment systems, break into networks through Wi-Fi light bulbs, melt down steel mills through cyberattacks, and so forth.

In addition to cybersecurity, there are very real problems with physical security. A data center uses many different techniques, fences, walls, mantraps, guards, surveillance cameras, burglar alarm systems, and the like. Dispersed fog nodes may be placed in server closets, regen huts, agricultural fields, mineshafts, etc., which can be either hard to lock down or prohibitively costly. 5G small-cell base stations will proliferate of necessity, due to their short effective transmission distance in the case of millimeter waves specifically. Moreover, these base stations may provide access to or in themselves be fog nodes. While this proliferation will provide low latency for 5G radio access network transmission, it also means that these stations will be located in numerous, easily accessible locations, driving numerous physical security challenges.

On a positive note, such proximity to users means that fog nodes can provide access control, support encryption, and offer contextual integrity and isolation. This can be beneficial, for example, precluding unauthorized network access and even preventing the node or base station or user device from being enrolled in a botnet supporting a distributed denial-of-service attack. Such nodes can also serve as aggregation, control, and encryption points for sensitive data or due to their flexibility perform select security functions as proxies for resource-constrained user devices or things.

On the other hand, because individual nodes or even groups of nodes have limited bandwidth and processing capability compared to a hyperscale data center, they are vulnerable not only to physical attack but also to denial of service attacks directed at the node(s). Moreover, openness of fog networks can promote flexibility and innovation but can also create vulnerabilities that can be exploited by untrustworthy or compromised devices or nodes.

5.6.3 Heterogeneity and Interoperability

One of the strengths of fog computing is that it comprises a heterogeneous abundance of components, products, and services from a global range of vendors and service providers. However, there is a danger that this becomes akin to the "Tower of Babel," involving a cacophony of protocols, containers, run-time environments, and signaling and control strategies. Therefore, there is a critical need to maximize interoperability and functional cooperation between heterogeneous resources, through a variety of mechanisms such as signaling, control, and data interfaces, standard ontologies [17], open APIs, service directories, interdomain agreements including possibly for pay-per-use resources, and operational interface standards.

5.6.4 Monitoring, Orchestration, and Management

Heterogeneity and dispersion also create problems with monitoring, orchestration, and management. Some simple examples: how can a message or data be passed from an endpoint such as a user device or thing, up to the cloud across a fog of multiple heterogeneous nodes, gateways, routers, elements, etc. spanning a variety of protocols? How can such messages be passed successfully – exactly once, no more and no less – in the presence of loss of nodes due to a variety of failures or lost packets? How can tasks successfully complete or successfully continue to run when some fog nodes may be mobile and intermittently connected, either due to loss of power, power-downs, or loss of connectivity as they go out of range of a base station, or switch carriers, or otherwise cross fog domains? How can we effectively predict user (device) or thing movement?

How do we monitor and manage a variety of quality of service (QoS) metrics and trade them off against other objectives such as network traffic caps or application processing delays or network costs?

Also, because various microservices can be resident at various points in the hierarchy, what is the optimal location for each service and its required data, given a variety of dynamic costs and states which may be time-varying, such as transport costs, delay costs, resource costs due to pay-per-use or other charging schemes, etc.

How can faults or errors be identified and localized, especially since a failed component in one place may cause a noticeable service issue somewhere else entirely?

5.7 Trade-Offs

In the preceding analysis, we saw that there are multiple dimensions to quantitatively and qualitatively assessing the benefits and challenges of fog computing economics: Strategic value through information excellence, solution leadership, collective intimacy, or accelerated innovation; bandwidth, latency, and response time trade-offs spanning network latency, server latency, total latency, and data traffic volume; the stochastic behavior of the trio of capacity requirements, utilization, and delivered resource costs; information value and service quality through the interrelationships of precision and accuracy and survivability, availability, and reliability; and data sovereignty, privacy, security, monitoring, orchestration, and management issues.

Sometimes benefits align nicely. For example, a highly dispersed solution can reduce latency while enhancing survivability. In other cases, trade-offs must be decided upon. For example, a heterogeneous solution eliminates the likelihood of a systemic outage through design or engineering errors of a single vendor but increases the complexity of system management. Unfortunately, there is no single best solution, only good solutions and best ones under certain circumstances.

Making matters even more complex, service providers that may offer elements of a total solution have a variety of pricing schemes. Compute resources may be owned and therefore usable without additional charges but incur costs even when not used. Their use may be free because their marginal cost is zero for fog computations because they are not congested. They may be charged for on a per-message basis, e.g. a status message from a sensor, or a per transaction and memory used, such as for cloud functions, or a pay-per-use or actually pay-per-allocation VM basis, on a billing increment of minutes or hours or months, or have a dynamic price as with spot instances or a long-term price such as with sustained-use discounts or an undefined price based on aftermarket trading of reserved instances.

5.8 Conclusion

Ultimately, fog architectures can contribute to reduced cost, increased revenues, enhanced performance and availability, increased customer satisfaction better compliance, etc. However, they are neither a panacea nor the only conceivable solution. Real-world architectures are likely to be a hybrid of cloud, fog, and edge; public and private; real-time, near-real-time, interactive, batch, and offline; consolidated and dispersed; and many other combinations that can offer the best of both worlds. Decisions on all these parameters will ultimately determine the constellation of benefits achieved and costs incurred to achieve them.

References

1 Weinman, J. (2012). *Cloudonomics: The Business Value of Cloud Computing*. Wiley.

2 Weinman, J. (2017). The 10 laws of Fogonomics. *IEEE Cloud Computing* 4 (6): 8–14.

3 Weinman, J. (2017). The economics of the hybrid multicloud fog. *IEEE Cloud Computing* 4 (1): 16–21.

4 Manyika, J., Chui, M., Bisson, P. et al. (2015). Unlocking the potential of the Internet of Things, https://www.mckinsey.com/business-functions/digital-mckinsey/our-insights/the-internet-of-things-the-value-of-digitizing-the-physical-world (accessed 19 August 2019)

5 Weinman, J. (2015). *Digital Disciplines: Attaining Market Leadership via the Cloud, Big Data, Social, Mobile, and the Internet of Things*. Wiley CIO.

6 Weinman, J. (2015). The cloud and the economics of the customer and user experience. *IEEE Cloud Computing* 2 (6): 74–78.

7 Weinman, J. (2011). As time goes by: the law of cloud response time. Working Paper. http://joeweinman.com/Resources/Joe_Weinman_As_Time_Goes_By.pdf (accessed 19 August 2019).

8 Böttger, T., Cuadrado, F., Tyson, G. et al. (2015). Open Connect everywhere: a glimpse at the internet ecosystem through the lens of the Netflix CDN. *ACM SIGCOMM Computer Communication Review* 48 (1): 28–34.

9 Yang, Y. (2019). Multi-tier computing networks for intelligent IoT. *Nature Electronics* 2: 4–5.

10 Weinman, J. (2017). The economics of computing workload aggregation: capacity, utilization, and cost implications. *IEEE Cloud Computing* 4 (5): 6–11.

11 Weinman, J. (2015). Intercloudonomics: quantifying the value of the intercloud. *IEEE Cloud Computing* 2 (5): 40–47.

12 Garey, M.R. and Johnson, D.S. (1979). *Computers and Intractability: A Guide to the Theory of NP-Completeness*. W. H. Freeman.

13 Yianilos, P.N. and Sobti, S. (2001). The evolving field of distributed storage. *IEEE Internet Computing* 5 (5): 35–39.

14 Lakshman, A. and Malik, P. (2010). Cassandra: a decentralized structured storage system. *ACM SIGOPS Operating Systems Review* 44 (2): 35–40.

15 Weinman, J. (2007). Is Metcalfe's Law way too optimistic? *Business Communications Review* 37 (8): 18–27.

16 Summary of the December 24, 2012 Amazon ELB Service Event in the US-East Region, https://aws.amazon.com/message/680587/ (accessed 19 August 2019).

17 Di Martino, B., Cretella, G., Esposito, A. et al. (2015). Towards an ontology-based intercloud resource catalogue – the IEEE P2302 intercloud approach for a semantic resource exchange. In: *2015 IEEE International Conference on Cloud Engineering*, 458–464. IEEE.

6

Incentive Schemes for User-Provided Fog Infrastructure

George Iosifidis[1], Lin Gao[2], Jianwei Huang[3], and Leandros Tassiulas[4]

[1] *School of Computer Science and Statistics, Trinity College Dublin, University of Dublin, Ireland*
[2] *Department of Electronic and Information Engineering, Harbin Institute of Technology, Shenzhen, China*
[3] *School of Science and Engineering, The Chinese University of Hong Kong, Shenzhen, China*
[4] *Department of Electrical Engineering, and Institute for Network Science, Yale University, New Haven, CT, USA*

6.1 Introduction

Today, we are witnessing two important socio-technological advances that herald the advent of a new era in wireless networks and mobile computing systems. First, the ever increasing needs of users for ubiquitous connectivity and real-time execution of computing tasks has created an unprecedented pressure to mobile networks and service providers. Second, we see nowadays the proliferation of user-owned equipment such as Wi-Fi access points (APs), advanced handheld devices, and various Internet of Things (IoT) devices with enhanced storage and computing capabilities. These devices not only can satisfy the communication and computing needs of their owners but can be also employed to provide related services to nearby users. In this context, each user is transformed to a *microservice provider* (a *host*) who may offer Internet access, storage, or computing resources to other users, giving rise to the so-called *user-provided infrastructures* (UPIs).

Such ideas were first studied in the context of network sharing [1], and interestingly, there are already related commercial applications. For instance, FON [2] is a Wi-Fi sharing service where users offer Internet connectivity through their residential Wi-Fi APs to other (mobile) users, in exchange for receiving such services when they need them. This model is based on earlier peer-to-peer Wi-Fi sharing models, which proposed pricing schemes for maximizing the number of served users [3]. Recent studies further discussed how the operator can maximize the overall network profit by optimizing the prices of different membership types under both complete and incomplete network information [4, 5]. Other similar models have been proposed where a user may host other users (clients) by acting as an Internet gateway, or even as a relay connecting them to user's gateways. This network sharing paradigm has attracted the focus of academia [6], and it has also inspired business models employed either by small startups or major network operators. New application scenarios such as crowd-sourced video streaming are enabled by such a new paradigm [7–9]. This interest is not surprising though since UPI-based services can yield substantial performance and economic benefits both for users and service providers.[1]

In the emerging era of fog computing, this architecture paradigm offers more opportunities but also becomes more challenging. Users can share network connectivity, storage resources, and computing power, or even their devices' battery energy. For example, a device that needs to execute a computation-heavy task can use the idle processor of a nearby device, or its storage resources for caching raw data. Hence, the dimension of the sharing problem increases from single-resource to multiresource problem, but there are more collaboration opportunities [10]. For instance, bandwidth can be exchanged with storage or computation power, and this creates further opportunities for synergistic interactions among users at the network edge. Several important tasks in fog computing require the orchestration of different resource types, and UPIs need to be adapted to go beyond only network sharing.

Clearly, the benefits of UPI systems that orchestrate different resources owned by end users can be unprecedented, and we can design UPI architectures suitable for different application domains. However, a bottleneck issue that hampers the successful deployment of such solutions is the participation of end users. This is a central issue in UPI solutions, as both the demand (clients' requests) and the provision (hosts' availability) depend on users' participation. Nevertheless, more often than not, the participants may have conflicting interests. For example, clients would prefer to receive resources and services at low costs, while hosts would prefer to charge high prices for their services or receive large amounts of resources in return. Aligning the interests and decisions of different types of users for heterogeneous resource sharing is certainly challenging.

This problem is further compounded by the decisions and interests of network operators, cloud computing providers, or cloud storage providers (we will use the term *providers* henceforth), which are directly affected when UPIs are in place [11]. In particular, a provider's incentive of supporting a fog computing UPI-based solution would be high if it can directly gain from such services. On the other hand, UPIs can not only reduce the revenue streams for the providers by satisfying the user demand in a peer-to-peer fashion but can even induce severe congestion when uncontrolled reallocation of edge resources takes place in massive scale.

---

1 Henceforth, we will use the term *provider* to refer to any type of service provider including network access provider (namely, a mobile operator or an ISP), cloud computing or cloud storage provider (e.g. Amazon), etc.

Therefore, it is of paramount importance to design incentive mechanisms for reconciling the objectives of all participants: users acting as clients, users acting as UPI hosts, and the service providers that are involved. Such mechanisms need to effectively tackle the following questions:

• How much a host needs to be compensated through payment (directly) or through service exchange (indirectly), for offering resources and services to clients in a UPI solution?
• Which bundles of charged prices and offered services render UPIs more attractive than conventional provider-offered computing or communication services?
• If the UPI service is enabled by a provider, how much it should charge the users clients and reimburse the users hosts?

Designing the proper UPI incentive schemes when the hosts are mobile, i.e. their devices are portable (such as a smartphones or moving IoT nodes), not attached to power sources and not having wireline Internet connections. First of all, the mobile Internet access cost is highly varying and often quite expensive. Moreover, mobile devices have tight energy budgets which lead to stringent energy consumption constraints for the UPIs. Additionally, the storage and computation capabilities of such small devices, albeit continuously improved, are often consumed by demanding mobile applications or the significant amounts of data that IoT applications need to collect. For all these reasons, we expect that users acting as hosts will be reluctant to share their resources unless they are adequately compensated.

These concerns are further perplexed due to the inherent volatility of the wireless medium that often results in varying Internet access performance. Similarly, the availability of storage and computation resources is often time-varying, as users change locations and their application requirements. In general, this problem has an inherent stochastic nature, where demands, resources, and sharing opportunities vary with time, often in a nonstationary and unpredictable fashion. Furthermore, in UPIs, there is no single entity having full access of the network information about all users' resources and needs. This hampers the derivation of an optimized resource orchestration policy and creates an additional level of inefficiency.

Despite these difficulties, UPIs are attracting growing interests and are expected to play an important role in realizing the vision of fog computing; and network sharing is already showing us the direction with several successful market paradigms. Two such interesting models have been proposed and implemented recently by the Open Garden [12] and Karma startups [13]. The former service enables mobile users to share their Internet access. The main idea is to exploit the diversity of users' needs and resources and crowd-source Internet connectivity by building an autonomous UPI service, i.e. without the intervention of network operators. The Karma mobile operator enables its subscribers to act as mobile Wi-Fi hotspots (MiFi) and serve nonsubscribers with proper compensations. In this case, the UPI service is controlled (or assisted) by the operator, who has to ensure the consensus of the hosts. Similar ideas have been also used for extending wired Internet coverage, with the Comcast Xfinity Wi-Fi service being a notable example [14].

However, these business models currently do not provide effective incentive mechanisms that enable long-term user participation, and hence do not address the previous three questions that we raised.[2] Our goal in this chapter is to analyze such incentive issues and propose potential solutions. We begin in Section 6.2 with an overview of UPIs by focusing on recently proposed models for network sharing, as this is the earlier and most established paradigm. We discuss the technical issues pertaining to resource allocation for these services and explain the importance of incorporating incentive schemes. Section 6.3 analyzes mobile UPI models that are inspired by Open Garden, where users collaborate in an autonomous fashion; and Section 6.4 discusses UPI models that are enabled by a central service provider, as in the case of Karma mobile operator. In both cases, we discuss the challenges in designing incentive mechanisms and present two novel solutions. Section 6.5 discusses a different collaboration mechanism that is fully decentralized and lightweight, with minimum signaling requirements, and hence suitable for large-scale fog computing architectures as those that expected to arise in IoT. Section 6.6 analyzes further key challenges for incentive mechanisms for mobile UPIs. We conclude in Section 6.7.

6.2 Technology and Economic Issues in UPIs

6.2.1 Overview of UPI models for Network Connectivity

We focus on UPIs that support network connectivity, as it is a prerequisite for sharing any other type of resource, i.e. computation or storage. The most prevalent example of a UPI network service is the Wi-Fi sharing model of FON, which was followed by other companies such as OpenSpark. In some cases, users need to purchase new equipment, e.g. customized Wi-Fi routers or install proprietary software, e.g. in Whisher [15]. The cooperation of users in these services is based on reciprocation, or in simple pricing rules where clients pay to gain instant Internet access. A slightly different model was proposed by Telefonica BeWiFi [16] where residential users in proximity create Wi-Fi mesh networks to increase their available bandwidth through resource pooling. On the other hand, Comcast Xfinity offers to mobile users access to residential Comcast APs [14].

---

2 The importance of such mechanisms is, perhaps, best exemplified by the class action lawsuit that was filed in California against Comcast for using the users' network and electricity resources to serve other mobile users without their agreement.

While these previous models involve mainly fixed hosts, the proliferation of mobile devices and the growth of mobile data have inspired the development of UPI models for network sharing that are only based on mobile hosts. For example, the mobile operator Karma equips each of its subscribers with a portable device (USB router), which operates as a MiFi hotspot and offers Internet access to other nonsubscribers (clients). Each subscriber pays a constant price per megabyte of data she consumes and earns a free quota of 100MB for every client she serves for the first time, at the expense of additional energy consumption and Internet access sharing (which reduces her Internet access speed). This provider-assisted mobile Internet sharing model, with the hybrid pricing-reimbursement scheme, enables the operator to increase the population of clients, and the subscribers to augment their data plans with free quotas.

A different model of mobile UPI is employed by Open Garden, which allows mobile users to create a mesh network and share their Internet connections. In this case, each user may act as a *client* node (consuming data), a *relay* node (relaying data to other nodes), or a *gateway* node connecting the mesh overlay to the Internet through a Wi-Fi or cellular connection. Each user may have multiple roles either simultaneously or sequentially during different time slots. Every user who cooperates and offers her services to others incurs an additional energy cost when acting as relay, and an additionally monetary cost when acting as gateway. In this autonomous UPI, the network operators do not have any control over the traffic exchanged among the users and transferred across heterogeneous networks in an ad hoc fashion.

From the aforementioned examples, it is clear that the diverse UPI models can be broadly classified as follows (see Figure 6.1):

- *Fixed host or mobile host*: One criterion for differentiating UPI services is whether they are offered by fixed hosts (e.g. FON) or mobile hosts (e.g. Karma).
- *Autonomous or provider-assisted*: Autonomous UPIs are transparent to the operators who do not intervene (e.g. Open Garden), while in provider-assisted UPIs, the operators may determine the pricing rules and/or the Internet bandwidth sharing policy.

Based on these criteria, each UPI model can be classified in one of the four categories as shown in Figure 6.2. One can further identify more subcategories, based on, for example, whether the sharing is realized over special equipment (e.g. FON) or software (e.g. Open Garden), and whether it is a service that involves only one type of resource (bandwidth, storage, or computation) or a combination of them. Furthermore, UPIs may be managed independently by each pair of host and client [17], by rules predetermined by users' communities [18], or based on policies that are set by the respective service provider.

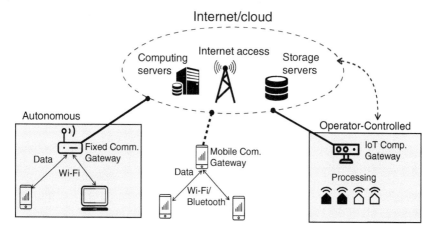

Figure 6.1 The architecture and possible operations of a UPI system. The operation of UPI may target extending network connectivity or availing computation resources to mobile handheld devices or offering computing or storage resources to IoT lightweight nodes. The services could be offered by fixed fog infrastructure or mobile fog nodes. Comm., Communications; Com., Communications; Comp., Computations.

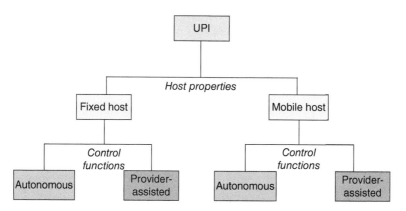

Figure 6.2 Taxonomy of UPI models according to two criteria are as follows: (i) fixed or mobile host, (ii) autonomous or provider-assisted servicing.

6.2.2 Technical Challenges of Resource Allocation

UPI solutions introduce challenging technical issues because the users undertake many of the tasks and services that otherwise were executed by the service providers. For example, one main concern in such fog computing solutions is security, and in particular regarding the potential threats that the hosting devices might face by malicious client devices. Fortunately some recent standards (e.g. Hotspot 2.0) address many of these problems, but clearly there are several open challenges here. For instance, when the UPI enables sharing of computation resources, the execution of malicious codes may significantly impact the hosting device. Another issue for mobile UPIs is the device discovery and coordination, which may cause significantly time overhead and consume scarce device resources such as bandwidth and energy. In this case, a provider maybe able to facilitate the user interactions due to the availability of more network information.

An important question that one may ask about UPI-based services is how much users can benefit from them. This has been studied in the context of data offloading where mobile users, instead of connecting to cellular networks, access the Internet through residential Wi-Fi APs of other users [19]. A related study from South Korea showed that Wi-Fi offloads 65% of the total mobile data traffic, and saves 55% of battery energy [20]. Moreover, UPIs must also consider the allocation of Internet access bandwidth so as to increase the performance of the services offered to clients, while ensuring a minimum service quality for the needs of the hosts [17].

This sharing problem is more challenging for mobile UPIs due to the stringent energy constraints of mobile devices. Paper [21] proposes an energy-prudent architecture that aggregates the cellular bandwidth of multiple hosts to build a MiFi hotspot and schedules the clients' transmissions so as to reduce the total energy consumption. Another salient feature of mobile UPIs is that the network load and availability are very dynamic, as both clients and mobile hosts are nomadic. Therefore, [22] presents a scheduling scheme that allows mobile hosts to dynamically admit client requests so as to maximize their revenue and ensure system stability. On the other hand, when each user can serve both as a client and as a gateway, a decision framework should assign these roles to users, considering their residual energy [23].

6.2.3 Incentive Issues

In all UPI models, where the devices are managed by different users, the hosts need to be offered proper incentives so as to agree to share their resources. This is very important, as network sharing may often induce a performance degradation for the hosts. For example, when a host shares her Wi-Fi AP, the additional traffic of the clients reduces the available bandwidth for the host's own communications. Similarly, when a host shares its memory or computation power, it will have less resources to store its own data or perform the computations required for its own tasks. Moreover, even if the host does not have such needs, supporting the demands of clients may induce costs in terms of data usage and energy consumption. The latter is particularly important for mobile hosts that often have tight energy budgets. The cost of data usage varies and can be negligible, e.g. as in the case of flat pricing, or very important, when usage-based data plans are in place. In sum, for any user device, supporting the communication, computation, or storage needs of other users can be very costly and even hinder its own performance. Therefore, the question that naturally arises is how to properly incentivize them in order to agree to act as UPI servers, despite these costs.

Similar incentive issues exist for peer-to-peer and ad hoc networks. However, the solutions proposed for these cases do not directly apply for UPIs since they do not account for users' different types of resources, nor for their data usage costs. Also, in UPIs, users can often access the Internet even without relying on other users and can execute their own computation tasks independently; while typically this is not the case for other autonomous systems. For example, in ad hoc networks, some nodes always rely on other nodes' help. This standalone (or, independent) operation possibility in UPI will serve as a benchmark for determining how the benefits brought by the UPI solution (i.e. the sharing gains) should be fairly dispersed among users.

In the sequel, we discuss the incentive mechanisms for UPIs, with emphasis on models with mobile hosts and clients for connectivity sharing. We consider two different classes of UPIs, namely autonomous and provider-assisted.

The incentive mechanisms for enabling cooperation among different users in autonomous networks can be classified as either reputation-based (or reciprocation-based) or credit-based. The former class allows only for bilateral service exchange, while the latter is more flexible and enables multilateral exchanges [18]. It is possible to combine both approaches in UPIs with a careful design [24]. A more sophisticated approach is to consider a cooperative (or, coalitional) game-theoretic framework and investigate the arising cooperative equilibriums. We present here one such solution that relies on Nash bargaining solution (NBS) [25], where the participation incentives are offered on the basis of a fair and efficient resource contribution and allocation of the service capacity. NBS is a prominent axiomatic criterion for resolving such "conflicts," and has been extensively used for sharing resources in wireless networks. Similarly, we also discuss the existence of a nonempty *core* on a coalitional game where the players correspond to the nodes of a network which decide if and how to collaborate with each other. A nonempty core means that all players can actually find a way to cooperate and exchange resources.

When the incentives are provided by the provider, it can explicitly reward the cooperating users. For example, authors of [26] proposed a resource allocation scheme for cellular networks that rewards cooperating users by allocating to them optimal spectrum chunks. Any incentive scheme for UPIs should take into consideration that these services exhibit at the same time positive network effects and negative congestion effects [27]. That is as more users join the service, both the numbers of hosts and clients increase, and the balance of demand and service provision depends on the resources and needs of the newcomers. We also present a novel provider-assisted UPI incentive mechanism, designed and implemented by the provider, aiming to maximize its revenue. Similar solutions apply for the case of sharing storage or computation resources, and we will discuss also these cases in the sequel.

6.3 Incentive Mechanisms for Autonomous Mobile UPIs

The autonomous mobile UPI connectivity service discussed here generalizes the Open Garden model by incorporating a carefully designed incentive mechanism. This mechanism determines how much resources each user needs to contribute, in terms of energy and Internet bandwidth, in order to maximize the overall service capacity, i.e. the aggregate amount of data delivered to users. Furthermore, it dictates how this capacity will be shared by the different clients, as each one of them should receive service commensurate to her contributions. In particular, this type of incentive schemes needs to address the following issues:

1) *Efficiency*: The mechanism should ensure the efficient allocation of the resources, i.e. maximize the service performance offered to users by taking into account the Internet access bandwidth, as well as the energy and monetary cost the user incurs. The satisfaction and costs of users will be reflected by their utility and cost functions, respectively.

2) *Fairness*: The mechanism should satisfy a fairness rule that accounts for the energy and bandwidth (and storage and computation power in the more general case) that each user contributes and consumes. Moreover, it should take into consideration the standalone (independent) performance of each user, i.e. the utility she has if not participating in the UPI solution and ensure that the UPI service will improve upon it for every user.

3) *Decentralized implementation*: The mechanism should be amenable to distributed execution for systems with a large number of nodes. Users have information only about their own needs and resource availability, and they should be able to decide independently their routing and Internet access strategies, based only on local information and minimum signaling from their one-hop neighbors.

4) *Indirect reciprocation*: A user being served by another user may not be able to return the favor immediately by offering similar services. Hence, a resourceful user may be reluctant to help other less resourceful users. The incentive mechanism needs to induce cooperation among users even for these cases.

5) *Future provision*: Some users may not have communication needs in a certain time period, and therefore may not be willing to participate. This may deteriorate the overall service performance. The mechanism should manage to encourage users to participate even if they currently have no communication needs.

In [28, 29], we proposed a mechanism that tackles the aforementioned challenges. Namely, the last two issues can be addressed by a virtual currency system. This way, users can pay and receive services even if they cannot reciprocate, while others with no communication needs are motivated to participate in the service so as to increase their currency budget. In this context, the payoff of each user is the performance from the service plus the virtual currency value that she collects by serving others.

We use the NBS to encourage users' efficient and fair contribution of resources and determine the corresponding service capacity allocation to each user. The scheme can work either for simple cases where there is no interference among the different pairs of collaborating users [28], or when the links interfere and proper channel assignment is needed [29]. The NBS is an axiomatic game theoretic concept suggesting how a group of players should share the surplus of their joint effort, in order to ensure that every participant will agree to cooperate. In particular, the NBS is derived by the solution of a convex optimization problem, where the objective function is based on the payoff functions of the players, by taking into consideration their standalone performance. The latter, which is known as the *disagreement point*, significantly affects how much each cooperating player hopes to receive through the cooperation [25]. For UPIs, NBS is attractive as it is Pareto optimal, i.e. no user can improve its payoff without decreasing another user's payoff. Moreover, it ensures that all users receive at least the payoffs they had if not participating in the collaborative service. This is a very important feature as it ensures that users will not be dissatisfied.

Finally, the NBS has a key practical advantage as it can be calculated in a distributed fashion by the users, thus enabling the decentralized implementation of the incentive mechanism. Although this latter property improves the scalability of the mechanism, it is clear that it induces additional overhead that may affect the system's performance. In particular, as it is explained in detail in [29], this NBS-based distributed incentive scheme has a message exchange overhead of $O(N^3)$, where N is the number of cooperating users. Moreover, the mechanism needs to recalculate the servicing policy each time a user joins or leaves the system, or when users' demands and/or their resources' availability change significantly. Dynamic optimization algorithms can be used however, e.g. see [30], to cope with this type of stochastic effects, where demand and resource availability change over time, possible in an unknown fashion.

The performance of this crowd-sourced UPI service increases with the diversity of the resources and needs of the participating users. This is depicted in Figure 6.3a, where we plot the data consumption and payoff functions for a group of 6 users, who are randomly located in a small area and cooperate under the proposed incentive mechanism. The energy consumptions and data costs are taken into account when deciding who will act as Internet gateway and

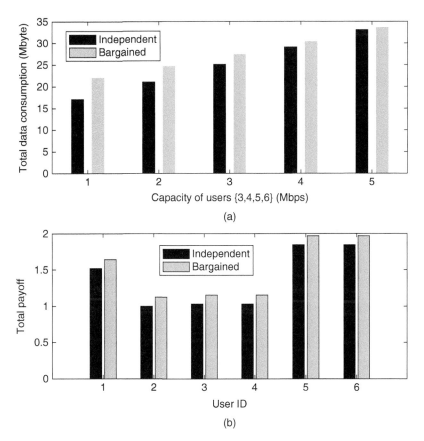

Figure 6.3 (a) Impact of capacity diversity on the service performance: $C_1 = C_2 = 12.7$Mbps, and capacities $C_3 = C_4 = C_5 = C_6$ are equal and change from 1 to 5Mbps (x-axis). Internet access prices and energy consumptions (per megabyte) are identical for all users. (b) Comparison of independent and bargained solution. For each user, the bargained solution ensures a higher performance than the independent (standalone) operation. Moreover, the aggregate total payoff improvement of the bargained solution (compared to the independent solution) is approximately 10%. More details can be found in [28].

who will relay data to the clients. Figure 6.3b shows that this scheme ensures that every user perceives higher performance comparing with her standalone (independent) operation.

Finally, Figure 6.4 presents an example of the independent operation and the bargained UPI operation for a downloading scenario. Namely, we consider a group of six mobile users in proximity, which have different Internet access capacities. When the users act in a standalone fashion, each user's data rate is limited by her Internet access capacity. In this case, some users may not be capable of accessing the Internet (e.g. user 2). On the other hand, in the UPI

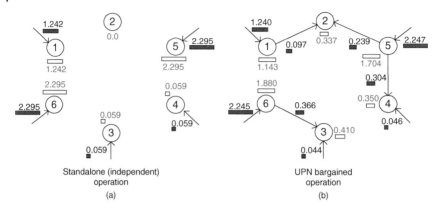

Figure 6.4 (a) Independent-standalone user performance (downloading scenario). Each user consumes (values associated with the white bars) only as much data as she downloads from the Internet (values associated with the black bars). (b) UPI performance based on NBS. Each user downloads and relays or receives data to/from others, as it is shown by the arrows.

bargained operation, the actual amount of data that each user will download and relay for others depends on their Internet access capacities, demands, and data usage costs. Through cooperation, a user who does not have direct Internet access may be able to receive data with the help of others. For simplicity, we have assumed that users have identical utility functions and data plans and that each user can only serve her one-hop neighbors. Note that when cooperation is enabled, the quantities of the downloaded data change, as there is now additional factors affecting the decisions of the users (e.g. the energy cost for exchanging data).

6.4 Incentive Mechanisms for Provider-assisted Mobile UPIs

The provider-assisted mobile UPI service discussed here is based on the Karma model [13] which, however, has certain limitations. First, each host is reimbursed with some free data quota when he shares the Internet connection with a client for the first time, independent of the actual amount of data that she routes for the clients. Later sharing with the same client will not bring any additional benefit to the host, hence does not encourage a long-term sustainable cooperation. Moreover, the reimbursement is identical for all hosts while they may have different bandwidth and energy constraints. Here, we consider a generalized provider-assisted UPI-based network service that allows for data price and quota reimbursement differentiation across users. This hybrid pricing scheme can be leveraged to increase the revenue of operator and the amount of data served by the hosts, by addressing the following issues:

1) *Quota-price balance*: This mechanism should select a proper combination of data prices and free quota rewards. Changing the price or the quota size has different impacts on the hosts' decisions to serve the clients and the operator's revenue.

2) *Effort-based reimbursement*: The free quota reimbursement should be effort-based, i.e. depends on the amount of data each host serves. This will motivate the hosts to increase the service that they offer to the clients.

3) *Users' demand-awareness*: The mechanism should consider whether the communication needs of the users are inelastic or elastic. Clearly, clients with elastic needs are less willing to pay high prices for Internet access. On the other hand, hosts with elastic needs are more willing to share their Internet access, even under small compensations.

4) *Price discrimination*: Finally, different hosts have different needs, energy limitations, and available cellular bandwidths (e.g. due to different locations and channel conditions). Hence, the incentives for routing traffic of the clients should be host-dependent.

In [31], we proposed a new incentive mechanism for such a provider-assisted mobile UPIs. Unlike the incentive mechanism for the autonomous UPI service, in this case, the mechanism is designed and applied by the operator. The latter determines the free quota reimbursement and the data price that is charged to each user (host or client), in order to achieve her goal. That is to increase the amount of served data and maximize her revenue. Based on the operator's decisions, each host determines accordingly how much traffic to admit and serve for other clients, and how much to consume for herself. If the charged prices are high and the reimbursements are low, it is possible that a host will not utilize all her available bandwidth. The operator anticipates the hosts' strategies and optimizes her decisions accordingly. This type of interaction can be modeled as a noncooperative Stackelberg two-stage game [31].

The particularly challenging aspect here is that the pricing and reimbursement decisions are intertwined and have different impacts on the hosts' data consumption and servicing policy. To illustrate this, we considered a scenario where 400 clients are uniformly assigned to 50 hosts. The demand of clients and the hosts is elastic. Figure 6.5a depicts the operator's optimal pricing-reimbursement strategy when the average user demand varies from $q = 0$ to $q = 1$ Mbit (per time period). Namely, each point represents the pair of charged price and free quota, which ensures the highest operator revenue for a given value of average user demand q. Observe that when data demand increases, the operator increases the charged prices and reduces the free quotas.

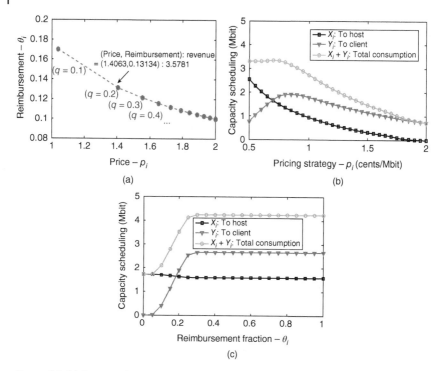

Figure 6.5 (a) Operators' optimal pricing–reimbursing strategy, (b) (average) host's servicing decisions for different data prices (fixed reimbursement), and (c) different reimbursement quotas (fixed charged prices). Network parameters: 400 clients uniformly assigned to 50 hosts, user elastic demand with respect to data prices, price variation from 0.5 to 2 (while $\theta_i = 0.2$), and free quota fraction from 0 to 1 (for $p_i = 0.7$). Further details can be found in [31].

In Figure 6.5b, we plot the total amount of data that a host i consumes on average, as a function of the charged data price. Moreover, we show how this data is allocated for the needs of the host and her clients. As it is anticipated, the aggregate consumed data $X_i + Y_i$ (Mbit) decreases with the price, since the demand of the users is considered elastic. What is interesting is that as the price increases, the host is motivated to allocate a larger portion of her servicing capacity to her clients (Y_i) and less for her own needs (X_i). This is because the reimbursement becomes relatively more beneficial for the host. However, for higher charged prices, the total transmitted data decreases, and for even higher prices, the host only downloads or uploads data to serve the needs of her clients. Finally, Figure 6.5c depicts the hosts' optimal downloading (uploading) and servicing policy as the operators reimbursement policy changes. The latter is described by the parameter $\theta_i \in [0, 1]$ that determines the fraction of the data served by a host i, that is returned as a reimbursement to her. For example, if

$\theta_i = 0.5$, for every 1 Mbit host i serves, she receives 0.5 Mbits of free data. We see that as the reimbursement increases, the amount of served data Y_i increases as well.

It is clear that the implementation of this hybrid pricing policy requires a monitoring and charging system so as to reward the hosts according to the traffic they serve. Such a system can be easily implemented as part of the accounting system that an operator already has. Finally, the design of this mechanism requires the collection and analysis of usage statistics so as to assess the average user demand and the elasticity of this demand. This will further introduce additional signaling overhead.

6.5 Incentive Mechanisms for Large-Scale Systems

In this section, we explore a fully decentralized and lightweight sharing mechanism for UPIs. Unlike the solutions in Sections 6.3 and 6.4, this system does not require any virtual currency scheme, exchange of messages among the participants, or the intervention of a central entity for aligning the hosts' and clients' actions. As the result, this mechanism has a good scalability property and is very suitable for large-scale systems such as those arising in the IoT. Indeed, in IoT thousands of devices will be interconnected in UPI platforms, and these devices often have limited computation capabilities and hence may not be able to support sharing protocols as those discussed in the previous sections.

The main idea of this resource sharing mechanism is surprisingly simple. Each user, at any given time, will have to serve with her spare resources (idle bandwidth, computation, or storage capacity) the neighboring user (with needs) that has been most generous with her in the past. As long as the users comply with this basic sharing rule, the entire network will be driven to an equilibrium which is lexicographically optimal, in terms of the resources each user contributes to the community and receives by it in return. The proof of this elegant result can be found in [32, 33].

Figure 6.6 presents a simple example of this idea. Assume there are $\mathcal{N}$ nodes, embedded in a network graph $G = (\mathcal{N}, \mathcal{E})$ that prescribes which nodes can collaborate directly with each other. It is assumed that two-hop collaboration is not possible, a practical technical limitation that commonly arises in such systems (e.g. mobile hotspots cannot connect users over two hops). The nodes share directly their spare resources, which can be network connectivity, storage, computation capacity, or actually any combination of these. A time-slotted operation is considered, where in each slot every node $i \in \mathcal{N}$ has a randomly generated amount of spare resource $D_i(t) \geq 0$, with $E[D_i(t)] = D_i$, that can share with its neighbors in $\mathcal{N}_i$. The allocation rule simply states that node i will allocate her entire resource $D_i(t)$ to the node $j \in \mathcal{N}_i$ with the currently lowest ratio $\rho_j = R_j(t)/D_j$, where $R_j(t)$ is the average

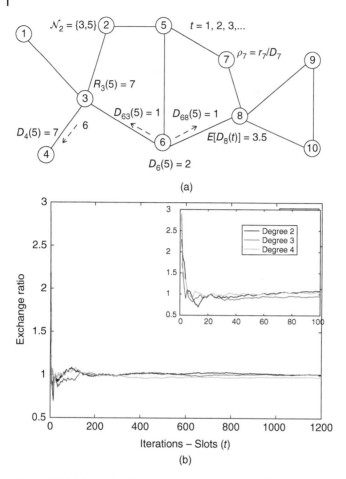

Figure 6.6 (a) A set of nodes exchange resources on a dynamic fashion. The system operation is time-slotted, and during each slot t a node i can have $D_i(t)$ spare resources that allocates to one or more of its neighboring nodes $j \in \mathcal{N}_i$, $D_{ij}(t) \le D_i(t)$. The performance criterion for this UPI model is the long-term exchange ratio r_i/D_i of the average received resources over the average resources it contributes to the UPI D_j. (b) The simple max-ratio allocation rule ensures that the vector of the exchange ratios $\rho = (\rho_1, \rho_2, \dots, \rho_N)$ will converge to its lexicographically optimal value.

(over time) amount of total resource j has received until slot t. Note that this node j is essentially the most generous neighbor of i since it allocates resource D_j (we assume all spare resources are allocated) and has received on average R_j resource (until slot t). In other words, node i will select the most generous of its neighbors and assign its entire spare resource in that slot.

Clearly, this is a very lightweight mechanism as it only requires each node to be aware of the current exchange ratio values of its one-hop neighbors. The nodes do not have to agree on pricing schemes, or run auctions, or bargain for the resources. Hence, the aforementioned mechanism is suitable for large-scale systems. At the same time, this sharing rule leads to quite impressive allocations. Namely, the resulting exchange ratio vector $\rho = (\rho_1, \rho_2, \ldots, \rho_N)$, which determines how much each node earns from the UPI in the long-run, is not only lexicographically optimal (hence Pareto efficient and fair) but also robust to strategic interactions. In other words, this algorithm is incentive-compatible in the sense that no single node or group of nodes can apply any other strategy that will maximize their received resources. This means that the achieved collaboration point is also a competitive equilibrium in a pertinent market game, where nodes allocate their resources in a selfish fashion. The practical implication of this result is that nodes have no reason not to comply with this simple sharing rule, and even a central system controller (who has the full control of the UPI) it would not be able to achieve a better overall system performance than this simple mechanism.

6.6 Open Challenges in Mobile UPI Incentive Mechanisms

Despite the recent efforts from industry and academia, there are still many open issues pertaining to incentive mechanisms for mobile UPI solutions. Here, we discuss several of them.

6.6.1 Autonomous Mobile UPIs

6.6.1.1 Consensus of the Service Provider

A particularly important aspect of the autonomous UPI services is that they bridge heterogeneous networks in an uncontrolled fashion. For instance, users participating in such services may transfer data from cellular to Wi-Fi networks (offloading) or from Wi-Fi to cellular networks (onloading). Similarly, when it comes to computation, users that have some local tasks to execute for, say, an Amazon-sponsored service might use the set-top boxes of another cloud provider. Furthermore, this is accomplished in an ad hoc basis, i.e. without the approval of the providers. Clearly, this could lead to additional servicing cost and even to significant congestion for some networks. Therefore, it is not surprising that providers have expressed their concerns regarding the adoption of these services.[3]

An important next step in designing incentive mechanisms for autonomous UPIs is to reconcile the conflicting objectives of providers and users. This will

---

3 For example, regarding the concerns raised by AT&T for Open Garden, please see: http://en .wikipedia.org/wiki/Open_Garden.

result in hybrid schemes where the mobile UPIs are not fully controlled by the provider nor are completely autonomous. To this end, a possible solution could be to design cost sharing policies that balance the benefits of users (for participating in such UPIs) and the cost of the operators for admitting traffic from nonsubscribers. A recent discussion on this issue based on the open-garden type of Internet sharing can be found [11].

6.6.1.2 Dynamic Setting

In these autonomous UPIs, user interactions are often spontaneous and short-term, and such network dynamics reduce the incentives of cooperation. For instance, we saw in Figure 6.6b that the exchange policy needs tens of iterations before converging to the fair point. Similar problems that appeared in peer-to-peer and ad hoc networks were addressed by reputation schemes, where the contribution of each user is quantified though accumulative metrics. However, in mobile UPIs, we need metrics beyond the simple ones in the literature to fully characterize the multiresource consumption and sharing history.

A possible solution for this problem could be the adoption of a virtual currency system, as we described in the previous section. Nevertheless, in such a dynamic environment, users are not able to predict the efficacy of this currency. This is because it is not guaranteed that, in the near future, they will find gateways or relays to serve them, even if they can pay for these services. Therefore, it is required to carefully tune the currency system's parameters (e.g. how much a relay service costs), based on extensive numerical simulations and field experiments. Other possible solutions include the design of dynamic policies that determine the resource sharing and servicing policy in an online fashion and without a priori knowledge of the demand or resource availability [30].

6.6.2 Provider-assisted Mobile UPIs

6.6.2.1 Modeling the Users

Our analyses rely on certain assumptions about the users. First, when acting as clients, it is assumed that they are always willing to accept services from the hosts, which the mechanism specifies. In a competitive environment with more than one hosting options (or even different types of UPI services), the clients will be able to select the most beneficial solution. In order to understand their strategic decisions, it is important to model the specific needs of the clients. For example, some users may have needs for delay-sensitive video streaming services, while others may prefer low-cost low-bandwidth basic Internet access. These different needs can be described, for example, with properly selected utility functions [17]. Moreover, it is important to understand how the decisions of clients affect the servicing cost of hosts and in turn the expenditures of operators. The coupling of these three decision planes renders the design of a good resource allocation mechanism particularly challenging, as it requires the coordination of many different entities.

When the users are acting as hosts, we have so far assumed that they make resource allocation decisions based on a utility model, i.e. they are driven by their expected benefits and costs. Nevertheless, recent studies conducted with actual users who were sharing network resources in a pertinent behavioral experiment [34], revealed that in practice, humans are more collaborative than expected. Namely, users were aiming to establish reciprocal relationships and exchange (almost) equal amounts of resource with their neighbors in the network graph, rather than solely maximizing their received resources. Clearly, UPI solutions need to account for such behavioral biases, especially for systems that actively include the humans in the decision loop.

6.6.2.2 Incomplete Market Information

Many times, the interaction among operators, hosts, and clients will be realized without complete network information to any of the parties involved. For example, the operator will not be aware of the actual communication needs of the hosts, which are essential to derive the proper charge and reimbursement rules. Similarly, a host may not know which client is willing to pay the highest price. For these cases it is necessary to leverage market mechanisms, such as auctions and contracts, in order to elicit this hidden information.

Ideally, one would like to design a provider-assisted UPI market where hosts and clients will reach agreements dynamically, based on their time-varying needs and the network congestion level. However, this mobile data market and the cloud computing market often have a hierarchical structure, with the hosts acting as intermediaries, that renders the design of such mechanisms challenging. For example, auctions, which are often used for clearing a market under incomplete information in networks, are only designed for two classes of market entities (sellers and buyers). This calls for new auction-based mechanisms tailored to such hierarchical settings. Similarly, it is worth to investigate how one can design optimal contracts for solving these problems, where a principle will propose suitable contract items, based on the scenario and the services, and the users will select the most preferable for their needs.

6.7 Conclusions

UPI-based services are expected to play a crucial role in the future network and computing systems [1], both for autonomous self-organizing user communities [7–9, 28, 29] and for provider-assisted services [4, 31]. Designing proper incentive schemes for these systems is a particularly challenging problem, as it requires us to bridge the interests of clients, hosts, and service providers. In this chapter, we considered both the technical and economic issues that arise in this context. Motivated by successful business models, we presented two novel mobile UPI models that focus on network sharing. For each one of them, we analyzed the salient features that a suitable incentive mechanism should have and outlined a potential solution. Furthermore, we presented a lightweight sharing algorithm that achieves a Pareto optimal and incentive-compatible sharing equilibrium, and hence is suitable for large-scale IoT systems. Finally, we discussed the remaining open challenges in this area along with possible directions for future research.

References

1 Sofia, R. and Mendes, P. (2008). User-provided networks: consumer as provider. *IEEE Communications Magazine* 46 (12): 86–91.

2 FON. www.FON.com (accessed 12 May 2018).

3 Antoniadis, P., Courcoubetis, C., Efstathiou, E.C., and Polyzos, G.C. (2003). Peer-to-peer wireless LAN consortia: economic modeling and architecture. Proceedings of the 3rd P2P Conference, pp. 198–199.

4 Ma, Q., Gao, L., Liu, Y.F., and Huang, J. (2017). Economic analysis of crowdsourced wireless community networks. *IEEE Transactions on Mobile Computing* 16 (7): 1856–1869.

5 Ma, Q., Gao, L., Liu, Y.F., and Huang, J. (2018). Incentivizing Wi-Fi network crowdsourcing: a contract theoretic approach. *IEEE/ACM Transactions on Networking* 26 (3): 10351048.

6 Iosifidis, G., Gao, L., Huang, J., and Tassiulas, L. (2014). Incentive mechanisms for user-provided networks. *IEEE Communications Magazine* 52 (9): 20–27.

7 Tang, M., Wang, S., Gao, L. et al. (2017). MOMD: A multi-object multi-dimensional auction for crowdsourced mobile video streaming. Proceedings of IEEE INFOCOM.

8 Tang, M., Pang, H., Wang, S. et al. (2018). Multi-dimensional auction mechanisms for crowdsourced mobile video streaming. *IEEE/ACM Transactions on Networking* 26 (5): 2062–2075.

9 Tang, M., Pang, H., Wang, S. et al. (2019). Multi-user cooperative mobile video streaming: performance analysis and online mechanism design. *IEEE Transactions on Mobile Computing* 18 (2): 376–389.

10 Tang, M., Gao, L., and Huang, J. (2017). A general framework for crowdsourcing mobile communication, computation, and caching. Proceedings of IEEE Globecom.

11 Zhang, M., Gao, L., Huang, J., and Honig, M.L. (2017). Cooperative and competitive operator pricing for mobile crowdsourced internet access. Proceedings of IEEE INFOCOM.

12 Open Garden Mobile App. www.opengarden.com (accessed 12 May 2018).

13 Karma Mobile Operator. https://yourkarma.com/wifi (accessed 12 May 2018).

14 Comcast xfinity. https://www.xfinity.com/learn/mobile-service (accessed 12 May 2018).

15 Whisher App. https://whisher.en.uptodown.com/ (accessed 12 May 2018).

16 Telefonica BeWifi. http://www.tid.es/research/areas/bewifi (accessed 12 May 2018).

17 Musacchio, J. and Walrand, J.C. (2006). Wi-Fi access point pricing as a dynamic game. *IEEE/ACM Transactions on Networking* 14 (2): 289–301.

18 Efstathiou, E., Frangoudis, P., and Polyzos, G. (2010). Controlled Wi-Fi sharing in cities: a decentralized approach relying on indirect reciprocity. *IEEE Transactions on Mobile Computing* 9 (8): 1147–1160.

19 Iosifidis, G., Gao, L., Huang, J., and Tassiulas, L. (2013). An iterative double auction for mobile data offloading. Proceedings of WiOpt, pp. 154–161.

20 Lee, K., Lee, J., Yi, Y. et al. (2013). Mobile data offloading: how much can Wi-Fi deliver?. *IEEE/ACM Transactions on Networking* 21 (2): 536–550.

21 Sharma, A., Navda, V., Ramjee, R. et al. (2010). Cool-tether: energy Efficient On-the-fly Wi-Fi hot-spots using mobile phones. Proceedings of ACM CoNEXT, pp. 109–120.

22 Do, N., Hsu, C., and Venkaramanian, N. (2012). CrowdMAC: a crowdsourcing system for mobile access. Proceedings of ACM Middleware, pp. 1–20.

23 Jung, E., Wang, Y., Prilepov, I. et al. (2010). User-profile-driven collaborative bandwidth sharing on mobile phones. Proceedings of ACM MCS.

24 Bogliolo, A., Polidori, P., Aldini, A. et al. (2012). Virtual currency and reputation-based cooperation incentives in user-centric networks. Proceedings of IWCMC, pp. 895–900.

25 Nash, J.F. (1950). The bargaining problem. *Econometrica: Journal of the Econometric Society* 18 (2): 155–162.

26 Haci, H., Zhu, H., and Wang, J. (2012). Resource allocation in user-centric wireless networks. Proceedings of IEEE VTC, pp. 1–5.

27 Afrasiabi, M.H. and Guerin, R. (2012). Pricing strategies for user-provided connectivity services. Proceedings of IEEE INFOCOM, pp. 2766–2770.

28 Iosifidis, G., Gao, L., Huang, J., and Tassiulas, L. (2014). Enabling crowd-sourced mobile internet access. Proceedings of IEEE INFOCOM.

29 Iosifidis, G., Gao, L., Huang, J., and Tassiulas, L. (2017). Efficient and fair collaborative mobile internet access. *IEEE/ACM Transactions on Networking* 25 (3): 1386–1400.

30 Giatsios, D., Iosifidis, G., and Tassiulas, L. (2016). Mobile edge-networking and control policies for 5G communication systems. Proceedings of WiOPT.

31 Gao, L., Iosifidis, G., Huang, J., and Tassiulas, L. (2014). Hybrid data pricing for network-assisted user-provided connectivity. Proceedings of IEEE INFOCOM.

32 Georgiadis, L., Iosifidis, G., and Tassiulas, L. (2015). Exchange of services in networks: competition, cooperation, and fairness. Proceedings of ACM SIG-METRICS.

33 Georgiadis, L., Iosifidis, G., and Tassiulas, L. (2014). Dynamic algorithms for cooperation in user-provided network services. Proceedings of NetGCoop.

34 Shirado, H., Iosifidis, G., Tassiulas, L., and Christakis, N.A. (2019). Resource sharing in technologically-defined social networks. *Nature Communications* 10: 1079.

7

Fog-Based Service Enablement Architecture

Nanxi Chen[1], Siobhán Clarke[2], and Shu Chen[3]

[1] *Chinese Academy of Sciences, Bio-vision Systems Laboratory, SIMIT, 865 Changning Road, 200050, Shanghai, China*
[2] *The University of Dublin, Distributed Systems Group, SCSS, Trinity College Dublin, College Green, Dublin 2, Dublin, Ireland*
[3] *IBM Ireland, Watson Client Solution, Dublin, Ireland*

7.1 Introduction

IoT empowers people's everyday environment by supporting their information and service requirements through interconnected devices. Nowadays, a large number of end users and devices have been substantially supported by IoT [1]. According to [2], by 2020, over 50 billion physical devices are expected to be connected by IoT. Many strategies for IoT application development adopts service-oriented computing architectures where resources (e.g. raw data generated by sensors, computation, and storage capability, and management configuration) are modeled as services and deployed in the cloud or at the edge [3]. To broadly serve the engagement of such resources, fog computing, as a paradigm extending cloud computing [4, 5], is proposed by Cisco in 2013 [6, 7]. Fog provides a novel mechanism to allow services being deployed in somewhere close to the end user or data sources to reduce the cloud's traffic demands and increase the response speed.

With the development of IoT and fog services deployment, smart cities have drawn a great interest in both the research and engineering fields [8]. In a smart city, massive end devices, including CCTV cameras, vehicles, mobile phones, and multimodal sensors, are connected to support functional services to maintain assets and resources in an urban area. Smart cities have triggered increasing demands for new IoT applications that are not only for users in a special vertical domain but also for those have cross-domain functionality requirements [9, 10]. In other words, the smart city system should collect and cooperate data from various end devices in different domains to make a more accurate decision to serve the urban residents.

Fog and Fogonomics: Challenges and Practices of Fog Computing, Communication, Networking, Strategy, and Economics, First Edition. Edited by Yang Yang, Jianwei Huang, Tao Zhang, and Joe Weinman.
© 2020 John Wiley & Sons, Inc. Published 2020 by John Wiley & Sons, Inc.

Taking an urban navigation system as an example, it reports real-time traffic status to drivers and to optimize the traffic routing with respect to traffic condition, this system uses not only real-time traffic volume data but also the arrangement of land buildings in urban areas, weather conditions, simultaneous alert on traffic emergencies, etc. The system also needs to bind different service providers to provide value-added services. For example, an urban fire alert system monitors the fire safety of living houses and needs to connect to various service providers including the fire stations, hospitals, and local buildings to guarantee an efficient and full functional reaction to a fire emergency.

Cross-domain data aggregation and services cooperation raise the requirement of flexible service provision [11]. Microservices has emerged as a promising solution to enable a flexible service provision and application development [12]. It decomposes application functions into a set of minimal self-contained entities and makes them accessible through a unified interface [13]. Such functionalities can include intelligent manufacturing, smart transportation, cellular networks, just to name a few. A microservice can be deployed on shared devices at the network edge (a.k.a., fog nodes) to support remote service invocation in an IoT environment. When a new application that has a complex business process is required, a software developer is allowed to find some microservices that match the required function and were previously deployed in the network. These microservices will then be composed to support a corresponding business process. However, microservice providers in a fog network are likely to be owned by heterogeneous vendors, such as the weather station and the traffic manager in the urban navigation example. Such heterogeneous vendors may follow different principles for creating service interfaces. In addition, cooperation among different fog nodes may be required to tackle a particular user requirement that is too complex or resource-intensive to be satisfied by microservices in a single device, and so a composition of microservices across multiple fog nodes will tackle such a requirement.

To enable microservices and service composition in fog computing networks, this chapter introduces a flexible, decentralized semantic-based service composition model named as FogSEA. FogSEA can span along with the cloud-to-things continuum to process cross-domain IoT requests. It manages microservices through a fog network that is formed by fog nodes. Fog nodes are the abstraction of devices that share their resource to the other devices in their vicinity. FogSEA utilizes microservices and allows them to be deployed on fog nodes. Each of the microservices can process a subtask of an IoT application.

7.1.1 Objectives and Challenges

FogSEA aims to address two important issues that are microservices organization and microservices disambiguation.

Microservices organization: Microservices of interest are generally contributed by different third-parties, so they are not supervised by one authority that can allocate services and maintain real-time service information. Asking all the available autonomous fog nodes to register their microservices on a central service provisioning system is not always possible because of the resource limitation in fog computing environments. In addition, a composition user's awareness of available microservices is limited by its communication range. An efficient service discovery and composition model is necessary to search services based on their real-time availability and compose possible services to achieve a composition user's request. FogSEA organizes various fog services providers and microservices through a decentralized service composition model. A cross-domain application can be formed by dynamically allocating a group of microservices that can cooperate to accomplish all the subtasks of this application. This mechanism reduces overhead caused by intercommunication between microservices and increases the robustness of service composition against those user requests composed of complex subtasks.

Microservices disambiguation: Microservices' ambiguation can be caused by heterogenous services interface. Such service interface specifies the microservice's functionality, the Input/Output (I/O) types, the Quality of Service (QoS), and any other attributes that relate to service invocation and execution. There is a lack of a unifying principle to specify such attributes for microservices from varied domains, which can cause semantic ambiguity when enabling service composition across multiple different application domains. Given the delay caused by communication between fog nodes, knowledge exchange in the network during service composition should reduce composition traffic. The semantic service description also needs to trade-off between including enough knowledge to support service matchmaking and minimizing the size of the knowledge base. FogSEA proposes a way to describe microservices according to the data-flow among them and adopts rapid semantic-based service matchmaking to bridge the semantic gap between the description of cross-domain applications and the functionalities of deployed microservices.

7.2 Ongoing Effort on FogSEA

FogSEA is a decentralized service enablement architecture that includes a two-layer fog infrastructure, as shown in Figure 7.1. The basic structure of FogSEA is first proposed in 2017 [14], which contains a hierarchical service management model that allows small entities like little companies or even an individual person to join a fog computing environment as a service provider. It supports service composition that combines services provided by different

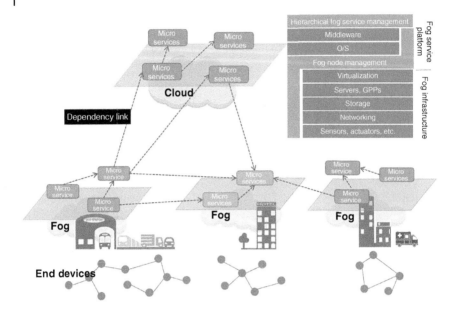

Figure 7.1 FogSEA service providers and the semantic dependency network.

entities to create a new value-added service. FogSEA has been extended with a flexible service deployment model [15] that grants software vendors to rent a set of fog infrastructure to host their services, and such services can be dynamically redeployed according to real-time service demand. This chapter will introduce FogSEA and propose a semantic microservice model to address the semantic heterogeneity issue.

Figure 7.1 illustrates a very basic structure of FogSEA. It mainly includes a service dependency overlay network built over the fog infrastructure (a.k.a., physical layer). In FogSEA, the fog infrastructure maintains various types of end devices connecting to the fog network, manages the connectivity among devices in this network, and implements visualization on computation capability, network topologies, and physical/visual storage resources. It adopts service enablement to allocate smart devices as fog nodes and to share the resources of these fog nodes in an IoT environment. The service dependency overlay network is managed by a fog service platform that is hidden from remote end users. It consists of node management modules, IoT middleware, and operating systems. Integrating these components constructs a horizontal computing platform across different systems, end devices, and networks to facilitate the development of cross-domain IoT applications over a distributed fog infrastructure. This platform contains two components, as a hierarchical fog service management module and a fog node management module. The former provides the functionalities to form cross-domain applications using deployed microservices and to manage real-time collaboration of various microservices to guarantee the fluency of executing those applications.

FogSEA organizes microservices according to their data dependency or logical dependency, to satisfy a specific functional requirement. That is to say, if one microservice is linked to another, this microservice whether has the similar function to another or its execution depends on the other one's execution result. In brief, FogSEA implements flexible service composition from microservices' perspectives to address the following issues in a microservice enablement fog network. FogSEA assumes that a fog-based application is formed by a series of domain-specific microservices. After receiving a service request, a fully functional service composition process decouples the request into a set of subtasks, and each of which is corresponding to a domain-specific microservice, rather than simply providing information query or file transmission services. Then, the fog networking searches for the corresponding microservices by which a specific service can be processed.

To serve semantic-based service composition, the fog infrastructure contains semantic models to do semantic disambiguation and allocation. This model indexes the predefined semantic dictionary and explains the subtasks in a user request using it. After task explanation, the semantic model searches for proper microservices which are semantically matched with these subtasks. FogSEA executes in the context of a component-based middleware structure, as illustrated in Figure 7.2. A fog node uses these middleware components to share its microservice. A fog service provider encapsulates its microservice's semantic interface as a probing message and uses this message to advertise to

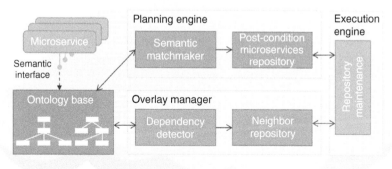

Fog service provider

Figure 7.2 Supporting architecture for fog services providers.

the network. Other fog nodes in the network receive the message and use the dependency detector and the ontology base to check if they have a microservice that matches the I/O parameters of the advertised microservices. The overlay manager's neighbor repository is used to hold those microservices it detects as a semantic dependency match and their dependency match values. During service composition, the planning engine processes a composition request, and the semantic matchmaker generates a match value for this request. If the value is bigger than a candidate threshold, the request's sender will be recognized as a post-condition service and held in the post-condition service repository. The service execution engine manages the post-condition service repository by setting a (configurable) lifetime for each composition request. If the time expires and a provider has not been invoked for the request, the repository maintenance module drops all the post-condition microservices associated with the request. This module also updates the neighbor repository when a neighbor can no longer be reached by this fog node.

7.2.1 FogSEA Service Description

The fog-based service platform provides microservice interfaces to allow a user request to access those interfaces through which subtasks in this request can be processed without knowing the implementation details of the corresponding interfaces. A microservice interface is defined by a third-party in terms of semantic annotations. It is described as a 3-tuple: $S = \langle f, \Omega_{in}, \Omega_{out} \rangle$, where f represents the functionality of S. Ω_{in} and Ω_{out} describe the set of input and output parameters of the service, respectively. Each input parameter $I_i \in \Omega_{in}$ is described as a 3-tuple $\langle IC, IS, IT \rangle$ that includes IC for the category of the input data, IS for its semantic name, and IT for its data type. Each output parameter $O_j \in \Omega_{out}$ is characterized as $\langle OC, OS, OT \rangle$. An example of a service specification is presented in Figure 7.3. The block illustrates the Semantic Service Interface (SSI) [16] of a RoutePlanning service and the annotation of this service from the provider's perspective.

In this chapter, we assume that some fog nodes in the open environment (e.g. crossroads, parks, shopping malls) agree to share the microservices on their devices and that there is no higher authority in the network to schedule services for service requesters. Consider a navigation task request from a pedestrian's mobile phone that includes a service workflow with three operations: Map Caching, Navigating, and Route Rendering. As shown in Figure 7.3, there are microservices in the network that are capable of supporting the full functionality of the user request, but their specifications do not semantically match that of the request, and the descriptions of I/O parameters of the data-dependent services also do not match. Without a central knowledge base that understands such differing specifications, or some other mechanism that can mediate between them, these operations cannot work together.

Requested workflow:

Map caching → Navigating → Route rendering

Available services:

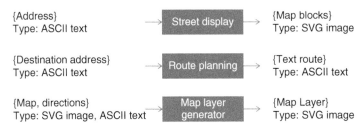

{Address}
Type: ASCII text
→ Street display →
{Map blocks}
Type: SVG image

{Destination address}
Type: ASCII text
→ Route planning →
{Text route}
Type: ASCII text

{Map, directions}
Type: SVG image, ASCII text
→ Map layer generator →
{Map Layer}
Type: SVG image

Figure 7.3 Example microservices.

We also assume that many of the devices in the environment are likely to be resource-constrained (in particular, bandwidth) or otherwise unwilling to invest a large proportion of their resources servicing a request.

Existing matchmakers for service composition usually employ a standard ontology for terms (e.g. WordNet [17]) or merge heterogeneous service ontologies in a service repository. Instead of using a full ontology for matchmaking microservices, generate ontology slices based on WordNet and hold these ontology slices for matchmaking.

We define an ontology slice for term a, $a.Ont_{id}$, where a represents an annotation, and Ont_{id} indicates an ontology slice id for a. To prepare ontology slices for matchmaking, a microservice can separately index all the terms used for annotating services in the WordNet's taxonomy, and then, extract hierarchy-based ontology slices for each term. An ontology slice is taxonomy-like, containing the indexed term's synonyms, sister terms, and all the direct hyponyms, hypernyms, as well as holonyms. Each set of synonyms corresponds to only one entity and can be represented by Syn (e.g. $goal \in destination.Syn$), and sister terms are captured as different entities in the same hierarchy.

For instance, one of the ontology slices for *DestinationAddress* can be characterized as *IS.Ont2: location{< address > {street address; mailing address; residence; abode; business address}}* as expressed in Figure 7.4. The *RoutePlanning* service's input *DestinationAddress* is described by two ontologies capturing features using different language constructs. Note that *DestinationAddress* is a composite term[1] capturing the entries for it in WordNet. As is common in

---

1 Bianchini defines the concept of a composite term in [18]: "A term is a simple term, denoted by st, if an entry for st exists in WordNet … A term is a composite term, denoted by $ct = \{st_1, st_2, …, st_n\}$, if ct is composed by more than one simple term st_i and an entry for ct does not exist in WordNet."

```
<xs: element name ="Route Planning" type="ASCII"
sawsdl : modelReference="RoutePlanning">...
<xs: element name ="current address" type="ASCII"
sawsdl : modelReference="Location.currentAddress">...
<xs: element name ="destination address" type="ASCII"
sawsdl : modelReference="Location.destinationAddress">...
<xs: element name ="text route" type="ASCII"
sawsdl : modelReference="Location.textRoute">...
...
<wsdl : interface name="RoutePlanning">
        <wsdl : operation name="RoutePlanning">
                <wsdl : input message="current_Address">
                <wsdl : input message="destination_Address
                <wsdl : output message="text_Route">
        </ wsdl : operation>
</ wsdl : interface>
...
```

-Semantic Service Interface-

$f_c = \{\text{Route Planning}\}$

$S_{in} : \{ IC_1 = \text{location},$

$IS_1 = \text{current address},$

$IT_1 = \text{ASCII text},$

$IC_2 = \text{location},$

$IS_2 = \text{destination address},$

$IT_2 = \text{ASCII text} \}$

$S_{out} : \{ OC_1 = \text{location},$

$OS_1 = \text{text route},$

$OT_1 = \text{ASCII text} \}$

Figure 7.4 Example of a service specification.

English, the central concept of a composite term is indicated by its rightmost simple term. So, the term *TextRoute* has a central concept Route, and the category of the concept Route represents the core category of the full term, which is Location as depicted in Figure 7.5. *Text* is the *TextRoute's* noncentral concept. This classification of the two kinds of concepts is used in matchmaking to determine terms' relevance.

7.2.2 Semantic Data Dependency Overlay Network

The semantic-based service composition in FogSEA differs from existing ontology-based service composition in its distributed maintenance of I/O parameter relations and its use of lightweight ontologies. We find semantically dependent relations rather than semantic similarities for services. Combined with the proposed microservice interface, this method can minimize the effort required to detect I/O parameter relations. We also extend a decentralized backward-chaining model [16] for service composition. To enable efficient detection of semantic I/O parameter relations during composition, we introduce a Semantic Data Dependency Overlay Network (SeDDON), which extends our previous Semantic Service Overlay Networks (SSONs) [16]. SSON introduced a semantic-based method to create a distributed service overlay network for service composition in mobile computing. Unlike traditional service overlay networks [19], where the data dependencies between services can be characterized as a directed acyclic graph (DAG) for global reasoning, SSON allows microservices to maintain a list of their semantic dependent neighbors to enable local reasoning. However, while SSON addressed semantic-based service composition, the construction of an SSON network relies on domain-specific ontologies, which are flexible for

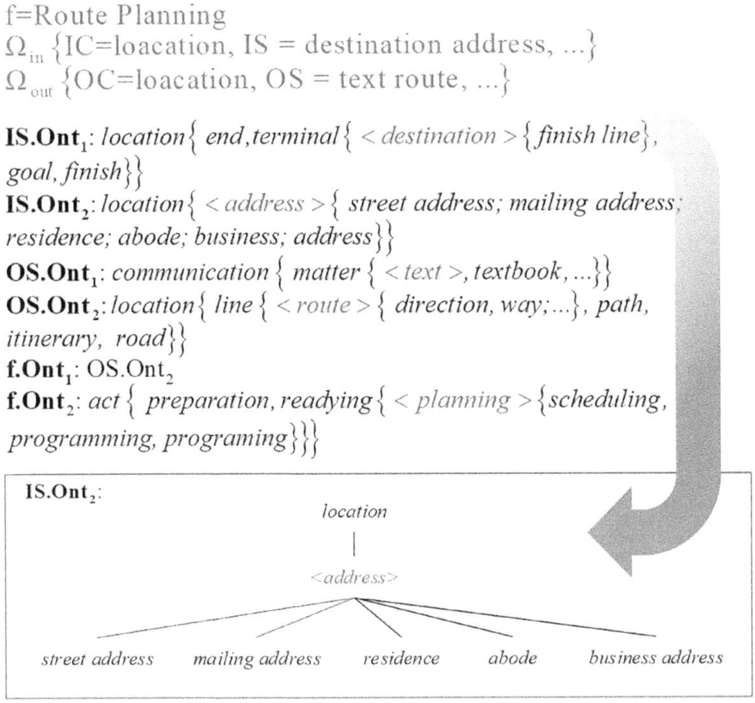

f=Route Planning

Ω_{in} {IC=loacation, IS = destination address, ...}

Ω_{out} {OC=loacation, OS = text route, ...}

IS.Ont$_1$: *location* { *end,terminal* { < *destination* > { *finish line* }, *goal, finish* } }

IS.Ont$_2$: *location* { < *address* > { *street address; mailing address; residence; abode; business; address* } }

OS.Ont$_1$: *communication* { *matter* { < *text* >, *textbook, ...* } }

OS.Ont$_2$: *location* { *line* { < *route* > { *direction, way;...* }, *path, itinerary, road* } }

f.Ont$_1$: OS.Ont$_2$

f.Ont$_2$: *act* { *preparation, readying* { < *planning* > { *scheduling, programming, programing* } } }

IS.Ont$_2$:

location

|

<address>

street address   mailing address   residence   abode   business address

Figure 7.5 Example for ontology slices (extracted from the example in Figure 7.4).

users with limited domain ontology knowledge. Moreover, the use of complete domain-specific ontologies and corresponding matchmaker is likely to require resource-rich microservices, which is inappropriate in fog environments with a large number of resource-constrained devices. In addition, SSON employs complex matchmaking that defines five different relations (depend, similar, share, support and exact) between microservices. This can cause an overhead because it requires microservices to maintain more links with their neighbors.

SeDDON proposed in this article improves the efficiency and scalability of both SSON and the semantic matchmaking process. It employs a lightweight SSI to capture the required information for dependence detection, an ontology generation scheme (Section 7.2.2.1), and matchmaking based on semantic distance (Section 7.2.2.2) for the creation of links between dependent microservices.

7.2.2.1 Creation and Maintenance

We define a semantic dependency match function *DepMatch* (S_1, S_2) between services S_1 and S_2. This approach extends the definition of semantic distance *d(a,b)* in EASY [20] with the concept of synonyms *(Syn.)* and applies it in Dep-Match. Our semantic distance function *d*(a_j, b.Ont_i)* for term $a_j \in a$ and *b* in ontology *b. Ont_i* is defined as:

$$d^*(a_j, b.Ont_i) =$$

$$\begin{cases} \infty(NULL) & \text{if } \forall a_j \in a, a_j \in b.Ont_i \\ 1 & \text{if } \forall a_j \in a, a_j \in b.Syn_i \\ pathLength(a_j, b.Ont_i) + 1 & \text{otherwise} \end{cases}$$

where the function *pathLength(a_j, b.Ont_i)* indicates the length of the shortest path from a_j to *b* in ontology tree *b.Ont_i*. For example, as shown in Figure 7.3, *d*(streetAddress, destinationAddress.Ont_2)* = 2. In addition, we define the semantic similarity function *SemSim(a, b)*∈*[0, 1]* between two terms a and b:

$$SemSim(a, b) = \frac{1}{n} \sum_{\substack{y \in a}} \max_{x \in b.Ont} ((d^*(y, x))^{-1} CategoryRank)$$

where *n* represents the number of ontologies that are used to describe *b*, and *CategoryRank* demonstrates the domain relevance of the two terms. In other words, if the core concept of the two terms is the same, their relevance is strong, otherwise it is weak. Therefore, function *DepMatch(S_1, S_2)* for acquiring the I/O parameter dependency between service S_1 and S_2 can be denoted as follows:

$$DepMatch(S_1, S_2) = \max_{x \in S_1.\Omega_{in}} \left(\max_{y \in S_2.\Omega_{out}} (SemSim(x, y)) \right)$$

If *DepMatch(S_1, S_2)* is above a dependency threshold value $\theta_{dep} \in [\ , 1]$ defined by the microservice according to its host device capabilities, we state that service S_1 depends on service S_2. As can be seen in Figure 7.2, we consider an output *{OC = location, OS = TextRoute, OT = Text}* for a navigation service as an instance. The ontologies for text and route can belong to two domain categories, Communication and Location, according to their semantics. Location is textroute's core concept. Route, with ontology *(location {line{<route> {direction, way; ...}, path, itinerary, road}})* matches OC and has stronger domain relevance than text with *(communication{matter {<text>,textbook, ...}})*. Consider a TextToAudio service using *IC=communication* and *IS=text* as input data. If we consider *CategoryRank* as set to 1 for terms with strong relevance and *CategoryRank* = 0.5 for terms with weak relevance, the dependency matching value for the microservice TextToAudio and the service Navigation is *DepMatch (TextToAudio,RoutePlanning)* = 0.25.

SeDDON takes advantage of these functions to allow participating fog node to deduce their microservices' I/O parameter relations. For example, let us' say a newcomer (fog node) *A* enters a network by broadcasting its semantic service profile, using a probe message. All the providers that can receive the message use the function *DepMatch* to detect if they depend on, or reverse-depend on, or have no relation with the newcomer. An existing node *B* receives this message and detects if the newcomer's output data semantically matches the dependent data it requires for the invocation of its microservices. If such a match is detected by *B*, a dependence link from *B* to *A* ($B \rightarrow A$) can be built and maintained by *B*. *A* can be regarded as *B*'s semantic dependent neighbor. *Node C* receives the message and finds it matches *A*'s input, not *A*'s output, which means the newcomer A relies on *C*'s execution result. In this case, the system creates a dependence link from *A* to *C* ($A \rightarrow C$) and stores it in *A*. Dependence links enable our distributed backward planning for service composition, reducing runtime communication overheads.

7.2.2.2 Semantic-Based Service Matchmarking

The previous Section 7.2.2.1 detailed how microservice interfaces are implemented, and FogSEA employs SeDDON to indicate semantic dependencies via the ontology generation scheme. The former introduces how FogSEA uses semantic annotation to determine microservices, and the latter is the foundation of semantic-based overlay network deployed across FogSEA. By combining these two mechanisms, FogSEA implements the service composition model which will be introduced next.

The service composition model in FogSEA (FogSEA-SCM) is based on the SeDDON network and is designed to support fast, lightweight service composition. To this end, SeDDON defines microservice interfaces by specifying its I/O parameters with ontologies, and in the perspective of microservices, a user request is specified as an abstract workflow with a sequence of subtasks.

Traditional distributed service composition models require systems to discover microservices for each of the ordered subtasks individually, and in the predefined sequence. That means each subtask's operation requires at least three communications with the subtask requester (as characterized in Figure 7.6a): first to discover microservices, second to bind the optimal one based on the fog topology and the status of fog nodes in usage, and last to invoke other essential microservices and send the required data. This mechanism has significant overhead for the network as network communication depletes a device's energy, which in turn would decrease the lifetime of the network overall [21].

FogSEA-SCM reduces the network communication required for a service composition process. First, the overall service delivery system sends out a composition message for a user request with the ultimate goal of obtaining the output required at the end of the task. The request is received and processed by microservices who decide if they will take part in the composition. A microservice will participate if it matches the last subtask in the request (i.e. those subtasks that can produce the final, required output). If so, this microservice updates the composition request by removing the ending subtask and forwards the updated request to its dependent neighbors. Figure 7.6b illustrates

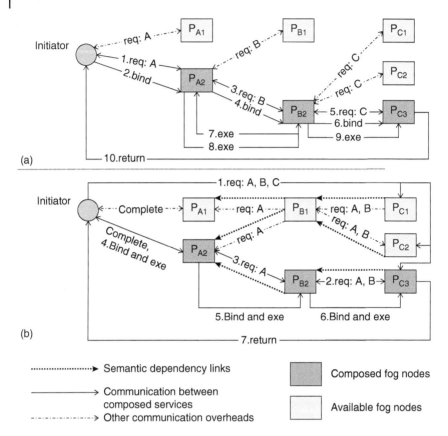

Figure 7.6 Service composition: (a) baseline and (b) backward (msg: message, req: request, exe: execute, P: fog node).

FogSEA-SCM, which may continue with the discovery of the next backward subtask before the current subtask-matchmaking process has collected all the matched results. A preliminary version of FogSEA-SCM was given in [16]. Results indicated a good success ratio for the preliminary FogSEA-SCM. In this article, we extend the work in [16] associated with the semantic similarity measures, increasing the likelihood that the final composition planning's output execution path better fits the user's requirement.

We define a complex user task (request) as $T = <T_{in}, T_{out}, \phi>$, where T_{in} expresses the input data provided by a requester, T_{out} represents the required results, and $\phi = \{R.f_1 \rightarrow R.f_2 \rightarrow R.f_3 \rightarrow ...\}$ indicates a series of subtasks structured as a workflow, and each of $R.f_i \in \phi$ indicates a set of function requirements for a subtask. Microservices match their published services S to a requested subtask $R.f_1$ using a semantic matching function to produce a match value for service binding:

$$Match(R.f_i, S) = \frac{1}{n} \sum_{x \in R.f_i} \max_{y \in S.f}(SemSim(x,y))$$

where n indicates the number of required functions in the subtask. If $Match(R.f_i, S)$ is above a candidate threshold value $\theta_{can} \in [\ , 1]$, we state that service S supports subtask $R.f_i$.

FogSEA-SCM is illustrated in Algorithms 7.1 and 7.2. A service composition request req for a complex user task T is defined as $req = <T, MV>$, where MV is the subsequent subtask's match value. In Algorithm 7.1, an initiator in the network sends a request $req = <T, MV>$ over the SeDDON network to start a service composition process and wait for composition participators' responses (Lines 1 through 3 in Algorithm 7.1). Note that the MV for the initiator is 0 as the subsequent task is unknown. Microservices in the SeDDON network receive the request and check if they match the ending subtask of the request. If a match is found by a microservice P, P stores the request sender as well as the match value MVi (Line 11 in 2), updates the request by eliminating a matched ending subtask and the corresponding output (Line 9), and modifies the MVh with MVi (Line 16 in 2). The request sender is P's post-condition service. If a microservice matches a subtask's entry (i.e. subtasks that directly receive data from the composition requester, or that are executed first in the workflow) in the request, this microservice sends a complete token that contains a match value to the requester (Lines 12 and 13 in Algorithm 7.2). The requester receives complete tokens and chooses microservices with the most appropriate match value to execute (Lines 6 through 11 in Algorithm 7.1). After processing the composition request, each microservice P participating in the composition maintains a group of post-condition services, whose invocation depends on P's execution result. During service execution, P invokes the provider with the highest match value in this post-condition group.

Algorithm 7.1   Initiator

```
1   send req
2   set timer C
3   /* Waiting */
4   receive token t=<T'in, T'out, φ', MV', P>
5   add t to TOK and add P as candidate
6   if C expires and ∩TOK φ'≠φ
7     then fail
8   if C expires and ∩TOK φ'=φ
9     then
10      choose K'⊂TOK such that ∩'K φ'=φ ∧ ∀K''⊂ TOK,
        ∩''Kφ''= φ, MV'>MV''
11  send input data to P in t such that t∈K'
```

Algorithm 7.2 Microservice provider

```
1   INPUT:  P : <f, Ωin, Ωout>
2           req : <T, MVh>
3           Candidate threshold θcan
4   /* Listening */
5   receive request req
6   foreach R.fi∈φ of T
7    if Match(R.fi, P)=MVi>θcan
8     φ\R.fi and ∃ outj∈Tout such that outj∈Ωout
9    then outj←Ωin
10     add sender D to repository R
11     store MVh
12     if φ has an empty branch
13       send t=<Tin, Tout, φ, MVi, P> to initiator
14     else
15       get semantic dependent neighbour list N
16       update req with the new T and the MVi
17       forward the updated req to N
```

7.3 Early Results

This section presents the evaluation on the performance of FogSEA in two parts: first, we measured the FogSEA service composition model's communication effort and response time, and second, we measured the communication effort. To determine a baseline, the evaluation hired the probe/discovery model [22] with distributed composition strategy [23]. This baseline binds all required services at the beginning and then executes the composed services. The probe/discovery model is based on semantic links in the network detected by probe messages, in which services with similar functionalities can be managed as semantic neighbors, so that service requesters can forward their requests via semantic neighbors and then locate the best matched services in the network. There are two forwarding policies for service requesting in the baseline: minimal, where the algorithm stops after finding one exactly matching service, and exhaustive, where the algorithm stops after finding all matching services from its overlay. In addition to comparisons of communication effort and response time, the experiment compared the traffic required to resolve user requests in FogSEA against the baselines introduced previously.

The experiments evaluated FogSEA and the baselines with the same scenario configuration, with multiple scenarios. Scenario configurations are defined by the different combinations of controlled or random variables that are shown in Table 7.1. Assume each service provider hosts one microservice. All the participating microservices first structured the corresponding logical overlay

Table 7.1 Simulation configuration.

General	
Simulator	NS-3
Clients	1
Communication	145 m
Random	
Node placement	
Service execute time	10–100 ms
Controlled	
Number of providers	50, 100, 150, 200 (Average number of providers for each subtask: 5, 10, 15, 20)
Composition mode (Evaluation A)	Baseline, backward
Overlays (Evaluation B)	Probe/discovery [22], FogSEA
Candidate threshold (for the backward)	0.5
Global similarity threshold (for the baseline)	0.5
Composition length	4, 5, 6, 7, 8, 9, 10

networks (i.e. SeDDON and the baseline semantic community overlay). Then, we simulated the service composition process to measure and record traffic (i.e. the number of messages sent) and the process's response time. Since the quantity of logical dependency links of these overlay networks can affect the results, we implemented a matchmaking control strategy for overlay construction, which ensures that the two evaluated overlay models have similar complexity.

7.3.1 Service Composition

(1) *Traffic*: Communication effort is measured by the number of communication messages sent (traffic). We measured service composition traffic with varying composition lengths (starting with 4 as smaller composition lengths did not provide significant results, and using 10 as the maximum) and with varying microservice densities of 50, 100, 150, and 200. Each round was run 10 times, and the results are shown in Figure 7.7. We first measured the two composition models with function-matching threshold of 0.5 and composition length of 5 (Figure 7.7a). The experiment investigated the baseline's forwarding policies, including baseline-minimal and baseline-exhaustive and illustrated the results of traffic for service

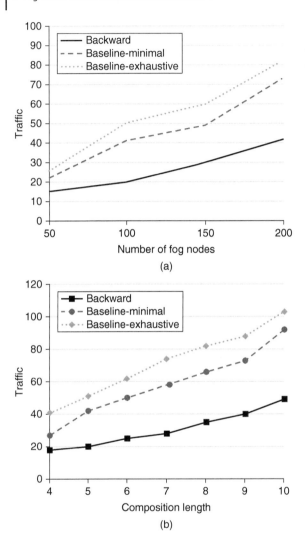

Figure 7.7 Traffic for service composition in varying (a) service densities and (b) composition complexity.

composition at different lengths of requests in Figure 7.7b. It shows that the message overheads for both models increase when increasing microservices' participation in the system, as high microservice densities indicate complex overlay networks. However, the results show that across the range of microservice numbers from 50 to 200, FogSEA produces about 42% less traffic compared to the better baseline approach (minimal forwarding policy). Another important point is that, from the perspective of an individual provider, the average cost of participation ranges from 1.2 to 1.9 messages when there are 100 microservices in the network, with

Figure 7.8 Response time for overlay construction in varying numbers of composition length and dense/sparse connectivity of the network.

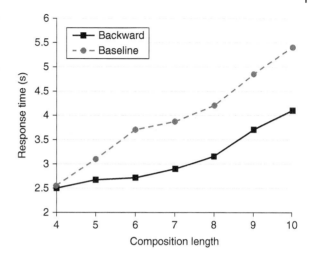

candidate threshold varying from 0.2 to 0.9. This does not vary significantly with higher numbers of providers or more complex requests.

(2) *Response time*: Figure 7.8 evaluated the response time for service composition from a composition requester's perspective. Response time is the period from dispatching a service composition request to getting the composed services' execution result. As shown in Table 7.1, we used a random value between 10 and 100 ms to represent execution time. Generally, as one would expect, the more complex the request, the more costly it is to discover and execute it, increasing the response time. The results illustrated in Figure 7.8 show that the planning-based service execution in FogSEA takes less time to return an execution result than the baseline approach. Note that Figure 7.8 illustrates the comparison against the better-performing baseline only.

7.3.1.1 SeDDON Creation in FogSEA

Figure 7.9 illustrated the results measuring the creation of SeDDON network in FogSEA against its communication effort, which is the number of communication messages sent (traffic in Figure 7.9). Generally, traffic increases as more microservices join the network. Figure 7.9 shows that SeDDON creation needs to send less messages across varying connectivity densities and microservice numbers. The main reason is that SeDDON allows microservices to advertise their services only once when they join a network, and only the microservice that detects the new node as a reverse-dependence neighbor needs to reply to it. Overall, across all microservice numbers from 50 to 200, SeDDON has about 19% less traffic.

After evaluating FogSEA, in this section, we discuss composition participants' resource constraints, which are intrinsic to distributed semantic-based

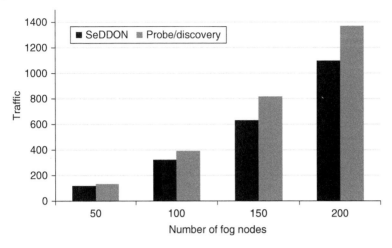

Figure 7.9 Traffic for overlay construction in varying numbers of microservices and dense/sparse connectivity of the network.

service composition to achieve composition-related tasks such as semantic matchmaking, interacting to reason about the best provider, maintaining composition information, and constructing semantic knowledge bases. In the early results of this chapter, we focused on measuring communication effort, which is a major energy consumer. The service composition model in FogSEA shows a reduction in communication over related approaches. With an average 1.1–1.8 messages for each participant, we conclude that reasonably small battery power is required for this task. In the worst case, there are two messages per node sent during a service response process: one to resolve composition requests and the other to bind and execute a subsequent service. Creating the SeDDON overlay also has less communication than other approaches, though does add some communication load. The numbers range from an average of 1.84 messages per provider for creating an overlay with 50 nodes, through to 3.15 : 100, 4.22 : 150, and 5.4 : 200 average message:nodes. When a node joins an existing overlay, the worst-case scenario is 1 message per node, if all existing nodes have a reverse dependency to the newcomer.

7.3.2 Related Work

Semantic Web technologies have emerged as a demonstrable solution to the semantic ambiguity problem in service-oriented computing [24, 25]. Specifically, systems can annotate services with semantic concepts that can be detected and matched with a service query at runtime. In general, however, dynamic matchmaking can be resource-consuming and reduce the overall performance of service composition, especially in service-rich environments such as those predominant in an IoT computing. Examples of such investigations applied to IoT computing environments are task computing [26], and the MEDUSA middleware [27], both of which use dynamic matchmaking for service discovery.

7.3.2.1 Semantic-Based Service Overlays

Semantic-based service overlays optimize dynamic matchmaking by moving a large part of the resource-consuming process associated with semantic matchmaking offline, thereby increasing service composition responses' efficiency as follows.

EASY [28] provides a lightweight semantic service discovery mechanism for distributed IoT environments. The main idea of this approach is to create a set of global DAGs to indicate related capabilities between published services. These DAGs define service profiles with three semantic elements: service inputs, outputs and categories, using the distance of a service's and the request's homologous concepts in ontologies to calculate the semantic similarity between them. CoSMoS [24, 29] is a dynamic service composition model for distributed component-based systems using semantic graphs and centralized ontologies. This approach maintains a graph indicating the semantic relations between components in a distributed system. Beyond their achievements, both EASY and CoSMoS have demonstrated that a centralized semantic graph is impractical in dynamic and resource constrained environments since it requires the presence of resource-rich nodes (e.g. central repository to organize such graph-based discovery).

Chord4S [30] presents a hierarchical service description based on domain taxonomies and equips distributed hash table (DHT)-based [31] overlay networks with the hierarchical service specification to promote data availability and service discovery efficiency. However, DHT supports only key-based query, thereby limiting the discovery to a single service. Pirro et al. [32] combine a semantic overlay network (SON) technique with DHT, increasing service discovery efficiency. In this approach, hash functions are used for service advertisement and semantic links for service grouping, where a group of services with similar functionalities can be routed quickly. These approaches [30, 32] are built on the assumption that services rely on concrete domain ontologies. Since there is no standard domain taxonomy for general commercial or industrial service domains, they have limited flexibility when matching services across domains as they expect a user (requester in an IoT environment) to have all required ontology knowledge. This would require each user to have ontologies covering all domains in question.

SECCO [33] proposes an ontology mapping method for structuring semantic links in P2P networks. It allows a system to describe services as ontologies. The system uses a matchmaking scheme between two ontologies to capture similar capabilities in corresponding services, and then a semantic link can be established between matching services, increasing efficiency in subsequent service queries. However, SECCO relies on a centralized ontology, which, as before, consumes significant execution resources.

To construct semantic links in P2P networks, P2P-SDSD [22] is proposed, which is a probe/discovery model for dynamic service discovery. In P2P-SDSD, a probing phase is applied before processing service discovery and then linking service providers with similar functionality. P2P-SDSD also proposes forwarding policies to reduce runtime traffic for service discovery.

ESTEEM [34] also uses semantic links and extends Bianchini's [18] matchmaking scheme. It proposes an overlay network called a service semantic community, which contains a group of services that have semantic relations. For example, when Service A extends Service B and intersects Service C, ESTEEM determines that A, B, and C have semantic relations to each other. However, neither P2P-SDSD nor ESTEEM takes implicit workflow descriptions into consideration during service composition, which is likely to reduce flexibility during service composition.

7.3.2.2 Goal-Driven Planning

Semantic-based goal-driven planning is an alternative solution to construct flexible service composition. For example, RodriguezMeier et al. propose [35] in which DAGs are employed to explore the relations between services in the network. This usage of DAGs differs from that in EASY since it maintains services' dependencies to underpin more flexible service planning. However, RodriguezMeier's approach shares a drawback with EASY as it also requires central entities for the graph-based service directory, which infers resource-rich nodes and also affects scalability.

WSPR [36] introduces a novel artificial intelligence (AI) planning-based algorithm for large-scale web services. This work is based on the analysis of complex networks by using two different matching schemes to support partial service matchmaking. The EQSQL-based planning algorithm [37] applies rank-based models to promote service composition efficiency. It employs WordNet [17] to resolve ontology heterogeneity. However, these approaches require central service repositories, and their composition engines are too heavyweight for distributed IoT environments.

Automatic bi-direction service composition has also been investigated to cope with the semantic heterogeneity problem [38, 39]. This composition model considers the implementation of a user request as a transition from an initial state (input data or executing preconditions) to the goals (output data or user requirements). The process resolves the request starting with a backward query, which explores the available services in the network from the request's goal to its initial state. Then, a forward-chaining process finds a solution. In [39], the authors model a web service as a conversion from an input state to an output state. It uses an OWLs ontology to maintain published services as a dependency graph. They employ a bi-direction planning algorithm to find a path with the smallest cost from the dependency graph. However, this approach relies on central repositories to maintain the dependency graph. Furno and Zimeo [38] advance fully distributed support for unstructured P2P networks. Bi-direction planning algorithms are used to deduce composition plans. Once a new service request is planned and bound to a composite service supported by a sequence of providers, these service providers elect a super-peer to record this composite service and group the providers by holding links to them. Thereafter, this super-peer responds quickly to service requests with the composite of services to process goals. If a task can be solved by a number of different super-peers, another super-peer is elected which maintains links to these. Thus, a tree-like overlay network is built over time, with super-peers as branches. By referring to super-peers, normal peers can locate provider compositions for their service requests if these requests have been resolved (or at least partly resolved) in the network. This work is the closest solution to our FogSEA approach, but it assumes the network has super-peers that are capable of managing network links. This assumption is not safe in dynamic environments, as super-peer-based groups may fail if any relevant peer leaves the network, or is no longer able to fulfill its role because of reduced capacity. In addition, this work assumes standard ontologies for goal matching, and as discussed in the previous Section 7.2.1, this is inappropriate considering environment heterogeneity. Finally, this work does not take dynamic service binding into consideration, as the service execution starts only when all the service providers have been bound. This has a communication overhead and is inflexible in dynamic environments [40].

WSMO [25] describes a single-direction planning models, which forward or backward chain service providers for user requests. PM4SWS [41]also introduces a model based on the same planning scheme. They do not consider the combination of the two directions of chaining, which differs from that of the bi-direction composition models. WSMO reasons about service execution plans and adapt composite services on the fly, addressing flexible service composition, but requiring central controllers to schedule services. PM4SWS is a distributed framework to discover and compose semantic web services. This service exploration process simply floods the network with query messages and applies dynamic matchmaking for service discovery. However, dynamic semantic service discovery is time-consuming for distributed IoT environments, especially where there are large numbers of services to be considered.

7.3.2.3 Service Discovery

To satisfy fast response and tailor IoT services for a specific requirement, researchers have been focusing on the service discovery associated with all the things in IoT networks. Investigations in [2, 42–45], although proposing various methodologies to address the service discovery, have their limitation, which those methodologies reply on a mobile management model provided by the vendors of IoT applications to solve the dynamic service discovery.

Reference [42] concludes that composition approaches based on Semantic Service Oriented Architecture (SSOA) fail to support dynamic user requests in heterogeneous IoT environments. CloudFit proposed in [2] adopts deploying tailored cloud services on top of a pervasive grid as an alternative to traditional cloud infrastructures. CloudFit advances a Platform as a Service (PaaS) middleware to create a private cloud and uses both computation capability over the generated pervasive networks. Reference [43] proposes an alternative solution of using pervasive service discovery, in which it defines a functioning prototype as a microservice and utilizes semi-lazy bandit algorithms to implement context-aware service discovery.

MI-FaaS [45] proposes anther service discovery mechanism to address dynamic cloud cooperation in mobile environments. To allocate services, MI-FaaS uses a hybrid algorithm, which defines a utility function recording the number of executed services and maximizes the historical record to predict the federation of local IoT clouds [45]. As a hybrid approach, MI-FaaS demonstrates its performance associated with the high density of IoT requests. To the cost, this mechanism requires significant historical records to achieve accurate predictions and uses a centralized orchestration which is increasing the traffic consumption of the overall network.

7.3.3   Open Issue and Future Work

This chapter introduced FogSEA to implement cross-domain application composition. FogSEA aims to solve the challenges of service provisioning in resource-limited IoT environments with respect to semantic-based service composition and seamless service providing in dynamic environments. Its architecture includes the fog infrastructure and the fog-based service platform. The fog infrastructure manages the cloud-thing devices and the corresponding functionalities in the fog network. The fog-based service platform hides the implementation details and provides various interfaces to serve user requests. FogSEA can dynamically organize a group of microservices to resolve cross IoT applications in which a set of subtasks are required to be executed sequentially. To ensure the ability of processing request dynamically, FogSEA adopts semantic-based dependency overlay network associated with ontology schemes and proxy fog nodes to allocate microservices. The chapter presented early experiments on FogSEA to demonstrate its performance regarding network traffic and response time.

To be applied in the physical world, FogSEA still confronts a number of challenges. We conclude a set of open issues for FogSEA as follows:

When processing real-time IoT requests, FogSEA should have the ability to decouple complex application, which could be absolutely new to its composition ability. During the service composition, another challenge is how to understand and disambiguate the semantic information of subtasks since the enthronement of deploying FogSEA, and the scenario of delivering such an application is dynamic at runtime. Last but not the least, the unpredictable user actions increase the complexity of constructing IoT applications across the network. For example, users could keep roaming and requesting different fog nodes to require one IoT application service. Based on this scenario, we conclude a series of challenges for FogSEA.

Complex application logic: A conceptual composite is an abstract model that states a specific functionality as a series of service requirements, each of which indicates a concrete service to be used. General service composition relies on previously generated conceptual composites and assigns microservices at runtime to recreate them. This chapter focused on requests with linear composite (or workflows). It is also important to cater for more composite types, such as parallel tasks, or iterative workflows. However, in an open environment, services of interest are independently deployed and maintained by different mobile devices so that it is not always possible to generate a conceptual composite for particular functionality. Composition users may provide the conceptual composite as a part of the service request, but such a user-defined composite is likely to be at variance since the operating environment is dynamic.

Unpredictable services availability: Mobile fog nodes, such as in-vehicle fog, may establish a wireless network in ad hoc ways. Any mobile fog node can arbitrarily offer and drop their services, as well as join and leave the network during execution. Such a dynamic mechanism can bring significant potential to improve the overall QoS by recomposing microservices including those participating in service execution. But the cost is increasing the dynamic of network topology and making the availability of executable microservices unpredictable at runtime. The former can cause an increasing consumption when recomposing microservices, and the later can lead to the network fails to construct some IoT applications due to the absence of essential microservices during application execution.

Limited system knowledge: Service discovery searches for and selects services that match the required functionality. A service provision system may need to dynamically reason about a functionality and find an appropriate combination of microservices that support it when an individual one is unavailable. Having a global system view of the computing environment is beneficial for such a reasoning-based service discovery since global service knowledge will facilitate reasoning processes and increase composition success. However, obtaining service knowledge from mobile devices leads to traffic overhead because it relies on multi-hop data transmission and when mobile devices have a limited communication range, there are likely to be a larger number of wireless transmission hops. In addition, maintaining a global system view can be expensive especially when the service and network topology are frequently changing.

Privacy and security are also important challenges for distributed service composition [46]. In general, hop-by-hop processing may expose the overall control and data flow to assigned microservices. To FogSEA, we believe that FogSEA-SCM goes some way toward addressing this issue by partitioning the data flow and the composition requester's goal. In particular, a subtask's microservice cannot see the whole data flow, and the requester's data are not visible to the full workflow's microservices.

Failure recovery has not been addressed in the current FogSEA. Composition failures may occur if the search result is empty, or if a participating provider drops out. We have previously outlined an adaptation model for dynamically detecting new candidate providers and updating invalid planning results, which goes some way toward preventing composition failures [16], but we still need a mechanism that can recover compositions when assigned providers drop out at very short notice.

References

1 Bandyopadhyay, D. and Sen, J. (2011). Internet of things: applications and challenges in technology and standardization. *Wireless Personal Communications* 58 (1): 49–69.

2 Steffenel, L.A. and Pinheiro, M.K. (2015). CloudFIT, a PaaS platform for IoT applications over pervasive networks. *European Conference on Service-Oriented and Cloud Computing*: 20–32.

3 Biswas, R. and Giaffreda, R. (2014). IoT and cloud convergence: opportunities and challenges. In: *IEEE World Forum on Internet of Things (WF-IoT)*, 375–376.

4 Chiang, M. and Zhang, T. (2016). Fog and IoT: an overview of research opportunities. *IEEE Internet of Things Journal* 3 (6): 854–864.

5 Mell, P. and Grance, T. (2010). The NIST definition of cloud computing. *Communications of the ACM* 53 (6): 50.

6 Cisco (2015). Fog computing and the Internet of Things: extend the cloud to where the things are. Cisco white paper.

7 Flavio Bonomi, Rodolfo Milito, Jiang Zhu, Sateesh Addepalli (2012). Fog computing and its role in the internet of things. *Proceedings of the first edition of the MCC workshop on Mobile cloud computing. ACM*, pp. 13–16.

8 Tang, B., Chen, Z., Hefferman, G. et al. (2015). A hierarchical distributed fog computing architecture for big data analysis in smart cities. In: *Proceedings of the ASE BigData & SocialInformatics*, 28. ACM.

9 Desai, P., Sheth, A., and Anantharam, P. (2015). Semantic gateway as a service architecture for IoT interoperability. *IEEE International Conference on Mobile Services*: 313–319, 2015.

10 Aloi, G., Caliciuri, G., Fortino, G. et al. (2017). Enabling IoT interoperability through opportunistic smartphonebased mobile gateways. *Journal of Network and Computer Applications* 81 (1): 74–84.

11 Chen, N., Cardozo, N., and Clarke, S. (2016). Goal-driven service composition in mobile and pervasive computing. *IEEE Transactions on Services Computing* 99: 1–1.

12 Wen, Z., Yang, R., Garraghan, P. et al. (2017). Fog orchestration for IoT services: issues, challenges and directions. *IEEE Internet Computing* 21 (2): 16–24.

13 Im, J., Kim, S., and Kim, D. (2013). IoT mashup as a service: cloudbased mashup service for the Internet of Things. *IEEE International Conference on Services Computing*. IEEE Computer Society, 462–469.

14 Chen, N., Yang, Y., Li, J., and Zhang, T. (2017). A fog-based service enablement architecture for cross-domain IoT applications. *IEEE Fog World Congress (FWC)*: 1–6.

15 Chen, N., Yang, Y., Zhang, T. et al. (2018). Fog as a service technology. *IEEE Communications Magazine* 56 (11): 95–101.

16 Chen, N. and Clarke, S. (2014). A dynamic service composition model for adaptive systems in mobile computing environments. In: *IEEE International Conference Service-Oriented Computation*, 93–107. Berlin, Heidelberg: Springer.

17 Miller, A. (1995). WordNet: a lexical database for English. *Communications of the ACM* 38 (11): 39–41.

18 Bianchini, D., Antonellis, V., and Melchiori, M. (2008). Flexible semantic-based service matchmaking and discovery. *World Wide Web* 11 (2).

19 Kalasapur, S., Kumar, M., and Shirazi, B.A. (2007). Dynamic service composition in pervasive computing. *IEEE Transactions on Parallel and Distributed Systems* 18 (7): 907–918.

20 Mokhtar, S., Kaul, A., and Georgantas, N. (2006). Efficient Semantic Service Discovery in Pervasive Computing Environments. In: *Proceedings of the ACM/IFIP/USENIX 2006 International Conference on Middleware*, 240–259. Springer-Verlag New York, Inc.

21 Groba, C. and Clarke, S. (2011). Opportunistic composition of sequentially-connected services in mobile computing environments. *IEEE International Conference on Web Services*: 17–24.

22 Bianchini, D., Antonellis, V., and Melchiori, M. (2010). P2P-SDSD: on-the-fly service-based collaboration in distributed systems. *International Journal of Metadata, Semantics and Ontologies* 5 (3): 222–237.

23 Yu, W. (2009). Scalable services orchestration with continuation-passing messaging. In: *International Conference on Intensive Applications and Services*, 59–64. IEEE.

24 Fujii, K. and Suda, T. (2005). Semantics-based dynamic service composition. *IEEE Journal on Selected Areas in Communications* 23 (12): 2361–2372.

25 Hibner, A. and Zielinski, K. (2007). Semantic-based dynamic service composition and adaptation. *IEEE Congress on Services*: 213–220.

26 Masuoka, R., Parsia, B., and Labrou, Y. (2003). Task computing: the semantic web meets pervasive computing. In: *International Semantic Web Conference*, vol. 2870, 866–881. Berlin, Heidelberg: Springer.

27 Davidyuk, O., Issarny, V., and Riekki, J. (2011). MEDUSA: middleware for end-user composition of ubiquitous applications. In: *Handbook of Research on Ambient Intelligence and Smart Environments: Trends and Perspectives*, vol. 11, 197–219. IGI Global.

28 Mokhtar, S., Preuveneers, D., Georgantas, N. et al. (2008). EASY: Efficient semAntic Service discoverY in pervasive computing environments with QoS and context support. *Journal of Systems and Software* 81 (5): 785–808.

29 Fujii, K. and Suda, T. (2004). Component service model with semantics (CoSMoS): a new component model for dynamic service composition. In: *International Symposium on Applications and the Internet Workshops*, 348–354.

30 He, Q., Yan, J., Yang, Y. et al. (2013). A decentralized service discovery approach on peer-to-peer networks. *IEEE Transactions on Services Computing* 6 (1): 64–75.

31 Harren, M., Hellerstein, J.M., Huebsch, R. et al. (2002). Complex queries in DHT-based peer-to-peer networks. *International Workshop on Peer-to-Peer Systems*: 242–250.

32 Pirro, G., Talia, D., and Trunfio, P. (2012). ADHT-based semantic overlay network for service discovery. *Future Generation Computer Systems* 4: 28, 689–707.

33 Pirro, G., Ruffolo, M., and Talia, D. (2009). SECCO: on building semantic links in peer-to-peer networks. *Journal on Data Semantics XII*, Springer, Berlin, Heidelberg: 1–36.

34 Montanelli, S., Bianchini, D., Aiello, C. et al. (2010). The ESTEEM platform: enabling P2P semantic collaboration through emerging collective knowledge. *Journal of Intelligent Information Systems* 2: 36.

35 Rodriguez-mier, P., Mucientes, M., and Lama, M. (2012). A dynamic QoS-aware semantic web service composition algorithm. In: *International Conference on Service-Oriented Computing*, vol. 7636, 623–630. Berlin, Heidelberg: Springer.

36 Oh, S., Lee, D., and Kumara, S. (2008). Effective web service composition in diverse and large-scale service networks. *IEEE Transactions on Services Computing* 1 (1): 15–32.

37 Ren, K., Xiao, N., and Chen, J. (2010). Building quick service query list using wordnet and multiple heterogeneous ontologies toward more realistic service composition. *IEEE Transactions on Services Computing* 4 (3): 216–229.

38 Furno, A. and Zimeo, E. (2014). Self-scaling cooperative discovery of service compositions in unstructured P2P networks. *Journal of Parallel and Distributed Computing* 74 (10): 2994–3025.

39 Ukey, N., Niyogi, R., Milani, A., and Singh, K. (2010). A bidirectional heuristic search technique for web service composition. *Computational Science and Its Applications*: 309–320.

40 Groba, C. and Clarke, S. (2010). Web services on embedded systems – a performance study. *8th IEEE International Conference on Pervasive Computing and Communications Workshops (PERCOM Workshops)*: 726–731.

41 Gharzouliand, M. and Boufaida, M. (2011). PM4SWS: AP2PModel for semantic web services discovery and composition. *Journal of Advances in Information Technology* 2 (1): 15–26.

42 Urbieta, A., Gonzalez-Beltran, A., Mokhtar, S.B. et al. (2017). Adaptive and context-aware service composition for IoT-based smart cities. *Future Generation Computer Systems* 76: 262–274.

43 Wanigasekara, N. (2015). A semi lazy bandit approach for intelligent service discovery in IoT applications. *Adjunct Proceedings of the ACM International Joint Conference on Pervasive and Ubiquitous Computing and Proceedings*: 503–508.

44 Ku, Y., Lin, D., Lee, C. et al. (2017). 5G radio access network design with fog paradigm: confluence of communications and computing. *IEEE Communications Magazine* 55 (4): 46–52.

45 Farris, I., Militano, L., Nitti, M. et al. (2016). Federated edge-assisted mobile clouds for service provisioning in heterogeneous IoT environments. *Internet of Things IEEE*: 591–596.

46 Buyya, R., Srirama, S.N., Casale, G. et al. (2018). A manifesto for future generation cloud computing: research directions for the next decade. *ACM Computing Surveys (CSUR)* 51 (5): 105.

8

Software-Defined Fog Orchestration for IoT Services

Renyu Yang[1,2], Zhenyu Wen[3], David McKee[1], Tao Lin[4], Jie Xu[1,2], and Peter Garraghan[5]

[1] School of Computing, University of Leeds, Leeds, UK
[2] Beijing Advanced Innovation Center for Big Data and Brain Computing (BDBC), Beihang University, Beijing, China
[3] School of Computing, Newcastle University upon Tyne, Newcastle, UK
[4] School of Computer and Communication Sciences, École Polytechnique Fédérale de Lausanne, Lausanne, Switzerland
[5] School of Computing and Communications, Lancaster University, Lancaster, UK

8.1   Introduction

The proliferation of the Internet and increasing integration of physical objects spanning sensors, vehicles, and buildings have resulted in the formation of Cyber-physical environments that encompass both physical and virtual objects. These objects are capable of interfacing and interacting with existing network infrastructure, allowing for computer-based systems to interact with the physical world, thereby enabling novel applications in areas such as smart cities, intelligent transportation, and autonomous vehicles. Explosive growth in global data generation across all industries has led to research focused on effective data extraction from objects to gain insights to support Cyber-physical system design. Internet of Things (IoT) services typically comprise a set of software components running over different geographical locations connected through networks (i.e. 4G, wireless LAN, Internet, etc.) that exhibit dynamic behavior in terms of workload internal properties and resource assumption. Systems such as datacenters and wireless sensor networks underpin data storage and compute resources required for the operation of these objects.

A new computing paradigm – fog computing – further evolves cloud computing by placing greater emphasis of computation and data storage at the edge of the network, allowing for reduced latency and response delay

Fog and Fogonomics: Challenges and Practices of Fog Computing, Communication, Networking, Strategy, and Economics, First Edition. Edited by Yang Yang, Jianwei Huang, Tao Zhang, and Joe Weinman.
© 2020 John Wiley & Sons, Inc. Published 2020 by John Wiley & Sons, Inc.

jitter for applications [1, 2]. These characteristics are particularly important for latency-sensitive applications such as gaming and video streaming. In this way, the data processing can be greatly decentralized by exploiting compute capacities from not only cloud infrastructures but also from the IoT network itself. In this environment, existing applications and massive physical devices can be leveraged as fundamental services and appliances, respectively. They are composed in a mash-up style (i.e. applications are developed using contents and services available online [3]) in order to control development cost and reduce maintenance overhead. IoT services which involve a great number of data-stream and control flows across different regions that require real-time processing and analytics are especially suitable to this style of construction and deployment. In this context, orchestration is a key concept within distributed systems, enabling the alignment of deployed applications with user business interests.

Let us see a motivating example. Smart cities aim to enhance the quality of urban life by using technology to improve the efficiency of services to meet the needs of residents. To this end, multiple information and communication technology (ICT) systems need to be integrated in a secure, efficient, and reliable way in order to manage city facilities effectively. Such systems consist of two major components: (i) sensors integrated with real-time monitoring systems and (ii) applications integrated with the collected sensor (or device) data. Currently, IoT services are rudimentary in nature and only integrate with specific sensor types. This is resultant of no existing universally agreed standards and protocols for IoT device communication and represents a challenge toward achieving a global ecosystem of interconnected things.

To address this problem, an alternative approach is to use an IoT service orchestration system to determine and select the best IoT appliances for dynamic composition of holistic workflows for more complex functions. As shown in Figure 8.1, the proposed orchestrator manages all layers of an IoT ecosystem to integrate different standalone appliances or service modules into a complex topology. An appropriate combination of these standalone IoT services can be used to facilitate more advanced functionality, allowing for reduced cost and improved user experience. For example, mobile health subsystems are capable of remote monitoring, real-time data analysis, emergency warning, etc. Data collected from wearable sensors that monitor patient vitals can be continuously sent to data aggregators, and in the event of detection of abnormal behavior, hospital personnel can be immediately notified in order to take appropriate measures.

While such functionality can be developed within a standalone application, this provides limited scalability and reliability. The implementation of new features leads to increased development efforts and risk of creating a monolithic application incapable of scaling effectively due to conflicting resource

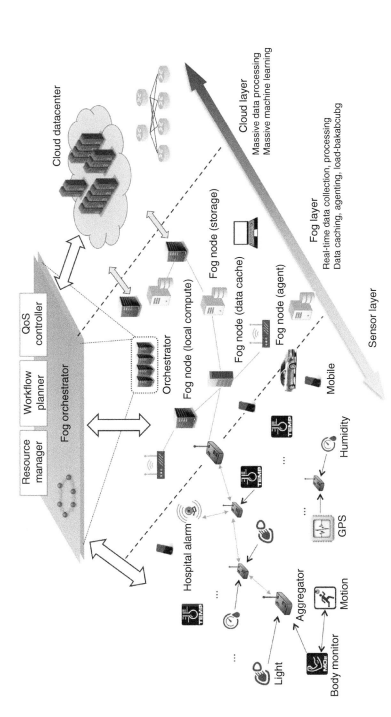

Figure 8.1 An orchestration scenario for an e-Health service. Different IoT appliances (diverse types of sensors and fog nodes) are orchestrated as a workflow across all layers of fog architecture. Several candidate objects can potentially provision similar functionality. The fog orchestrator acts as a controller deployed on a workstation or cloud datacenter and across all organization layers based on global information. Its primary responsibility is to select resources and deploy the overall service workflow according to data security, reliability, and system efficiency requirements. It is noteworthy that the orchestrator is a centralized controller only at a conceptual level and might be implemented in a distributed and fault-tolerant fashion, without introducing a single point of failure.

requirements for effective operation. For reliability, increased application complexity leads to tedious, time-consuming debugging. The use of orchestration allows for more flexible formation of application functionality to scale, and it also decreases the probability of failure correlation between application components.

At present, orchestration within cloud computing environments predominantly address issues of automated interconnection and interaction in terms of deployment efficiency and resource satisfaction from the perspective of the cloud provider [4–8]. However, these works do not consider the effects of network transmission characteristics outside the operational boundary of the datacenter. In reality, the heterogeneity, mobility, and the dynamicity introduced by edge devices within fog environments are greater than those found within cloud environments. Additionally, the emergence of 4G or 5G techniques are still far from mature in terms of response latency and energy efficiency. This has resulted in increasing network uncertainty that may incur tailed execution and security hazards. In this context, it is significantly important to take all these factors into account within the automated resource provisioning and service delivery. Therefore, the fog orchestrator should provide (i) a centralized arrangement of the resource pool, mapping applications with specific requests, and an automated workflow to physical resources in terms of deployment and scheduling; (ii) workload execution management with runtime Quality of Service (QoS) control such as latency and bandwidth usage; and (iii) time-efficient directive operations to manipulate specific objects.

In this chapter, we propose a scalable software-defined orchestration architecture to intelligently compose and orchestrate thousands of heterogeneous Fog appliances (devices, servers). Specifically, we provide a resource filtering-based resource assignment mechanism to optimize the resource utilization and fair resource sharing among multitenant IoT applications. Additionally, we propose a component selection and placement mechanism for containerized IoT microservices to minimize the latency by harnessing the network uncertainty and security while considering different applications' requirement and capabilities. We describe a fog simulation scheme to simulate the aforementioned procedure by modeling the entities, their attributes, and actions. We then introduce the results of our practical experiences on the orchestration and simulation.

8.2 Scenario and Application

8.2.1 Concept Definition

Prior to discussing technical details of orchestration, we first introduce a number of basic terms and concepts.

Appliance: Appliance is the fundamental entity in the fog environment. Appliances include fog things, fog nodes, and cloud servers. Things are defined as networked devices including sensors and devices with built-in sensors which can monitor and generate huge amount of data. Cloud servers store the data and provide parallelized capability of computation. It is noteworthy that a fog node is defined as a particular equipment or middleware residing within the midst of edge things and the remote cloud. It serves as an agent that collects data from a set of sensors, which is then transmitted to a centralized computing system that locally caches data and performs load balancing.

IoT microservice: It is a software unit that provisions a specific type of functionality. For instance, there are a number of demands for data collecting, data streaming gateway or routing, data preprocessing, user data caching, load balancing, firewall services, etc. These functionalities are independently executed, encapsulated into a container, and then placed onto an appliance (except for sensors that simply generate data). Additionally, several candidate objects potentially provision similar functionality and one of them will be eventually selected and deployed as the running instance.

IoT service (IoT application): A complete IoT application typically consists of a group of IoT microservices. All microservices are interconnected to form a function chain that best serve user's requirements. Formally, an IoT application can be depicted as a directed acyclic graph (DAG) workflow, where each node within the workflow represents a microservice. An example is illustrated in Figure 8.2, where the aforementioned e-Health application can be divided into many independent but jointly working microservices.

Fog orchestration: The orchestration is a procedure that enables the alignment of deployed IoT services with users' business interests. Fog orchestration manages the resource pool; provides and underpins the automated workflow with specific requests of IoT service satisfied; and conducts the workload execution management with runtime QoS control. A full discussion of this concept can be found within Section 8.4.

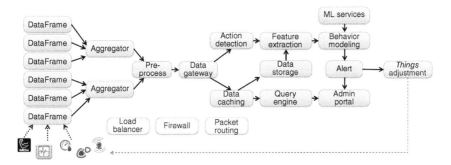

Figure 8.2 E-Health system workflow and containerized microservices in the workflow.

Table 8.1 Comparison between web-based application and fog-enabled IoT application.

Attributes	Web-based	Fog-enabled
Architecture	Cloud + devices	Cloud + fog + things
Communication	Centralized	Hybrid
Interfaces	WSDL/SOAP protocol web service	MQTT protocol [9] Lightweight API
Interoperability	Loosely decoupled	Extremely loosely decoupled
Reliability	Medium	Low

8.2.2 Fog-enabled IoT Application

Traditional Web-based service applications are deployed on servers within cloud datacenters that are accessed by end devices such as tablets, smart phones, and desktop PCs. Similar to Web-based service applications, the cloud provisions centralized resource pools (compute, storage) in order to analyze collected data and automatically trigger subsequent decisions based on a predefined system logic. The most significant difference, however, is the use of fog nodes that transmit data to cloud datacenters. For example, the vast majority of wearable sensor data is collected and preprocessed by smart phones or adjacent workstations. This can either significantly reduce transmission rates or improve their reliability.

We summarize the main differences between Web-based and IoT applications in Table 8.1.

First, IoT communication is performed using a hybrid centralized–decentralized approach depending on context. Most message exchanges between sensors or between a sensor and the cloud are performed using fog nodes. Purely centralized environments are ill-suited for applications that have soft and hard real-time requirements. For example, neighboring smart vehicles need to transfer data between other vehicles and traffic infrastructure to prevent collisions. Such a system was piloted in New York City using Wi-Fi to enable real-time interactions to assist drivers in navigating congestion and to communicate with pedestrians or oncoming vehicles [10]. Furthermore, given the huge number of connected devices, the data volume generated and exchanged over an IoT network is predicted to become many orders of magnitude greater than that of conventional Web-based services, resulting in significant scalability challenges.

Interoperability is another aspect where Web-based and IoT applications diverge. Software-defined networking technologies enable the decoupling of software control and heterogeneous hardware operations. This approach provides an opportunity to dynamically achieve different quality levels for different IoT applications in heterogeneous environments [11]. Moreover, application-level interoperability benefits from Web technologies, such as the RESTful architecture, which provide a high level of interoperability. Using these technologies and the MQTT messaging protocol [9], an abundance of programming Application Programming Interfaces (APIs) can be distributed across entire fog domains and utilized to increase the flexibility of loosely coupled management [12]. Lightweight APIs, such as RESTful interfaces, result in agile development and simplified orchestration with enhanced scalability when composing complex distributed workflows.

A third aspect is reliability. Physical systems make up a significant part of IoT applications; thus, the assumptions that can be made regarding fault and failure modes are weaker than those for Web-based applications. IoT applications experience crash and timing failures stemming from low-sensor battery power, high network latency, environmental damage, etc. [13, 14]. Furthermore, the uncertainty of potentially unstable and mobile objects increases difficulties in predicting and capturing system operation. Therefore, an IoT application workflow's reliability needs to be measured and enhanced in more elaborate ways.

8.2.3 Characteristics and Open Challenges

The diversity among fog nodes is a key issue – location, configuration, and served functionalities of fog nodes all dramatically increase this diversity. This raises an interesting research challenge, namely how to optimize the process of determining and selecting the best software components onto fog appliances to compose an application workflow while meeting nonfunctional requirements such as network latency, QoS, etc. We outline and elaborate on these specific challenges as follows:

Scale and complexity: With the increase of IoT manufacturers developing heterogeneous sensors and smart devices, selecting optimal objects becomes increasingly complicated when considering customized hardware configurations and personalized requirements. For example, some applications can only operate with specific hardware architectures (e.g. ARM, Intel) or operating systems, while applications with high security requirements might require specific hardware and protocols to function. Not only does orchestration cater to such functional requirements but it must also do so in the face of increasingly larger workflows that change dynamically. The orchestrator must determine whether the assembled systems comprising of cloud resources, sensors, and fog nodes coupled with geographic distributions and constraints are capable of provisioning complex services correctly and efficiently. In particular, the orchestrator must be able to automatically predict, detect, and resolve issues pertaining to scalability bottlenecks, which may arise from increased application scale.

Security criticality: In the IoT environment, multiple sensors, computer chips, and communication devices are integrated to enable the overall communication. A specific service might be composed of a multitude of objects, each deployed within different geographic locations, resulting in an increased attack vector of such objects. Fog nodes are the data and traffic gateway that is particularly vulnerable to such attacks. This is especially true in the context of network-enabled IoT systems, whose attack vectors can range from human-caused sabotage of network infrastructure, malicious programs provoking data leakage, or even physical access to devices. A large body of research focuses on cryptography and authentication toward enhancing network security to protect against cyberattacks [15]. Furthermore, in systems comprising of hundreds of thousands electronic devices, how to effectively and accurately evaluate the security and measure its risks is critically important in order to present a holistic security and risk assessment [16]. This becomes challenging when workflows are capable of changing and adapting at runtime. For these reasons, we believe that approaches capable of dynamically evaluating the security of dynamic IoT application orchestration will become increasingly critical for secure data placement and processing.

Dynamicity: Another significant characteristic and challenge for IoT services is their ability to evolve and dynamically change their workflow composition. This is a particular problem in the context of software upgrades through fog nodes or the frequent join-leave behavior of network objects, which will change its internal properties and performance, potentially altering the overall workflow execution pattern [8]. Moreover, handheld devices inevitably suffer from software and hardware aging, which will invariably result in changing workflow behavior and its properties. For example, low-battery devices will degrade the data transmission rate; and unexpected slowdown of read/write operations will manifest due to long-time disk abrasions. Finally, the performance of applications will change due to their transient and/or short-lived behavior within the system, including spikes in resource consumption or data generation [13]. This leads to a strong requirement for automatic and intelligent reconfiguration of the topological structure and assigned resources within the workflow, and importantly, that of Fog nodes.

Fault diagnosis and tolerance: The scale of a fog system results in increased failure probability. Some rare-case software bugs or hardware faults that do not manifest at small-scale or testing environments have a debilitating effect on system performance and reliability. For instance, the straggler problem [17] occurs when a small proportion of these tasks experience abnormally longer execution compared with other sibling tasks from the same parallel job, leading to extended job completion time. At the scale, heterogeneity, and complexity we are anticipating, it is very likely that different types of fault combinations will occur [18]. To address these, redundant replications and user-transparent fault-tolerant deployment and execution techniques should be considered in orchestration design.

8.2.4 Orchestration Requirements

According to the discussed user cases within fog environments, a user firstly provides a specification of their requirement that explicitly describes the basic topological workflow (e.g. from the data collection to the final monitoring system) and the detailed requirements for each workflow node in terms of data locality, response latency, reliability tolerance level, minimum security satisfactory level, etc. In this context, the ultimate objective of the fog orchestration is to transform the logical workflow design from the user perspective into the physically executable workflow over different resources of fog appliances. In this procedure, the requirements that should be at least satisfied can be primarily summarized as follows:

(1) *Exploit fog appliance heterogeneity*: The orchestrator should recognize the diversity of edge devices, fog nodes, and cloud servers and fully exploit the capabilities of CPU, memory, network, and storage resources over the fog layers. At present, neither the conventional cluster management systems [19]–[20] nor the container management frameworks [1, 21, 22] can efficiently detect and leverage the edge resources due to the deficient design of current inter-action protocol and state management mechanism.

(2) *Enable IoT appliance and application operation*: Unawareness of resource availability and IoT application status makes it unfeasible to manipulate any instructions of resource allocation or parameter tuning at runtime. This is also a fundamental step for realizing the interoperations among different appliances in the workflow. Loosely coupled functions or APIs should be designed and accessed via predefined interfaces over the network, which enables the reuse and composition to form a chain of functions.

(3) *Conduct workflow planning optimization and network latency-aware container placement*: For general purposes, the orchestrator is expected to support topology-based orchestration standard TOSCA [23]. Afterward, according to the topological workflow, how to choose the most suitable microservice from the candidates and how to choose the most suitable fog appliances for hosting the selected containerized microservices are two research problems. Due to physically widespread in a local or wide area network of fog appliances, the software services are ideally deployed close to the data sources or data storage in order to reduce the transmission latency. With other factors considered, the orchestrator must support

a comprehensive placement strategy while being aware of appliances' characteristics such as physical capabilities, locations, etc.

(4) *Leverage real-time data and learning techniques for optimization and simulation*: Performance-centric and QoS-aware learning can significantly steer the effectiveness and efficiency of resource allocation, container placement, and the holistic orchestration. This is highly dependent of data-driven approach and machine-learning techniques.

8.3 Architecture: A Software-Defined Perspective

8.3.1 Solution Overview

To fulfill the aforementioned requirements, the initial steps are resource allocation and microservice-level planning before those microservices are deployed and launched. An exemplified construction problem is to first find a suitable microservice instance into container, and then find a physical entity with adequate resources to host those containers. Namely, after obtaining a candidate i that can serve the functionality from a candidate list I for a specific type of microservice t (which is the node within the whole topology T of IoT application), we deploy the selected instance into a container which is hosted by a physical machine or portable device r from the resource set R. The objective is to maximize a utility function (utilFunc) that describes the direction of resource selection and container placement (such as minimizing the performance interference while maximizing the security and reliability) under QoS and capacity constraints. The entire procedure can be outlined in Figure 8.3.

---

maximize:

$$\sum_{t \in T} utilFunc(i, r), i \in C_t, r \in R$$

subject to: $QoS(i, r)$, $i \in C_t$, $r \in R$
$Cap(r)$, $r \in R$

To satisfy the application-specific needs with hard or soft constraints, and the platform-level fairness of allocations among different IoT applications, it is highly preferable to accurately sort out the appliances that can best serve each IoT application. Also, for online decision-making, real time or sometimes faster than real time is urgently required. In some cases, orchestration would be typically considered computationally intensive, as it is extremely time-consuming to perform combination calculation considering all specified constraints and objectives.

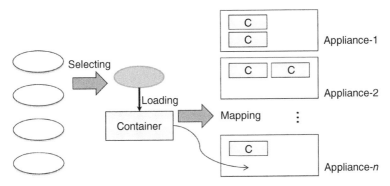

Figure 8.3 Mapping between microservice candidate, containerized microservice instance, and the physical appliances.

After resource selection and allocation, we can obtain an optimal or near-optimal placement scheme based on current system status before the application deployment. After the IoT application is deployed, workload running status and system states should be timely monitored and collected to realize dynamic orchestration at run-time with QoS guaranteed. Meanwhile, with the huge amount of data generated, data-driven optimization and learning-based tuning can facilitate and drive the orchestration intelligence.

8.3.2 Software-Defined Architecture

A significant advantage of a software-defined solution [24] is the decoupling of the software-based control policies and the dependencies on heterogeneous hardware. On one hand, along with the rapid development of mobile and embedded operating system, the programming API and virtualization techniques can be greatly utilized to increase the flexibility of manipulation and management. Virtualization, through the use of containerization, can provide minimized granularity of resource abstraction and isolated execution environment. Resource operations are exposed as interoperable system APIs and accessible to upper frameworks or administrators. On the other hand, the orchestration controls the software-defined architecture. In order to mitigate the overloaded functionalities of control plane in previous architecture, the information plane is decoupled from the control plane. The independent information plane can therefore provision more intelligent ingredients into the orchestration and resource management by integrating pluggable libraries or learning frameworks. Additionally, we adopt container technology to encapsulate each task or IoT microservice. Containers can ensure the isolation of running microservices and create a development and operation environment that is immune to system version revision, submodule updates.

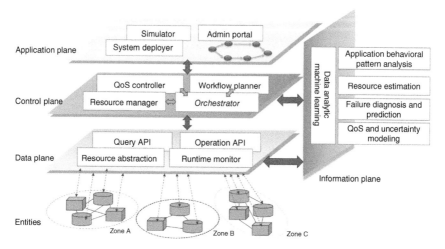

Figure 8.4 Fog orchestration architecture.

As shown in Figure 8.4, the fog orchestration framework is incorporated with the emerging networking and resource management technologies. We design the layered architecture according to the popular SDN reference architecture [24, 25]. The main components are described as follows:

Data plane: The first responsibility of data plane is to regulate and abstract the resources from heterogeneous fog entities. It also provisions an easily accessed API for resource management and application runtime control across the entire fog system. Furthermore, the monitored system states such as resource usage, application-specific metrics, etc. are collected and maintained in the data plane. The build-in query APIs are provided for visualization or administration.

Control plane: The control plane is the decision-making plane that works on the basis of control logic in the overall architecture. It dominates both data flow and control flow, and interconnects with the deployment module and operations of underlying entities and running appliances. The orchestrator mainly takes charge of resource management, workflow planning, and runtime QoS control:

- *Resource manager*: The resource manager is responsible for the resource prefiltering according to the basic demands and constraints in the requests and available resources in the fog environments. In addition, after the final decision made by the planning step, the resource manager also takes the responsibility of resource binding and isolation against other applications. It also takes charge of the elastic resource provisioning during the appliances' execution. They are depicted in Section 8.4.1.

- *Workflow planner*: The planner calculates the optimal mapping of candidate microservices, containerized appliances and the hosting entities. We will detail the relevant techniques in Section 8.4.2.
- *QoS controller*: The controller dynamically tunes the allocated resource, the orchestration strategy at run-time with the QoS guaranteed. They are detailed in Section 8.4.3.

The control module can be implemented in distributed (with each suborchestrator managing its own resource partitions without global knowledge) or centralized (with all resource statuses in the central orchestrator), or a hybrid way for the consideration of scalability and dependability.

Information plane: The information plane lends itself as a vertical within this architecture, provisioning data-driven supporting and intelligent solutions. By exploiting the stored sensed information and system real-time statuses, the data analytic and machine learning (ML) submodule can abstract and analyze the application's behavior pattern and give more accurate resource estimation and location preference in the resource allocation. Also, with the aid of big data analytics, this module can build the performance model based on the QoS and network uncertainty modeling and diagnose system failures preventing them from the regular orchestration.

Application plane: The application-level plane firstly contains an administration portal that aggregates and demonstrates the collected data and allows for visualized interaction. Additionally, a containerized deployer is integrated in this plane, providing cost-effective fog service deployment. It automatically deploys the planned IoT application or services into the infrastructure and continently upgrades current services. The simulation module by leveraging the collected data, modeling the user and appliance's characteristics, resource allocation and placement policies, and the fault patterns, etc.

8.4 Orchestration

In this section, we discuss the detailed research subtopics that we believe are key to tackling the challenges outlined previously. As shown in Figure 8.5, within the life-cycle management, these include the resource filtering and assignment in resource allocation phase; the optimal selection and placement in planning phase; dynamic QoS monitoring and guarantees at runtime through incremental processing and replanning; and big data driven analytics and optimization approaches that leverage adaptive learning such as ML or deep learning to improve orchestration quality and accelerate the optimization for problem-solving. The functionality decomposition based on the life-cycle perspective is orthogonal to the software-defined architecture. In particular, the construction and execution part are mainly implemented in the control

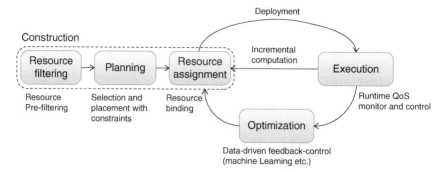

Figure 8.5 Orchestration within the life-cycle management. Main functional elements in our fog orchestrator: resource allocation for filtering and assigning the most suitable resources to launch appliances; the planning step for selection and placement; runtime monitoring and control during execution; and the optimization step to make data-driven decision based on adaptive learning techniques.

plane, and all functionalities are underpinned by the information plane and data plane. The data-driven optimization is associated with the information plane. The application deployment and overall administration will manifest in the application plane.

8.4.1 Resource Filtering and Assignment

Fog infrastructures typically consist of heterogeneous devices with diverse capacity of compute, memory, and storage size. Therefore, resource allocation is a very fundamental procedure for system entities to be launched and executed. One of the responsibilities of Fog orchestrator is to optimize the use of both cloud and fog resources on top of Fog applications. There are two main tasks in the fog ecosystems: containers which encapsulate the microservices and run across tiers in Fog ecosystem and computation-intensive tasks that run in parallel to process the huge volume of data. The resource requests proposed by both sides needs to be timely dispatched and handled in the resource manager. Meanwhile, the resource manager will trigger new iterations of resource allocation by leveraging recently aggregated resources (such as CPU, memory, disk, network bandwidth, etc.). Allocated resources will be guaranteed and reserved for the requesting application. Additionally, the resource manager keeps track of the task progress and monitors the service status.

In essence, the procedure of resource allocation is the matchmaking or mapping between the requirements from the applications that are waiting for execution and the available resources that are dispersed over the fog environment. Therefore, the resource allocation subsystem should fully exploit the diversities and dynamicity of computing clusters at massive scale to improve throughput of computation jobs and reduce the negative impact of unexpected latencies stemming from the jitter of network and occurrence of ineffective queuing. Only through recognizing the accurate targets for placement can the scheduler mitigate the computation straggler or promote resource utility and compaction. Considering heterogeneity in fog [8] is extremely important when conducting the mapping, and such heterogeneity leads to different resource capacities and unique machine characteristics. We need to find out the machines that are determined to be most suitable for specific purposes. In cloud datacenter, this can be typically done through a multistep filtering process that comprehensively considers estimated load, correlative workload performance, and queue states. Similarly, fog resource allocation should be conducted from the following three aspects:

Label-based resource representation: In the fog environment, there is a considerably growing trend of the resource heterogeneity stemming from the rapid development of IoT devices and new types of hardware. This growth provisions more choices for upper applications. For example, hardware such as GPU, FPGA, TPU (Tensorflow Process Unit, NVM, etc.) makes it possible to accelerate the computations and data processing in deep learning, vision recognition, etc. Moreover, for those applications that involve a great many of geo-distributed data access and processing, it is preferable to require the affinities between tasks and the stored raw data. Therefore, consideration of such data and resource affinity is extremely meaningful especially for latency-sensitive applications. In the procedure of resource filtering, we should firstly sort out the collections of destinations that have sufficient available resources and satisfy all specific requirements. To this end, we can adopt the label-based matchmaking between the requests and resources. Formally, the request can be expressed as an n-tuple: $\mathbf{ResReq_i} = (\mathbf{Req_i},$ *LatBound, LocPref*) where $\mathbf{Req_i} = \{Req_i^1, \ldots, Req_i^d\}$ represents the requested resource amount of different labels. The label represents a specific description of resource dimension or a certain constraint. Latency requirement *LatBound* specifies the detailed acceptance level of the latency and response time, and the *LocPref* indicates the preferable execution locations according to the data distribution and processing requirements. On the other hand, the fog resources existing in an fog appliance e can be described as $\mathbf{Res_e} = \{Res_e^1, \ldots, Res_e^d, Priority\}$, where Res_e^i represents the value of ith label, and Priority attribute implies the prioritized level according to the appliance type.

Candidate filtering and resource assignment: Combined all requests with available resources from all active entities, the resource manager tries to rank the candidate fog appliances according to system metrics such as resource status, device load, queuing states, etc. An intuitive latency-aware resource allocation strategy is to firstly allocate resources of edge devices to microservices requiring lower-delay, and microservices with lower level of delay requirement are then allowed onto entities such as fog node or cloud resources.

The resultant collection of candidate entities will be further considered in the next phase; and the final resource binding is conducted once the component selection and placement is determined. It is worth noting that the prefiltering and candidate selection can dramatically reduce the aimless and unnecessary calculations. Therefore, this step cannot be ignored for fog orchestration.

Node and executor management: Node manager in conventional cluster management systems is an agent that runs on each individual node and serves two purposes: the resource monitoring and the executor (worker process, virtual machine [VM], or Docker container) control. The latter is made possible through the aid of process launch, monitor, isolation, etc. Compared with the clusters in cloud data centers, the network condition, vulnerability of physical devices, and communication stability are entirely different, resulting in the disability to directly apply all current methods of node management in the resource management. For example, to handle frequent variations of node status, the resource manager should reserve the allocated resources instead of directly killing all running processes in face of frequent node joining-in or departing and the node anomaly stemming from temporary network delays or transient process crashes, etc. Additionally, the high-rate data exchange between the cloud and edge devices is fundamental to underpin the IoT applications. Long-thin connections between mobile users and remote cloud have to be long-lived, maintained, and isolated for the sake of network resource reservation.

8.4.2 Component Selection and Placement

The recent trend in composing cloud applications is driven by connecting heterogeneous services deployed across multiple datacenters. Similarly, such a distributed deployment helps improve IoT application reliability and performance. However, it also exposes appliances and microservices to new security risks and network uncertainty. Ensuring higher levels of dependability is a considerable challenge. Numerous efforts [26, 27] have focused on QoS-aware composition of native VM-based cloud application components but neglect the proliferation of uncertain execution and security risks among interactive and interdependent components within the DAG workflow of an IoT application.

Cost model: As we discussed in Section 8.3.1, the composition and placement of components can be regarded as an optimization problem. To be precise, the optimization includes two main factors in order to capture the increasing characteristics in fog environment – the network uncertainties and service dependability (such as security and reliability risks). We assume that the uncertainty and security of microservice s_i are defined as Unc_i and Sec_i, respectively. Importantly, there are parameters to represent the dependency relation between two adjacently chained microservices. For example, $DSec_{ij}$ represents the risk level of interconnecting s_i and s_j. Similarly, the uncertainty level between s_i and s_j is described as $DUnc_{ij}$. Thus, the optimization objectives can be formalized as:

$$\begin{cases} argmax_{s_i \in S, s_j \in S} \sum Sec_i + \sum DSec_{ij} \\ argmin_{s_i \in S, s_j \in S} \sum Unc_i + \sum DUnc_{ij} \end{cases}$$

As the equation shows that we need to maximize the security while minimizing the impact of uncertainties on the services, there are two ways to solve this problem: (i) Optimize a utility function which includes both objectives with different weights; (ii) set a constraint for one of the objective and then optimize the other one. For some multiobjective problems, it is unlikely to find a solution that has optimal values for all objective functions simultaneously. Alternatively, a feasible solution is the Pareto optimal [28] where none of the objectives can be improved without degrading an objective. Therefore, IoT service composition is to find a Pareto optimal solution which meets users' constraints.

Parallel computation algorithm: Optimization algorithms or graph-based approaches are typically both time-consuming and resource-consuming when applied into a large-scale scenario and necessitate parallel approaches to accelerate the optimization process. Recent work [29] provides possible solutions to leverage an in-memory computing framework to execute tasks in a cloud infrastructure in parallel. However, how to realize dynamic graph generation and partitioning at runtime to adapt to the shifting space of possible solutions stemming from the scale and dynamicity of IoT services remains unsolved.

Late calibration: To ensure near-real-time intervention during IoT application development, a potential approach could be correction mechanisms that could be applied even when suboptimal solutions are deployed initially. For example, in some cases, if the orchestrator finds a candidate solution that approximately satisfies the reliability and data transmission requirements, it can temporarily suspend the search for further optimal solutions. At runtime, the orchestrator can then continue the improvement of decision results with new information and a re-evaluation of constraints and make use of task and data migration approaches to realize workflow redeployment.

8.4.3 Dynamic Orchestration with Runtime QoS

Apart from the initial placement, the workflow dynamically changes due to internal transformations or abnormal system behavior. IoT applications are exposed to uncertain environments where variations in execution are commonplace. Due to the degradation of consumable devices and sensors, capabilities such as security and reliability that initially were guaranteed will vary accordingly, resulting in the initial workflow being no longer optimal or even totally invalid. Furthermore, the structural topology might change in accordance to the task execution progress (i.e. a computation task is finished or evicted) or will be affected by the evolution of the execution environment. Abnormalities might occur due to the variability of combinations of hardware and software crashes or data skew across different management domains of devices due to abnormal data and request bursting. This will result in unbalanced data communication and subsequent reduction of application reliability. Therefore, it is essential to dynamically orchestrate task execution and resource reallocation.

QoS-aware control and monitoring: To capture the dynamic evolution and variables (such as dynamic evolution, state transition, new operations of IoT, etc.), we should predefine the quantitative criteria and measuring approach of dynamic QoS thresholds in terms of latency, availability, throughput, etc. These thresholds usually dictate upper and lower bounds on the metrics as desired at runtime. Complex QoS information processing methods such as hyper-scale matrix update and calculation would give rise to many scalability issues in our setting.

Event streaming and messaging: Such performance metric variables or significant state transitions can be depicted as system events, and event streaming is processed in the orchestration framework through an event messaging bus, real-time publish-subscribe mechanism or high-throughput messaging systems (e.g. Apache Kafka [30]), therefore significantly reducing the communication overheads and ensuring responsiveness. Subsequent actions could be automatically triggered and driven by cloud engine (e.g. Amazon Lambda service [31]).

Proactive recognition: Localized regions of self-updates become ubiquitous within fog environments. The orchestrator should record staged states and data produced by fog appliances periodically or in an event-based manner. This information will form a set of time series of graphs and facilitate the analysis and proactive recognition of anomalous events to dynamically determine such hotspots [32]. The data and event streams should be efficiently transmitted among fog appliances, so that system outage, appliance failure, or load spikes will rapidly feedback to the central orchestrator for decision-making.

8.4.4 Systematic Data-Driven Optimization

IoT applications include numerous geographically distributed devices that produce multidimensional, high-volume data requiring different levels of real-time analytics and data aggregation. Therefore, data-driven optimization and planning should have a place in the orchestration of complex IoT services.

Holistic cross-layer optimization: As researchers or developers select and distribute applications across different layers in the fog environment, they should consider the optimization of all overlapping, interconnected layers. The orchestrator has a global view of all resource abstractions, from edge resources on the mobile side to compute and storage resources on the cloud data center side. Pipelining the stream of data processing and the database services within the same network domain could reduce data transmission. Similar to the data-locality principle, we can also distribute or reschedule the computation tasks of fog nodes near the sensors rather than frequently move data, thereby reducing latency. Another potential optimization is to customize data relevant parameters such as the data-generation rate or data-compression ratio to adapt to the performance and assigned resources to strike a balance between data quality and specified response-time targets.

Online tuning and history-based optimization (HBO): A major challenge is that decision operators are still computationally time-consuming. To tackle this problem, online ML can provision several online training (such as classification and clustering) and prediction models to capture the constant evolutionary behavior of each system element, producing time series of trends to intelligently predict the required system resource usage, failure occurrence, and straggler compute tasks, all of which can be learned from historical data and a HBO procedure. Researchers or developers should investigate these smart techniques, with corresponding heuristics applied in an existing decision-making framework to create a continuous feedback loop. Cloud ML offers analysts a set of data exploration tools and a variety of choices for using ML models and algorithms [33].

8.4.5 Machine-Learning for Orchestration

Although current deployment of orchestration has been explored by human experts and optimized by some hand-crafted heuristics algorithms, it is still far from meeting the challenge of automated management and optimization. Learning-based methods, or more precisely, ML, open a new door to tackle the challenges raised from IoT orchestration. ML approaches automatically learn underlying system patterns from historical data and explore the latent space of representation. It not only significantly reduces human labor and time but is also capable of dealing with multi-dimension and multi-variety data in dynamic or uncertain environments.

Metric learning: The current evaluation of a given workflow normally involves the knowledge of human experts as well as the numerical characteristic, quality of hardware, etc. However, the dynamicity within heterogeneous environments makes it infeasible and inaccurate to handcraft standard metrics for the evaluation over different orchestrations. Instead, metric learning [34] aims to automatically learn the metric from data (e.g. hardware configuration, historical records, runtime logs), providing convenient proxies to evaluate the distance among objects for better complex objects manipulation. Regarding orchestration scenarios, it is interesting if the algorithm can consider the topology layout of data during the learning.

Graph representation learning: Connecting Metric Learning with the graph structure provides an orthogonal direction for current methodologies of resource filtering and resource allocation. However, traditional orchestration approaches normally use user-defined heuristics to explore the optimal deployment over the original graph with structural information. Those summary statistics again significantly involve hand-engineered features that are inflexible during learning process and design phase. By using Graph Representation Learning (GRL), we can represent or encode the complex structural information of a given workflow [35]. Furthermore, we can either use it for better exploitation of the ML models or provide more powerful workflow metrics for better orchestration. For example, the current label-based resource representation may easily encounter the issue of sparse one-hot representation, and it would be more efficient to represent different hardware/services in a low- and dense-latent space [36].

Reinforcement learning: Design of good heuristic or approximation algorithms for NP-hard combinatorial optimization problems often requires significant specialized domain knowledge. However, traditional algorithms are often insufficient in such knowledge when extreme complicated IoT applications are orchestrated. Given the efficient representation of the workflow, graph embedding [37] shows the potentiality of using neural network with reinforcement learning (RL) methods to incrementally construct an optimal solution in dynamic environments. There are a great number of research opportunities since current deep RL solutions of combinatorial optimization merely focus on the standard traveling salesman problem (TSP), whose scenario is much simpler than the IoT application orchestration.

8.5 Fog Simulation

8.5.1 Overview

Simulation is an integral part of the process of design and analyzing systems in engineering and manufacturing domains. There is also a growing trend to analyze distributed computing systems using technologies such as CloudSim [38] or SEED [39], for example to study resource scheduling or analyze the thermodynamic behavior of a data center [40]. In these contexts, it is essential to understand the categories of simulation [41]:

- Discrete event simulation (DES) [42, 43] in which the system events are modeled as a discrete sequence.
- Continuous simulations [41] that are typically constructed based on ordinary differential equations (ODEs), which represent properties of physical systems.
- Stochastic simulation [44] such as Monte-Carlo methods.

Live, Virtual, and Constructive (LVC) simulation providing interactive simulations often supported by technologies as the IEEE HLA 1516 [45].

The LVC category of simulation introduces the concept of co-simulation, whereby two or more simulations are run concurrently to explore interactions and complex emergent behaviors. In the domain of engineering, co-simulation is typically limited to a handful of simulations due to the complexity of integrating simulations with differing time-steps and simulations of differing types. With the rise of IoT, Industry 4.0, and the Internet of Simulation (IoS) paradigm [46, 47], there is a growing trend to explore the use of simulation as a technology for facilitating online decision-making. There are several key factors that must be considered in this context:

- Inter-tool compatibility between simulations and also between simulations and other tools/systems.
- Performance and scalability in terms of the size of the simulations that need to be run and the time needed to do so. For example the use cases for simulation with IoT may require simulation to perform in near real-time.
- Understanding the complexity of the models involved in the simulations and understanding the trade-offs between complexity and abstraction [48].

The remainder of this section explores the application of simulation across two key areas: Simulation for IoT and also online analysis as part of a decision-making system such as an orchestrator.

8.5.2 Simulation for IoT Application in Fog

Within the traditional IoT sector, there are two broad situation categories in which simulation is typically used:

(a) Design and analysis of devices which may include control systems or 3D modeling during the design phases of the engineering process. This may occur at the component, system, or system of systems level. During the early stages, the simulations are typically abstract concepts of functional behavior, which are then iteratively, ideally using co-simulation, expanded upon to provide detailed insight into specific behaviors and emergent interactions. This process in the context of traditional engineering life cycles is

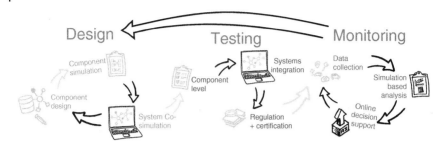

Figure 8.6 The workflow of system simulation.

depicted in Figure 8.6 where the traditional V-model of component and system-level design is integrated with component and systems level testing.
(b) Analysis of data as described by the Industry 4.0 movement [49]. In this context, data are collected by IoT sensors and systems and fed into analysis systems that increasingly involve simulation. An example is the automotive industry where data are collected from vehicles as they are used and fed back to the manufacturer. In the automotive space, this is typically limited to periodic data collection during servicing, but there is a growing trend with connected vehicles to provide more frequent or even a continuous data feedback to the manufacturer. As Figure 8.6 shows, data can be collected from deployed systems or devices and fed into simulations for further analysis, which may or may not occur at the design phase of another engineering life cycle iteration.

Figure 8.7 depicts the abstract layers of IoS: virtual or federated cloud, traditional service layers (IaaS, PaaS, SaaS, FaaS, etc.). On top of this are the simulation layers with virtual things deployed as simulations (SIMaaS) and then

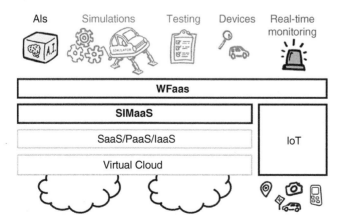

Figure 8.7 The architecture of simulation as a service.

virtual things becoming virtual systems at the workflow service layer (WFaaS). IoT lends itself as another vertical within this model providing physical things that can be connected to the virtual system workflows.

In order to precisely model and build the fog simulator, we need to depict the attributes and behaviors of fog appliances and the services. For a specific appliance, we include the appliance type, physical capacities such as CPU cores, RAM, storage, up/downlink bandwidths, and the connection status such as which appliances it is interconnected with and the latency information of connections among different appliances. The attributes also contain the hardware specifications (i.e. GPU, FPGA, TPU, etc.), software specifications (i.e. OS version, libraries, etc.), and other machine attributes that comprehensively described in [50]. All these are implemented as different labels. To simulate the service, we should provide the interfaces to define the IoT service DAG topology and the dependencies among different microservices. The detailed resource requests and other requirements of each microservice (the vertex in the DAG) are also determined as inputs within the simulator. For example, the main attributes of action detection microservice in Figure 8.2 can be depicted as follows:

```
{ "ActionDetection": {
    "Resource": {
    "CPU": 2 vcores,
    "RAM": 2GB,
    "DISK": 10GB
    },
    "Priority": "Medium",
    "Security": "High",
    "Computation": "Medium",
    "Latency": "Low"
}}
```

8.5.3 Simulation for Fog Orchestration

Moving away from the traditional IoT sector and the common uses of simulation, there are two growing trends for simulation adoption as part of online decision-making systems. The first trend is the automated parameterization and deployment of simulations based on data to provide immediate data analytics to decision makers. Second, the use of real-time simulation is in-the-loop with other systems. There are two key challenges with both trends, which are the need for timeliness while dealing with the scale of the systems being modeled.

An example in the context of orchestration is the use of simulation as part of both the optimization and planning phases. During system execution, the collected data are used to update relevant simulation models in terms of system behavior (this could include network latencies, server performance, etc.). The optimization process is able to use the simulation as a data representation for the ML algorithms, which in turn feeds into the planning phase. For example, using the genetic algorithm (GA)-based approaches used in Section 8.6, simulations can be run with each individual and generation to provide a more detailed and informed fitness function. Although this has the ability to significantly increase the capability of the system, there remains a significant trade-off in deciding the complexity of the simulation versus the performance that is required.

8.6 Early Experience

8.6.1 Simulation-Based Orchestration

Design overview: Based on the design philosophy and methods discussed, we propose a framework that can efficiently orchestrate fog computing environments. As demonstrated in Figure 8.8, in order to enable planning and adaptive optimization, a preliminary attempt was made to manage the composition of applications in parallel under a broad range of constraints. We implement a novel parallel genetic algorithm (GA-Par)-based framework on Spark to handle orchestration scenarios, where a large set of IoT microservices are composed. More specifically, in our GA-based algorithm, each chromosome represents a solution of the composed workflow, and the

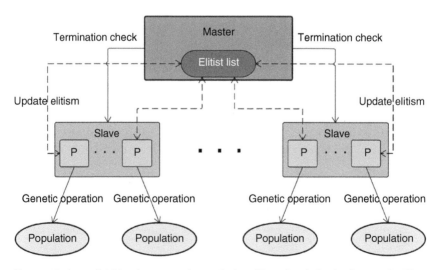

Figure 8.8 A parallel GA solver to accelerate the handling of optimization issues raised in the planning and optimization phase.

gene segments of each chromosome represent the IoT microservices. We normalize the utility of security and network QoS of IoT appliances into an objective fitness function within GA-Par to minimize the security risks and performance degradation.

To strike a balance between accuracy and time efficiency, we separate the total individual population into parallel compute partitions dispersed over different compute nodes. In order to maximize parallelism, we set up and adjust the partition configuration dynamically to make partitions fully parallelized while considering data shuffling and communication cost with the topology change. To guarantee optimal results can be gradually obtained, we dynamically merge several partitions into a single partition and then repartition it based on runtime status and monitored QoS. Furthermore, the quality of each solution generation can be also maintained by applying an elitist method, where the local elite results of each partition will be collected and synthesized into global elite. The centralized GA-Par master will aggregate the full information at the end of each iteration, and then broadcasts the list to all partitions to increase the probability of finding a globally optimal solution.

Experiment setup: To address data skew issues, we also conduct a joint data-compute optimization to repartition the data and reschedule computation tasks. We perform some initial experiments on 30 servers hosted on Amazon Web Services (AWS) as the cloud datacenter for the fog environment. Each server is hosted as an r3.2xlarge instance with 2.5 GHz Intel Xeon E5-2670v2 CPUs, 61 GB RAM, and 160 GB disk storage. We use simulated data below to illustrate the effectiveness of composition given IoT requirements. For this, we randomly select four types of orchestration graphs with 50, 100, 150, and 200 workflow nodes, respectively. For each node within a workflow, we stochastically prepare 100 available IoT appliances as simulated agents. The security levels and network QoS levels are randomly assigned to each candidate agent. We compare our GA-Par with a standalone genetic algorithm (SGA). The metrics quality, execution time, and fitness score (with lower values indicating better results) are used to evaluate SGA and GA-Par.

Evaluation: As can be observed in Figure 8.9, GA-Par outperforms SGA. The time consumption of GA-Par has been significantly reduced to nearly 50% of that of SGA, while the quality of appliance selection in GA-Par is always at least 30% higher than that of SGA. However, the scalability of our current approach is still slightly affected by increasing numbers of components and requests, indicating that we still need to explore opportunities for incremental replanning and on-line tuning to improve both time-efficiency and effectiveness of IoT orchestration.

Figure 8.10 demonstrates the experimental results under different workflow size and candidate number of microservice by using GA-Par. We can observe

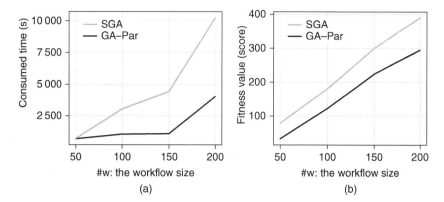

Figure 8.9 Initial results demonstrate that the proposed approach can outperform a standalone genetic algorithm in terms of both time (a) and quality (b) aspects.

that with the increment of workflow size, the time consumption increases accordingly. The linear increase demonstrates that the growth of task number in the workflow will augment the searching range to find optimal solution, thereby taking longer time to finish the overall computation. In Figure 8.10a, the number of microservice candidate number is not an obvious factor that influence on the time consumption. The consumed time slightly fluctuates when the topology and size of the workflow is determined. Apparently, given the workflow size w and each node in the orchestrated workflow has s candidates, the searching space is s^w. Thus, the impact of s on the consumed time will not be as significant as that of w. Likewise, a similar phenomenon can be observed in terms of the fitness calculation. In particular, the increased workflow size will naturally degrade the optimization effectiveness given the fixed setting of the total population. Compared with a smaller scale workflow, larger workflows with soaring number are less likely to converge and obtain the optimal result once the population is set up.

Discussion: IoT services are choreographed through workflow or task graphs to assemble different IoT microservices. Therefore, it is very worthwhile if we intend to obtain a precise decision and deploy IoT services in a QoS guaranteed, context- and cost-aware manner in spite of the magnitude of consumed time. In the context of pre-execution planning, static models and methods can deal with the submitted requests when the application workload and parallel tasks are known at design time. In contrast, in some domains, the orchestration is supplied with a plethora of candidate devices with different geographical locations and attributes. In the presence of variations and disturbances, orchestration methods should typically rely on incremental orchestration at runtime (rather than straightforward complete recalculation by rerunning static methods) to decrease unnecessary computation and minimize the consumed time.

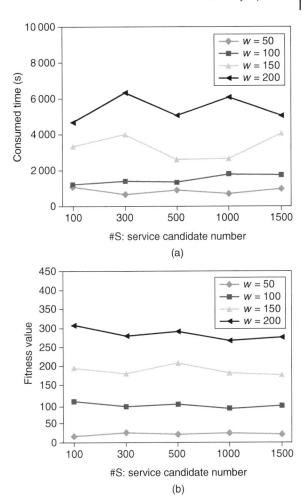

Figure 8.10 Initial results of GA-Par in terms of both time (a) and quality (b) aspects.

Based on the time series of graphs, the similarities and dependencies between successive graph snapshots should be comprehensively studied to determine the feasibility of incremental computation. Approaches such as memorization, self-adjusting computation, and semantic analysis could cache and reuse portions of dynamic dependency graphs to avoid unnecessary recomputation in the event of input changes. Intermediate data or results should also be inherited as far as possible, and the allocated resources that have been allocated to the tasks should also be reused rather than be requested repeatedly. Through graph analysis, operators can determine which subgraphs changes within the whole topology by using subgraph partitioning and matching as an automated process that can significantly reduce overall orchestration time.

Another future work is to further parallelize the simulation and steer the complexity of GA-Par to achieve better scalability over large-scale infrastructures. Potential means include using heuristic algorithm or approximated computing into some key procedures of algorithm execution and value estimation.

8.6.2 Orchestration in Container-Based Systems

There are numerous research efforts and system works that address the orchestration functionality in fog computing infrastructures. Most of them are based on the open source container orchestration tools such as Docker Swarm [1], Kubernetes [21], and Apache Mesos marathon [22]. For instance, Refs. [51, 52] gave an illustrative implementation of a Fog-oriented framework that can deliver containerized application onto datacenter nodes and edge devices such as Raspberry Pi. Reference [53] comprehensively compared how these tools can meet the basic requirement to run IoT applications and pointed out Docker Swarm is the best fit to seamlessly running IoT containers across different fog layers. Docker Swarm provisions robust scheduling that can spin up containers on hosts provisioned by using Docker-native syntax.

Based on the evaluation conclusions drawn by [53], we developed the proposed orchestrator as a standalone module that can be integrated with the existing Docker Swarm built-in modules (Figure 8.11). The orchestrator will overwrite the Swarm Scheduler and take over the responsibility of orchestrating containerized IoT services. Other sensors or Raspberry Pi devices are regarded as Swarm workers (Swarm Agents) and managed by the Scheduler. We integrate the proposed techniques such as label-based resource filtering and allocation, microservice placement strategy, and parallelized optimization solver GA-Par with the provided filter and strategy mechanisms in Swarm Scheduler. As a result, whenever a new deployment request from the client is received,

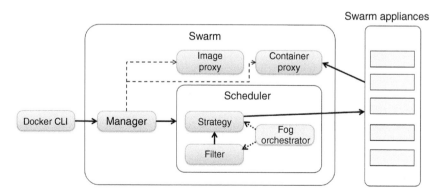

Figure 8.11 Fog orchestrator with Docker Swarm.

the Swarm Manager will forward it to the orchestrator, and the planner in the orchestrator will try to find the optimal placement to place and run containers on the suitable appliances. For the scalability and adaptability, we also design the planner as a pluggable module that can be easily substituted by different policies.

8.7 Discussion

The emergent of fog computing is one of particular interest within computer science. Within the coming decades, the concept of the exascale system will become increasingly commonplace, interconnecting billions and tens of billions of different devices, objects, and things across a vast number of industries, which will likely co-exist in some form of fog ecosystem. The challenges pertaining to security, reliability, and scalability will continue to play a critical concern for designing these systems, as well as a number of additional characteristics:

Emergent behavior: Systems operating at scale have begun to show increasingly operational characteristics not envisioned at system design-time conception. This is particularly true due to the massive heterogeneity and diversity of orchestrating various IoT services in tandem. Such emergent behavior encompasses not only positive aspects such as emergent software [54] but also encompasses failures [17, 55–58] unforeseen at design time. As a result, we will likely see increased use of meta-learning in order to dynamically adapt workflow orchestration in response to user demand and adversarial system conditions.

Energy usage: 10% of global electricity usage stems from ICT [59], and coupled with technological innovations and massive demands for services will likely see this electricity demand grow in both quantity and proportion. Given that these systems will operate services which produce vast quantities of emissions and economic cost, the environmental impact of these services executing within IoT will likely become increasingly important in the coming years. This is particularly true if legal measures are created and enforced to control and manage such power demand and carbon emissions.

Centralization versus decentralization: Within the past two decades, distributed systems have seen paradigms spanning clusters, web services, grid computing, cloud computing, and fog computing. It is noticeable that these paradigms appear to pivot between centralized [19, 60–62] and decentralized architectures [20, 63, 64] in response to technological breakthroughs, combined with demands for new types of applications. We foresee that this pattern will continue to evolve and potentially see the realization of massive-scale fog ecosystems that are capable of both centralized and decentralized architectures combined together in response to demand.

8.8 Conclusion

Most recent research related to fog computing explores architectures within massive infrastructures [2]. Although such work advances our understanding of the possible computing architectures and challenges of new computing paradigms, there are presently no studies of composability and concrete methodologies for developing orchestration systems that support composition in the development of novel IoT applications. In this chapter, we have outlined numerous difficulties and challenges to develop an orchestration framework across all layers within the Fog resource stack and have described a prototype orchestration system that makes use of some of the most promising mechanisms to tackle these challenges.

Acknowledgment

This work is supported by the National Key Research and Development Program (2016YFB1000103), EPSRC (EP/P031617/1), National Natural Science Foundation of China (61421003), and the Beijing Advanced Innovation Center for Big Data and Brain Computing (BDBC).

References

1 Docker Swarm. https://www.oasis-open.org/committees/tosca/ (accessed 28 August 2019).

2 Yi, S., Li, C., and Li, Q. (2015). A survey of fog computing: concepts, applications, and issues. In: *Proceedings of the 2015 Workshop on Mobile Big Data (June 2015)*, 37–42. ACM.

3 Yu, J., Benatallah, B., Casati, F., and Daniel, F. (2008). Understanding mashup development. *IEEE Internet Computing* 12 (5): 44–52.

4 Manh, P., Tchana, A., Donsez, D. et al. (2015). Roboconf: a hybrid cloud orchestrator to deploy complex applications. In: *2015 IEEE 8th International Conference on Cloud Computing*, 365–372. IEEE.

5 Wang, X., Liu, Z., Qi, Y., and Li, J. (2012). Livecloud: a lucid orchestrator for cloud datacenters. In: *4th IEEE International Conference on Cloud Computing Technology and Science Proceedings (December 2012)*, 341–348. IEEE.

6 Liu, C., Loo, B., and Mao, Y. (2011). Declarative automated cloud resource orchestration. In: *Proceedings of the 2nd ACM Symposium on Cloud Computing (October 2011)*, 26. ACM.

7 Ranjan, R., Benatallah, B., Dustdar, S., and Papazoglou, M. (2015). Cloud resource orchestration programming: overview, issues, and directions. *IEEE Internet Computing* 19 (5): 46–56.

8 Bonomi, F., Milito, R., Zhu, J., and Addepalli, S. (2012). Fog computing and its role in the Internet of things. In: *Proceedings of the First Edition of the MCC Workshop on Mobile Cloud Computing (August 2012)*, 13, 13–16, 16. ACM.

9 Message Queuing Telemetry Transport (MQTT) (2014) [Online]. Available: http://mqtt.org/ (accessed 29 August 2019).

10 Dikaiakos, K.M.D., Florides, A., Nadeem, T., and Iftode, L. (2007). Location aware services over vehicular ad-hoc networks using car-to-car communication. *IEEE Journal on Selected Areas in Communications* 25 (8): 1590–1602.

11 Qin, Z., Denker, G., Giannelli, C. et al. (2014). A software defined networking architecture for the internet-of things. In: *2014 IEEE Network Operations and Management Symposium (NOMS)*, 1–9.

12 Nastic, S., Sehic, S., Le, D.H. et al. (2014). Provisioning software-defined IoT cloud systems. In: *2014 International Conference on Future Internet of Things and Cloud*, 288–295. IEEE.

13 Madsen, H., Burtschy, B., Albeanu, G., and Vladicescu, F. (2013). Reliability in the utility computing era: towards reliable fog computing. In: *2013 20th International Conference on Systems, Signals and Image Processing (IWSSIP)*, 43–46. IEEE.

14 Atzori, L., Iera, A., and Morabito, G. (2010). The Internet of things: a survey. *Computer Networks* 54 (15): 2787–2805.

15 Roman, R., Najera, P., and Lopez, J. (2011). Securing the Internet of things. *IEEE Computer* 44 (9): 51–58.

16 Riahi, A., Challal, Y., Natalizio, E. et al. (2013). A systemic approach for IoT security. In: *2013 IEEE International Conference on Distributed Computing in Sensor Systems*, 351–355. IEEE.

17 Garraghan, P., Ouyang, X., Yang, R. et al. (2019). Straggler root-cause and impact analysis for massive-scale virtualized cloud datacenters. *IEEE Transactions on Services Computing* 12 (1): 91–104.

18 Yang, R., Zhang, Y., Garraghan, P. et al. (2016). Reliable computing service in massive-scale systems through rapid low-cost failover. *IEEE Transactions on Services Computing* 10 (6): 969–983.

19 Zhang, Z., Li, C., Tao, Y. et al. (2014). Fuxi: a fault-tolerant resource management and job scheduling system at internet scale. *Proceedings of the VLDB Endowment* 7 (13): 1393–1404.

20 Karanasos, K., Rao, S. et al. (2015). Mercury: hybrid centralized and distributed scheduling in large shared clusters. In: *2015 USENIX Annual Technical Conference (USENIX ATC 15)*, 485–497.

21 Google Kubernetes. https://kubernetes.io (accessed 28 August 2019).

22 Mesos Marathon. http://mesos.apache.org (accessed 28 August 2019).

23 TOSCA. http://docs.oasis-open.org/tosca/TOSCA/v1.0/os/TOSCA-v1.0-os .pdf (accessed 28 August 2019).

24 Kreutz, D., Ramos, F., Verissimo, P. et al. (2015). Software-defined networking: a comprehensive survey. *Proceedings of the IEEE* 103 (1): 14–76.

25 Jarraya, Y., Madi, T., and Debbabi, M. (2014). A survey and a layered taxonomy of software-defined networking. *IEEE Communication Surveys and Tutorials* 16 (4): 1955–1980.

26 Zheng, Z., Zhang, Y., and Lyu, M.R. (2014). Investigating QoS of real-world web services. *IEEE TSC* 7 (1): 32–39.

27 Wen, Z., Cala, J., Watson, P., and Romanovsky, A. (2016). Cost effective, reliable and secure workflow deployment over federated clouds. *IEEE Transactions on Services Computing* 10 (6): 929–941.

28 Deb, K., Pratap, A., Agarwal, S., and Meyarivan, T. (2002). A fast and elitist multiobjective genetic algorithm: NSGA-II. *IEEE Transactions on Evolutionary Computation* 6 (2): 182. https://doi.org/10.1109/4235.996017.

29 Gonzalez, J.E., Xin, R.S., Dave, A. et al. (2014). GraphX: graph processing in a distributed dataflow framework. In: *11th USENIX Symposium on Operating Systems Design and Implementation (OSDI 14) (November 2014)*, 599–613. USENIX.

30 Apache Kafka. http://kafka.apache.org (accessed 28 August 2019).

31 AWS Lambda. https://aws.amazon.com/lambda (accessed 28 August 2019).

32 Yamanishi, K. and ichi Takeuchi, J. (2002). A unifying framework for detecting outliers and change points from non-stationary time series data. In: *Proceedings of the Eighth ACM SIGKDD International Conference on Knowledge Discovery and Data Mining (July 2002)*, 676–681. ACM.

33 InfoWorld (2016). Cloud machine learning [Online]. http://www.infoworld.com/article/3068519/artificialintellegence/review-6-machine-learning-clouds.html (accessed 29 August 2019).

34 A. Bellet, A. Habrard, and M. Sebban (2013). A survey on metric learning for feature vectors and structured data. *arXiv Preprint arXiv:1306.6709*. [Online]. Available: https://arxiv.org/pdf/1306.6709.pdf.

35 W. Hamilton, R. Ying, and J. Leskovec (2017). Representation learning on graphs: methods and applications *arXiv Preprint arXiv:1709.05584*. [Online]. Available: https://arxiv.org/pdf/1709.05584.pdf.

36 Mikolov, T., Sutskever, I., Chen, K. et al. (2013). Distributed representations of words and phrases and their compositionality. In: *Advances in Neural Information Processing Systems (December 2013)*, 3111–3119.

37 I. Bello, H. Pham, Q. Le, M. Norouzi, and S. Bengio (2016). Neural combinatorial optimization with reinforcement learning *arXiv Preprint arXiv:1611.09940*. [Online]. Available: https://arxiv.org/pdf/1611.09940.pdf.

38 Calheiros, R.N. et al. (2011). CloudSim: a toolkit for modeling and simulation of cloud computing environments and evaluation of resource provisioning algorithms. *Software: Practice and Experience* 41 (1): 23–50.

39 Garraghan, P., McKee, D., Ouyang, X. et al. (2016). SEED: a scalable approach for cyber-physical system simulation. *IEEE Transactions on Services Computing* 9 (2): 199–212.

40 Clement, S., McKee, D., and Xu, J. (2017). A service-oriented co-simulation: holistic data center modelling using thermal, power and computational simulations. In: *IEEE/ACM UCC*.

41 Law, A.M. and Kelton, W.D. (2007). *Simulation Modeling and Analysis*, vol. 3. New York: McGraw-Hill.

42 Fujimoto, R.M. (1990). Parallel discrete event simulation. *Communications of the ACM* 33 (10): 30–53.

43 Varga, A. (2001). Discrete event simulation system. European Simulation Multiconference (ESM'2001).

44 Ripley, B.D. (2009). *Stochastic Simulation*, vol. 316. Wiley.

45 Martinez-Salio, J.-R., Lopez-Rodriguez, J.-M., Gregory, D., and Corsaro, A. (2012). A comparison of simulation and operational architectures. In: *2012 Fall Simulation Interoperability Workshop (SIW)*. Simulation Interoperability Standards Organization (SISO).

46 McKee, D., Clement, S., Ouyang, X. et al. (2017). The Internet of simulation, a specialisation of the Internet of things with simulation and workflow as a service (sim/wfaas). In: *2017 IEEE Symposium on Service-Oriented System Engineering (SOSE)*, 47–56. IEEE.

47 Clement, S., McKee, D., Romano, R. et al. (2017). The Internet of simulation: enabling agile model based systems engineering for cyber-physical systems. In: *2017 12th System of Systems Engineering Conference (SoSE)*, 1–6. IEEE.

48 Singer, M. (2007). Modelling both complexity and abstraction: a paradox? In: *Modelling and Applications in Mathematics Education*, 233–240. Boston, MA: Springer.

49 Lasi, H. et al. (2014). Industry 4.0. *Business & Information Systems Engineering* 6 (4): 239–242.

50 Google Cluster Trace. https://github.com/google/cluster-data (accessed 28 August 2019).

51 Pahl, C., Helmer, S., Miori, L., Sanin, J., and Lee, B. (2016). A container-based edge cloud PaaS architecture based on Raspberry Pi clusters. IEEE FiCloud.

52 Bellavista, P. and Zanni, A. (2017). Feasibility of fog computing deployment based on docker containerization over Raspberry Pi. In: *Proceedings of the 18th International Conference on Distributed Computing and Networking (January 2017)*, 16. ACM.

53 Hoque, S., Brito, M., Willner, A. et al. (2017). Towards container orchestration in fog computing infrastructures. In: *2017 IEEE 41st Annual Computer Software and Applications Conference (COMPSAC)*, vol. 2, 294–299. IEEE.

54 Porter, B., Grieves, M., Filho, R.V.R., and Leslie, D. (2016). REX: A development platform and online learning approach for runtime emergent software systems. In: *12th USENIX Symposium on Operating Systems Design and Implementation (OSDI 16)*, 333–348.

55 Dean, J. and Barroso, L.A. (2013). The tail at scale. *Communications of the ACM* 56.

56 Schroeder, B. and Gibson, G.A. (2010). A large-scale study of failures in high-performance computing systems. *IEEE Transactions on Dependable and Secure Computing*.

57 Garraghan, P., Yang, R., Wen, Z. et al. (2018). Emergent failures: rethinking cloud reliability at scale. *IEEE Cloud Computing* 5 (5): 12–21.

58 Ouyang, X., Garraghan, P., Yang, R. et al. (2016). Reducing late-timing failure at scale: Straggler root-cause analysis in cloud datacenters. In: *46th Annual IEEE/IFIP International Conference on Dependable Systems and Networks*. DSN.

59 Mills, M.P. (2013). *The Cloud Begins with Coal*. Washington, DC, USA: Digital Power Group Tech. Rep. [Online]. Available: http://www.tech-pundit.com/wp-content/uploads/2013/07/Cloud_Begins_With_Coal.pdf.

60 Vavilapalli, V.K., Murthy, A.C., Douglas, C. et al. (2013). Apache Hadoop yarn: yet another resource negotiator. In: *Proceedings of the 4th Annual Symposium on Cloud Computing (October 2013)*, 5. ACM.

61 Hindman, B., Konwinski, A., Zaharia, M. et al. (2011). MESOS: a platform for fine-grained resource sharing in the data center. In: *NSDI (March 2011)*, vol. 11, 22–22.

62 Verma, A., Pedrosa, L., Korupolu, M. et al. (2015). Large-scale cluster management at Google with Borg. In: *Proceedings of the Tenth European Conference on Computer Systems (April 2015)*, 18. ACM.

63 Boutin, E., Ekanayake, J., Lin, W. et al. (2014). Apollo: scalable and coordinated scheduling for cloud-scale computing. In: *11th USENIX Symposium on Operating Systems Design and Implementation (OSDI 14) (October 2014)*, 285–300.

64 Sun, X., Hu, C., Yang, R. et al. (2018). ROSE: cluster resource scheduling via speculative over-subscription. In: *2018 IEEE 38th International Conference on Distributed Computing Systems (ICDCS)*, 949–960. IEEE.

9

A Decentralized Adaptation System for QoS Optimization

Nanxi Chen[1], Fan Li[2], Gary White[2], Siobhán Clarke[2], and Yang Yang[3]

[1] Bio-vision Systems Laboratory, SIMIT, Chinese Academy of Sciences, 865 Changning Road, 200050 Shanghai, China
[2] Distributed Systems Group, SCSS, Trinity College Dublin, The University of Dublin, College Green, Dublin, Dublin 2, Ireland
[3] Shanghai Institute of Fog Computing Technology (SHIFT), ShanghaiTech University, Shanghai, China

9.1 Introduction

In the Internet of Things (IoTs), billions of cyber-physical objects are connected to collect and share data, providing a near real-time state of the physical world. The functionalities provided by these objects can be offered as a service [1], which is achieved by providing a well-defined interface that hides any complexity emerging from device heterogeneity. IoT services are referred to as software components running over different spatial locations connected through various networks such as 4G, WLAN, and Internet. Different from traditional Web services, IoT services exhibit dynamic behavior in terms of availability and mobility of resources, unpredictable workload, and unstable wireless network condition.

Considering the resource-constrained capability of heterogeneous objects, the IoT and cloud are merged together to benefit from the cloud's unlimited capabilities in terms of storage and processing power [2]. The IoT data sourced in different spatial locations are sent to a centralized cloud for processing, and then delivered to the distributed users who request the data or other IoT devices. However, on one hand, some mission-critical IoT applications do not fit the Cloud paradigm due to the requested low latency, location awareness, and mobility support [3]. On the other hand, the growth rate of cloud-based data centers falls behind that of the data aggregation and processing requirement, especially the requirement contributed by IoT [4]. Fog computing has emerged as a new paradigm to resolve such issues, which extends both computation and storage services in Cloud to the network edge

Fog and Fogonomics: Challenges and Practices of Fog Computing, Communication, Networking, Strategy, and Economics, First Edition. Edited by Yang Yang, Jianwei Huang, Tao Zhang, and Joe Weinman.
© 2020 John Wiley & Sons, Inc. Published 2020 by John Wiley & Sons, Inc.

to satisfy the increasing data processing requirement and reduce latency and response delay jitter for latency-sensitive applications.

Fog computing allows its infrastructures to span not only the network perimeter but also the cloud-to-things continuum. A fog infrastructure provider can be considered as a fog service provider that resides anywhere along this continuum and shares its compute, storage, and network resources to the other entities in its vicinity. A fog service provider can be an individual fog node or a fog network that consists of linked fog nodes. Fog nodes can differ in their capacity and specificity, and resources on them are not restricted to any particular application domain. A fog network can pool resources and can deploy its service in the cloud, at the edge or on the things. It changes the way IoT services are delivered to customers. IoT applications can be deployed in a fog service provider or a set of linked fog service providers that reside close to the customers or the data sources and are accessed by the customers through service interfaces in the fog network. As a fog node does not have to be resource-rich and is meant to be widely spread, it allows small companies and even an individual person to deploy and operate computing, storage, and control services at different scales as a fog service provider in an IoT environment. This will inspire new business models and require reasonable mechanisms to deal with the service provisioning process.

The fog computing business model liberates the market from only large organizations who can afford to build and operate powerful servers and huge data centers to small companies and even an individual person, which makes the fog a promising solution to the IoT. Microservices technology and flexible service-oriented computing models give fog computing more possibility to support complex application logic rather than simply providing storage services or content caching services [5]. However, the features of fog bring new problems to IoT service provisioning in terms of quality of service (QoS) and trust. Challenges of supporting complex application logic in fog environment emerge from three aspects:

Dynamics and spatial distribution: Many IoT end devices and sensors are designed to be connected via wireless communication channel and located in a widely dispersed network. The network topology and the service requirement in a particular area are likely to change with mobility or the availability of the devices. To efficiently provide services to IoT customers, fog service providers have to be scattered and reconfigurable. Sometimes they even need to be replaced to where the corresponding resource is in shortage. In such a fog computing environment that is dynamic and widely distributed, it is difficult for IoT customers to always access the same fog service provider when they request for an IoT service. A fog computing environment should be able to provide services with acceptable QoS for each of IoT application request.

Resource-limitation: Fog computing involves many heterogeneous nodes that have different resource capabilities. It is inefficient and removes the possibility of obtaining a better QoS or value-added services from other service providers if all the services of an IoT application are deployed on one single fog service provider. To satisfy various application demands, the composition technique is required to compose the services deployed in different fog service providers. The QoS and resource demand are usually application dependent, making the service consumer to negotiate Service Level Agreement (SLA) with multiple fog service providers when composing services.

Trustworthiness: IoT is an ultra-large-scale network containing a billion or even trillion nodes [6]. A large number of services providing the same or similar functionalities create a competitive market that allows different third-party fog service providers to join a service composition, which bring trustworthiness issues. A malicious fog service provider may tamper the services deployed on it, making a false execution result returned to the service consumer or a reduction of QoS. To enable a reliable service composition, a reputation system is needed to track a service's reliability and provides a basis for service selection during the service execution stage.

To address the aforementioned challenges, a service management is necessary for fog computing environments. Considering the scale of IoT network and the resource limitation in the target environment, a centralized service management will be infeasible. Current work in service composition does not propose much decentralized methods. Some distributed solutions require frequent communication among different fog service providers, which is heavyweight and can increase latency to fog network. Existing solution GoCoMo [7] provides a time-efficient decentralized composition of services on mobile devices by considering task resolution in a dynamic environment and adaptation of the combination of services as appropriate to the service providers' and the environment's changing factor. Service execution paths' reliability and QoS issues are considered when selecting services for invocation to reduce the possibility of execution failures. However, this approach assumes that all the service providers in the environment are honest and trustworthy, and there is a lack of a reputation model to fully support QoS-aware composition.

For mission-critical applications such as transportation, health care, and emergency response, the stringent demand on QoS indicates that the requested service should be delivered with a stable quality level [8], we refer it as the hard QoS requirements. In cloud computing, the service level is specified in the SLA, which is an agreement between a service consumer and a provider in the context of a particular service provisioning. It specifies QoS properties that must be maintained by the service provider during service operation time. These properties are referred to as Service Level Objectives (SLOs) [9], which are evaluated using measurable data by the SLA monitor during the runtime. Any violation against SLOs would trigger the penalty clauses specified in the SLA, which may consists of a financial sanction, a decrease in the reputation of the provider, and an adaptation mechanism such as SLA termination and renegotiation. In other words, the SLA-supported services are more competitive in mission-critical applications than best effort services, since the agreed service level can be maintained by providers and supervised by quality assurance techniques such as runtime QoS monitoring.

Fog computing targets the latency issue and the mission-critical applications. Considering the dynamics of IoT environments, maintaining the requested service level is a long way to go. Service-level degradations may derive from the unstable wireless network conditions, fluctuating of service workload, and unpredictable events such as the sudden damage of an outdoor fog device by weather. To deal with changes in the IoT environment, the dynamic SLA renegotiation is a useful adaptation mechanism, which improves the trust of a service and maintains the service continuity. Given the participation of multiple third-party fog service providers in an IoT application and the lack of centralized nodes at the network edge, SLA renegotiation model for fog computing is required to deal with fog-to-fog (F2F) renegotiation in the case of a service level degradation.

To recover a composition from a service-level degradation after renegotiation, we need QoS-aware composition. Accurate QoS values are essential when supporting reliable QoS-aware composition, which are provided by the runtime QoS monitor. In a fog-based IoT, there is a trade-off between the extra load on the network caused by the monitoring and the accuracy of the values. To minimize the amount of actual invocations needed for the composition, we use a combination of prediction and probing. When selecting possible candidate services in the environment to switch to, we use the IoTPredict algorithm [10], which uses other similar users in the environment to predict QoS values instead of having to invoke the actual service. When monitoring services at runtime, we use a forecasting algorithm to identify if a service may be about to fail [11]. This allows us to make proactive recomposition actions to deal with any services that may be about to fail. We combine the prediction approaches with a distributed probing mechanism that allows for both soft QoS requirements such as a route planning application to find the best way through a smart city as well as more accurate monitoring values for applications with hard QoS requirements, such as monitoring patients in a hospital or collision avoidance in a self-driving car [12]. The combination of prediction and probing allows us to deal with a range of IoT applications, while also reducing the load on the network.

This chapter introduces AdaptFog, a distributed reputation system for QoS-aware service choreography in a fog environment. It includes a runtime QoS monitoring component, a F2F dynamic SLA renegotiation component, and a reputation assessment component. A reputation model based on direct graph is proposed to trace a service's reputation value, and a cross-cutting verification mechanism is introduced to verify service performance. Section 9.2 discussed the state-of-the-art solutions toward QoS-aware service choreography, SLA negotiation, and service monitoring. Section 9.3 introduces the system model and presents the detail of AdaptFog. Section 9.4 illustrates the open issues and concludes this chapter.

9.2 State of the Art

9.2.1 QoS-aware Service Composition

Dynamic composition planning uses real-time service availability information to reason about a service workflow. The solution includes open-service discovery and goal-oriented service discovery. Generally, goal-driven service planning is more flexible, but most existing solutions still require a centralized planning engine to support complex service workflows. Open service discovery uses predefined function flows, limiting the diversity of composition results.

Reference [13] takes nonfunctional properties of services into consideration when selecting service providers in pervasive computing environments. In this work, service providers predict their own QoS before service execution and enclose this QoS information into their service advertisement messages. The approach uses a set of QoS metrics to quantify QoS information, which includes a probability value for service availability and a value that indicates execution latency. This work uses probability values to define service availability, which only caters for when a service is withdrawn by providers and not when the provider moves out of range. Reference [14] predicts the quality of entire service composition instead of individual service provider. They model service compositions' cost and reliability, defining six cost behaviors and four workflow reduction models for different service flows. However, they do not consider the reliability and the cost of communication among successive services. HOSSON [15] maintains a user perspective service overlay network (PSON) for all the candidate service providers. The PSON is graph-based and performs service selection using multiple criteria decision-making to find the optimal service execution path with minimal service execution cost and good reputation.[1] Reference [16] also considers the quality of overall service execution paths when selecting service providers. They adopt path measurement including reliability, execution latency, and network conditions. However, the aforementioned approaches assume that the QoS information is predefined by service providers, and as network conditions are likely to change, this assumption is not safe in a dynamic environment.

---

1 Reputation is a ranking value that is given by previous service users to represent a service's dependability [15].

Reference [17] selects services at an early stage in a service discovery process, depending on the strength of service links. During service advertisement, if a service provider announces its service specification, a neighboring node caches the service specification and also maintains information relating to the strength of the path to the service provider. When a node receives a service request, it will choose the service with the strongest path to itself from a list of functionally equivalent services. When a selected destination is unreachable, a backtracking mechanism can be performed to recover service binding. Though path strengths are involved in binding decision-making, and a binding recovery mechanism is proposed for failed routing, this approach is still unsafe as beforehand cached path strength information may be out of date, and the backtracking based recovery is expensive itself. Surrogate Models are used to facilitate service composition in mobile ad hoc networks [18].

ProAdapt [19] monitors operating environments and adapts service composites to a list of context changes, including response time changes, availability of services, and availability of service providers. FTSSF [20] applies a monitoring and fault handling process for service provisioning in pervasive computing. Monitoring is concurrent with a service delivery process, allowing for service reselection if the execution crashes. Reference [21] allocates a service provider to the best performing candidate in terms of the QoS while keeping a group of backup providers. A backup provider can replace a previously bound provider if its execution fails. This model is more efficient than ProAdapt and FTSSF, as it monitors the service composite only when it is executing, which means there is no need for a service composition system to continuously monitor the operating system throughout the composition process.

Reference [22] models service composition as a problem of finding a service provider for each abstract service in a predefined conceptual composite. Service binding adaptation in this approach is based on automatic QoS prediction. Specifically, a service composition system firstly allocates a set of services to form an execution path, the QoS of which conforms to the QoS constraints that are defined in the composition request. The system then predicts the failure probability of the path and finally finalizes the service composite according to the prediction result. If necessary, reselection for providers can be performed. This solution regards the service execution path's reliability as an important criterion for service selection and allocation instead of considering each of the service providers independently. However, the approach relies on a centralized composition handler to perform its prediction.

OSIRIS [23] selects service providers on demand at execution time. This selection of service providers depends on run-time device load and dynamic invocation cost. Service discovery is performed offline to find candidate service providers, but repeated to detect new service providers. However, the approach depends on a central repository to store service specifications, which is limited in dynamic systems.

Opportunistic service composition [24–26] also proposes an on-demand service binding mechanism. Unlike OSIRIS, opportunistic service composition does not use central repositories to keep service specification. Instead, it discovers service providers on-the-fly relying on request flooding. After service providers are located, the approach asks for permissions to lock service provider's resources, and then invokes the service. This model ensures an available service provider will be invoked for execution, but refinements are required to further consider result routing when binding a service provider, at the same time reducing the cost of its flooding-based service discovery and increasing the flexibility of composition planning.

QoS-based service binding addresses dynamic environments by self-describing services' runtime properties (e.g. availability, reliability, response time, etc.). Existing approaches [13, 15, 17] depend on QoS descriptions provided by service providers that predict their own service performance. Although this mechanism is lightweight for a service composition system as no monitoring effort is required, such QoS descriptions are likely to be inaccurate. Moreover, mobile service providers may have to frequently update their QoS description. Adaptable binding detects changes on operating environments and adapts a service composite accordingly. Detecting changes or failures requires different levels of monitoring. Execution time monitoring is the most efficient, but failures can only be detected after they occur. Recovering a composition from emerged failures can introduce additional time and communication cost. On-demand binding selects service providers using up-to-date service information, requiring no extra environment monitoring or infrastructure maintaining efforts. It would be interesting to improve the existing on-demand service binding mechanisms to support flexible as well as reliable service composition.

Nanxi et al. proposed a novel Goal-driven service Composition model (GoCoMo) for Mobile and pervasive computing environments [7], which performs service discovery, planning, binding, invocation, and adaptation in a fully decentralized manner. The heuristic service discovery which supports complex composites such as parallel or hybrid service flows is achieved by performing a dynamic backward planning algorithm. The discovered potentially diverse service flows contain all the reached usable composite participants, allowing for dynamic selection among them during the execution time.

9.2.2 SLA (Re-)negotiation

The SLA management has been widely discussed in the Web services provisioning, cloud computing, and grid computing, which can be seen in the projects such as the OPTIMIS [27], Brein [28], SLA@SOI [29], etc. However, SLA management has not been fully considered in the IoT middlewares [30, 31]. For example, Gaillard et al. [32] outlined a centralized SLA management component in the wireless sensor network (WSN) to improve the service quality level. However, this framework relies on human intervention to finish the negotiation process, which is not practical due to the scale of the IoT network. Misura et al. [33] proposed a cloud-based mediator platform that connects IoT devices and the applications, where automatic negotiation is performed to find the conditions of data provision that are acceptable to both parties. They assume the negotiation strategy, such as minprice, will be specified when a device is registered in the mediator platform. However, this may not be suitable for a highly dynamic environment where workload, supply, and demand are continuously changing. Also, the strategy for a device owner may change as well according to their cost-profit model. Mingozzi et al. presented an SLA negotiation framework for M2M application [34]. The negotiation is performed in a cross-layer manner where an application negotiates with services and service negotiates with thing services. They provide a QoS model with three types of services: real time, assured services, and best effort, but the chapter does not specify the QoS parameters in different layers. Generally, three broad areas need to be considered when building a negotiation component: the negotiation object, negotiation protocol, and negotiation strategy [35]. The negotiation object is the set of issues over which agreement must be reached. Kim et al. [36] represented an ontology model for describing sensing data as service-oriented properties, which outlines users' expectations rather than focusing on the physical sensor information. The SENSEI project [37, 38] provides an architecture for wireless sensor and actuator networks that enables real-world entities accessible through services by modeling the entities as resources. The semantic ontology associated with the resource includes resource type, location, temporal availability, semantic operation description (e.g. input, output, preconditions, postconditions), observation area, quality, and cost. For WSNs, accuracy, delay, coverage, packet loss, throughput, number of active sensors, sample rate, and energy consumption are the general QoS requirements [32, 39–41]. In M2M applications, availability, response time, and request rate are used as example QoS parameters in the negotiation template [34].

Negotiation protocol defines the type and format of information exchanged during a negotiation process and describes the interaction rules when receiving negotiation information [35]. For instance, a set of messages were designed for a negotiation procedure that aimed to create the connection with demanded QoS parameters when delivering a composite service [42]. Karl Czajkowski et al. [43] presented the Service Negotiation and Acquisition Protocol (SNAP) for negotiating access to, and usage of, resources from resource providers in a distributed system. However, it is too complex and not flexible enough for automatic negotiation. Nabila et al. [44] illustrated the generic alternating offers protocol proposed by Rubinstein, for bargaining between agents. FIPA Contract Net Interaction Protocol [45] is another commonly used negotiation protocol, which supports recursive negotiation to find a compromise [46]. The WS-Agreement Negotiation language is the extended negotiation layer on top of a standard Web service SLA specification WS-Agreement, which proposed a (re-)negotiation protocol that describes the possible negotiation states, the actions that trigger the state transitions, and the corresponding actions in a particular state [47]. Based on the negotiation language, Mobach et al. proposed a (re-)negotiation framework to handle auctioning by using knowledge bases and predefined workflows [48].

Negotiation strategy is a mathematical model that is used to evaluate the proposal and generate a counter proposal. Yao et al. adopted a concession strategy that takes time, resources, the priority of attributes, and currently available counterparts into consideration [35]. Sim et al. combined Bayesian learning with genetic algorithms to search for the optimal strategy when negotiating with incomplete information [49]. However, this approach is not suitable for negotiating multiple QoS parameters, and the success rate is unsatisfactory when the deadline range is lower than 30 negotiation rounds. Zheng et al. [50] mixed the concession and trade-off tactics to balance the success rate and QoS utility, but the approach cannot guarantee to find a solution when the solution exists.

9.2.3 Service Monitoring

A key research question in the IoT is how to increase the reliability of applications that use services provided in the environment [51]. A number of approaches have been proposed for dynamic service composition [7], service selection [52], run-time QoS evaluation [53], and service ranking prediction [54]. These approaches assume that the QoS values are already known, but, in reality, user side QoS may vary significantly, given unpredictable communication links, mobile service providers, and resource constrained devices. Service monitoring is important in IoT to ensure that SLAs are enforced and that users can have retribution if a service provider fails to adhere to the agreement. The fog environment provides a number of challenges that monitoring approaches in other domains such as cloud do not have to deal with such as resource constrained devices. To create a suitable monitoring approach for a fog environment, we have to balance the accuracy of the monitoring approaches with the additional load they can enforce on the network. To achieve this, we use a combination of active monitoring probing for hard QoS situations where there is a need for strict QoS guarantees and prediction-based approaches when we require soft QoS guarantees [8]. In this section, we first introduce the prediction-based approaches that can be used to make predictions for candidate services during service composition and a forecasting approach that can be used to identify that a currently executing service is about to fail.

Collaborative filtering can be used to make predictions for candidate services that could be used in the service composition based on other similar users in the environment. It has been used in other domains such as recommender systems [55–57]. There are two main collaborative filtering models, which can typically be classified as either memory or model-based. Memory-based approaches store the training data in memory and in the prediction phase, similar users are sorted based on the current user. There are a number of approaches that use neighborhood-based collaborative filtering, including user-based approaches [58], item-based approaches [59], and their combination [56]. VSS [58] and PCC [55] are often used in similarity computation methods.

Model-based approaches, which employ a machine learning technique to train a predefined model from the training data sets, have become increasingly popular. Several approaches have been studied, including clustering models [60], latent factor models [61], and aspect models [62]. Latent factor models create a low-dimensional factor model, on the premise that there are only a small number of factors influencing the QoS [55, 63]. These are usually factors such as the location of the service or the time of day, which the service is usually invoked.

Both approaches can be combined to predict missing QoS values by using global information from matrix factorization and local information from similar users and items [64, 65]. Previous approaches have focused on Web services deployed in the cloud [64, 66–69]. However, these make up only one part of the IoT, and we need to collect QoS information and make predictions for all service types to create a reliable application that use all the services provided in the IoT. Approaches such as IoTPredict can deal with the increased variability of IoT services, but further work needs to be done on the distribution and reduction of training time for fog nodes [10].

The addition of QoS forecasting to traditional monitoring approaches during service execution allows for adaptive actions to take place before a service fails. A number of techniques have previously been used for QoS attribute forecasting. A recent study conducted a detailed empirical analysis of time series forecasting for the dynamic quality of Web services [70]. Time series forecasting techniques can be categorized into two groups: classical methods based on statistical/mathematical concepts and modern heuristic methods based on artificial intelligence algorithms [71]. The former includes exponential smoothing models, regression models, autoregressive integrated moving average (ARIMA) models, threshold models, and generalized autoregressive conditional heteroskedasticity (GARCH) models. The latter includes artificial neural networks, long short-term memorys (LSTMs) and evolutionary computing-based approaches.

Classical QoS forecasting methods have included traditional augmented reality (AR)-based methods/variations, such as self-exciting threshold autoregressive (SETAR), ARIMA, and GARCH models can be used to measure time varying QoS parameters at different layers of the IoT network [8]. Two time-series methods ARIMA and GARCH have been combined to produce more accurate QoS forecasting results than traditional ARIMA but requiring extra processing time [72]. Two common time series methods ARIMA and self exciting threshold autoregressive moving average (SETARMA) were also considered, but the method used depends on the linearity of the predicted QoS time series training data. The incoming time series is tested to detect whether it is linear, in which case the ARIMA method is used otherwise the SETARMA method is adopted. The traditional ARMA model has also been used without the integrated (I) component used in ARIMA [73, 74]. Other more simplistic methods such as exponential smoothing have also been used where the forecast is given as a weighted moving average of recent time series observations. The weights assigned decrease exponentially as the observations get older.

The approaches based on traditional AI algorithms have focused on artificial neural networks (ANNs) but have not compared their approach against any established method [75–77]. Recurrent neural networks (RNNs) have developed in recent years as the most used neural network for time-series forecasting as they can utilize past data through a feedback path. Recent experiments using the LSTM architecture have shown promising results in other domains such as event forecasting [78], acoustic modeling [79], and sequence tagging [80]. They have also recently been used for QoS forecasting and have shown improved forecasting accuracy against other state-of-the-art approaches [11] and can be deployed at the edge of the network by using a deep edge architecture [81, 82]. This can allow for a number of interesting approaches to interacting with services that were not possible in a traditional cloud-based environment such as augmented reality, which can take advantage of the reduced response time and jitter at the edge [83] as well as real-time quantified-self analytics [84].

In order to enforce SLAs we need an active probing component that provides a greater level of monitoring accuracy, which can be used to resolve disagreements between service consumers and providers. Monitoring tools such as SALMon can be used to monitor Web service SLA violations [85]. The tool focuses on the monitoring of availability and the time behavior characteristics of the services. SALMon's architecture contains three components monitors, analyzers, and decision-makers. The monitors define the quality attributes, the analyzers perform the task to the SLA rules, and the decision-makers perform corrective actions in order to satisfy the SLA rules. Other approaches have focused on key metrics such as response time [86].

This method uses a proxy connected to the required services, which is able to calculate the response time. These approaches have mostly focused on SLA monitoring in traditional environments such as the cloud, in our system model, we describe a distributed SLA monitoring approach that is more suitable for a fog architecture.

9.3   Fog Service Delivery Model and AdaptFog

This chapter focuses on the process of delivering the services that have previously deployed on fog nodes by IoT application providers. AdaptFog assumes that a set of resource-intensive fog nodes (e.g. gateways, WLAN switch, base station, etc.) reside close to the IoT devices to support real-time data processing and low-latency response for mission-critical applications. The service delivery model in fog computing makes a reusable IoT service available to many fog service consumers in the environment. A fog service consumer can be an end user or another service, as shown in Figure 9.1. Fog service consumers can rent and use services according to the provisions of an SLA to achieve their different goals.

To achieve service composition for IoT applications, we assume fog nodes rely on a service composition middleware to work in a distributed manner. Such a middleware supports adaptable service management functionalities such as service discovery, service choreography, QoS monitoring, and SLA renegotiation. These fog nodes communicate with each other, composing a large-scale dynamic fog network that manages IoT services and transmits data. In other words, when a request comes, fog nodes are allowed to discover the needed services, manage service composition, and invoke services based on the services' availability, reputation, and QoS. In this case, a fog node will have to deliver services to another fog node, and so the middleware will need to support F2F SLA renegotiation. This section introduces an AdaptFog architecture that enables a reputation mode to facilitate decentralized service composition and underpins a F2F SLA renegotiation.

9.3.1   AdaptFog Architecture

AdaptFog employs a graph-based service aggregation method to realize a decentralized service composition. It targets flexible planning by modeling services and their I/O parameters in an aggregation graph based on service

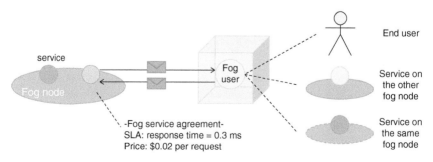

Figure 9.1 Fog service delivery model.

dependency in the services. Service dependency includes I/O parameters' relations and the services' invocation conditions. For example, two services have a dependency relation if one service's invocation relies on the other service's output data or any other execution results such as a change on the system state. AdaptFog composes services using a distributed overlay network called self-organising service overlay network (SSON) [87], which is built over P2P networks based on such service dependency and enables self-organizing service composition through management of the network. Complex IoT applications are resolved according to predefined workflows that consist of a set of abstract services. Service composition models will dynamically compose services to finalize an abstract service when no single service can support it independently. A composite service is found if the aggregation graph contains a path to link the abstract service's output parameter to its input parameter.

Figure 9.2 shows the distributed architecture of the envisioned fog computing network, which can be abstracted as a three-layer structure. The lowest layer is the device layer, which manages physical IoT devices such as sensors, actuators, smart phones, etc. The middle layer is the service layer, consisting of services with annotated SLA, which specifies a service's functionality, inputs, outputs, QoS, renegotiation information, and other concepts taken from a common standard ontology. The semantic of SLA is used to generate a service dependency network which is an overlay network building the semantical dependency of deployed IoT services for dynamic composition [7]. The highest layer is the cloud service layer, which provides a scalable way to manage all aspects of IoT deployment including data storage, security, data analysis, etc.

IoT devices and IoT services are assumed to be randomly distributed across multiple fog nodes. Each fog node stores a portion of SSON, which maintains information about all the local deployed services and the links to their semantically related services. A complete SSON is backed up in the cloud. For example, Node 1 hosts two services A2 and B1, thus it stores a service dependency network including A2, E2, B2, B3 (i.e. A2 and A2's semantical dependency) and B1, A1, C1 (i.e. B1 and B1's semantical dependency).

When a request comes, a goal-driven backward planning algorithm [7] is performed to discover a possible execution path. Each service in the service dependency network can adapt composition requests and generate potential execution fragments through a service match-making process. After the discovery stage, the bound service (compositor) selects the best matched configuration fragment to send its execution result. The fragment selection is based on the monitored QoS values and the reputation of services within the fragment. If a service has no monitored data (i.e. the service has not been invoked yet), the QoS values are predicted using the IoTPredict algorithm [10]. During the process, new services may be discovered to replace any that result in composition failures or service outages, composing them into the current execution. After the execution, if a service does not fulfill its promise defined in the

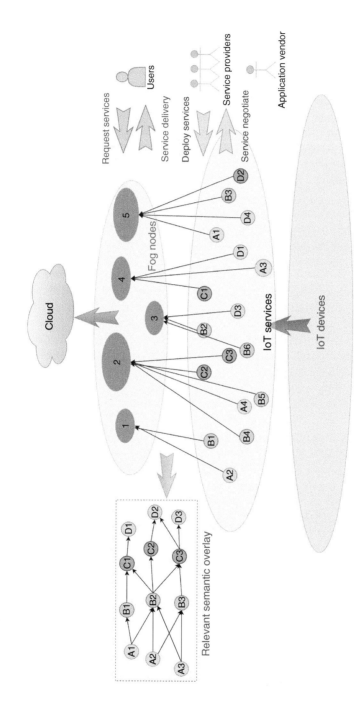

Figure 9.2 Fog computing architecture.

SLA, a renegotiation process is conducted to reconfirm the service level with its provider.

The fog service delivery will require a F2F service renegotiation as the fog computing environment lacks of a centralized controller node, and the participation of a lot of third-party fog nodes increases the potential of including malicious services and malicious fog nodes. Malicious services may report false information about its dependent services or do not provide the promised QoS. For example, an abstract workflow A, B, C, D is specified for a user task[2], its relevant dependency overlay is showed in Figure 9.2. The detail of creating the service dependency links is described in [87]. During the discovery stage, the goal-driven backward planning algorithm brings the issue that C2 sends a message to B2, which includes the service information of C2 and D2. If C2 is a malicious node that overstates the performance of C2 or D2 to increase the chance of being selected during the execution stage, which may result in a service level degradation. Thus, we equipped the composition mechanism with a trust system to monitor the real-time performance of a service and record its reputation that assists the service user to make an informed decision about selecting the most optimized solution.

9.3.2 Service Performance Validation

The trust system adopted a self-policing mechanism to address the threat of compromised entities, which consists of three modules: runtime QoS monitoring module, SLA renegotiation module, and reputation assessment module (Figure 9.3). The system architecture is shown in Figure 9.4. Inspired by the transaction validation in IoTA [88], we design an SLA validation mechanism that a service is validated by two of its prior alternative services in SSON, or its competitors' prior services, which does not contain its prior execution service if the service does not have multiple prior alternative services. If the validated service is the compositor, the validation is done by the user. For example, the SSON in Figure 9.2 indicates a set of possible execution paths that can satisfy a user's functional requirement. When the user receives a discovered plan from composers A1, A2, and A3, the current optimized solution A2, B2, C2, and D2 are selected for execution. After execution, A2 is validated by the user, B2 is validated by A1 and A3, C2 is validated by B1 and B3, D2 is validated by C1 and C3. The validation is conducted by the distributed monitoring engine of fog nodes where A1 and A3 are deployed, the detail is described in Section 9.3.3.

The reason to design such validation mechanism is that each node in the SSON can measure the execution risk of its links so that when next discovery request comes from its subsequent nodes, it can choose not to cooperate with

---

2 Each capital letter represents a collection of micro-services which provide the same functionality.)

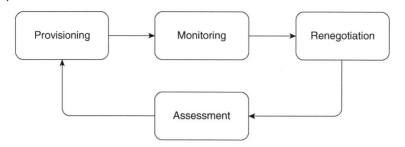

Figure 9.3 Reputation assessment life cycle.

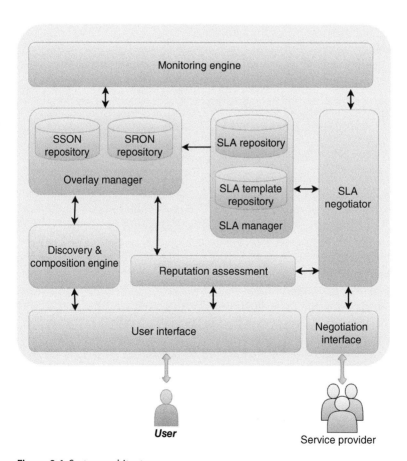

Figure 9.4 System architecture.

the malicious dependent services even though there may have a potential to match the functional requirement to maintain its reputation and the global QoS. The malicious nodes in SSON can be identified and isolated from the execution paths as time passes. Since the services in SSON are neighboring logically, and we assume the services are randomly deployed in different fog nodes, the possibility of group cheating during the validation stage can be neglected. The detail of service validation is described in Section 9.3.2.

Based on the validation mechanism, we defined an SSON-based service reputation overlay network (SRON). The SRON uses a directed graph to store the reputation assessment data, which can be regarded as a selection criterion during the executed time. The nodes in SRON represent validating services and validated services. The line between two nodes has a direction that points from the validating service to the validated services. Considering the validation is time-sensitive in the dynamic environment, we assume the validation has a lifetime from time t_0 to time t_{max}, which means the link disappears after the deadline t_{max}. The time t can be normalized to $\tilde{t}$ based on Eq. (9.1).

$$\tilde{t} = \frac{t - t_0}{t_{max} - t_0} \qquad (9.1)$$

The link pointed from service s_k to service s_i represents that s_i has been validated by s_k. The validation result is modeled by validation scoring function, which evaporates as time passes once a validation has been performed. If multiple validation from s_k to s_i has been performed before the deadline, a new link would be generated to replace the old one. Figure 9.5 shows an example of SRON that corresponds to the SSON in Figure 9.2. Each dashed line representing a validating relationship between two services. We define a scoring function $V_{ki}(\tilde{t}, T_{s_i})$ to compute the weight of each directional link: $V_{ki}(\tilde{t}, T_{s_i}) = P_f \times P_i \times f(\tilde{t}, T_{s_i})$, where P_f is the performance of the execution fragment that starts from the validated service to the last executed service, P_i is the performance of

Figure 9.5 An example of the SRON.

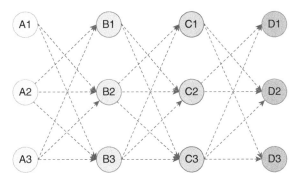

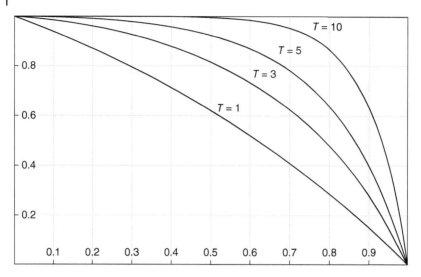

Figure 9.6 Evaporation function.

the validated service s_i, $f(\tilde{t}, T_{s_i})$ is the evaporation function, which is defined as:

$$f(\tilde{t}, T_{s_i}) = \frac{e^{T_{s_i}(\tilde{t}-1)} - 1}{e^{-T_{s_i}} - 1} \tag{9.2}$$

where T_{s_i} ($T_{s_i} \in [1, 10]$) is the trust value measuring the degree to which s_i can be trusted. When a new service join in the network, it has an initial trust value that equals to 10. This value can be dynamically changed based on the service performance and the renegotiation result of the service. Figure 9.6 shows how the trust value controls the evaporation rate.

The performance of service s_i is measured by the utility function, which can be modeled as the weighed sum of all the requested QoS utility:

$$U(s_i) = \sum_{j=1}^{n} w_j u(q_{ij}) \tag{9.3}$$

$$\sum_{j=1}^{n} w_j = 1 \tag{9.4}$$

where n is the number of requested QoS attributes, w_j is the weight of the j-th attribute specified by the user, $u(q_{ij})$ is the utility of j-th requested QoS attribute. Considering the QoS attribute can be classified as a "lower-the-better" attribute such as response time, or a "higher-the-better" attribute such as availability.

We assume the SLA uses a threshold value to specify the SLO (e.g. response time <10 ms, or availability greater than 98%), the utility of a QoS attribute can be defined with a simple step function:

$$u(q_{ij}) = \begin{cases} 1, & \text{if } q_{ij} \leq Q_{ij} \\ 0, & \text{if } q_{ij} > Q_{ij} \end{cases} \tag{9.5}$$

$$u(q_{ij}) = \begin{cases} 1, & \text{if } q_{ij} \geq Q_{ij} \\ 0, & \text{if } q_{ij} < Q_{ij} \end{cases} \tag{9.6}$$

where Q_{ij} is the threshold value of j-th QoS attribute specified in the SLA, q_{ij} is the monitored value provided by the monitoring engine. Equation (9.5) is used for "lower-the-better" attributes, and Eq. (9.6) is used for "higher-the-better" attributes.

The reputation assigned by service s_k to service s_i is defined as the reputation of s_k times its validating score to s_i (i.e. the weight of the link). The overall reputation of service s_i is computed during the runtime by adding the weights of all the existing links, which is defined as $R(s_i)$:

$$R(s_i) = \sum_{k=1}^{m} \rho(s_k)R(s_k)V_{ki} \tag{9.7}$$

$$\sum_{k=1}^{m} \rho(s_k) = 1 \tag{9.8}$$

where m is the number of links that point from validating service s_k to the validated service s_i, $\rho(s_k)$ is the importance factor of s_k, which is used to diminish the risk of malicious validation. For example, A1 and A3 are semantically related to B2, reporting a malicious validation result may affect their future reputation or decrease the chance of being executed. However, B1 has no semantic dependency with C2. It may report an unreal validation result of C2 so that C1 may be more competitive than C2, which further increases the chance of itself to be executed with C1. Thus, we assume that the semantically related service is more likely to report an unbiased validation result, and its importance factor is two times higher than the services that have no semantic dependencies. It should be noted that maliciously reporting an unreal validation result may damage the long-run benefit of the validating service. For instance, if B1 and B3 underestimate the performance of C2, the low reputation value of C2 may affect the reputation of D1 and D3, which may decrease the chance of executing a configuration path that contains D1 and D3. In other words, maliciously validation does not guarantee a business benefit. Also, the random service deployment decreases the possibility of group cheating from different fog nodes.

9.3.3   Runtime QoS Monitoring

To address the requirements of monitoring in a fog environment where devices may be resource constrained and there is a limited amount of bandwidth available on the network, we propose a distributed monitoring approach that makes use of both probing and predictions to provide accurate QoS monitoring while reducing the additional bandwidth on the network. We first describe our distributed monitoring approach, which is used for applications with hard QoS requirements, we then show how we can use prediction-based approaches to reduce the load on the network while still providing accurate QoS values for soft QoS applications [8].

With the large-scale size of the IoT in terms of services and requesters, a centralized approach cannot be used due to complexity and scalability issues. There are a number of QoS parameters that can be measured and recorded as part of an SLA either at the provider or client side. We use a hybrid approach where the monitor is not located on the client or server but in the middle distributed in the fog nodes. This allows for a number of parameters such as availability, which can be specified in the SLA to be probed by our monitoring framework. All of these monitoring operations are driven by SLAs, which include both the objectives and the quality of the monitoring probes. Due to often overlapping objectives, the probes are optimized by combining them. The monitoring approach is distributed across all the fog nodes in the network to allow for total coverage of the services in the environment. The most recent monitored values are stored in a distributed database, and after a certain time period they are compressed and sent to the cloud. This allows for a quick response at the edge, while also having a long-term back-up for any disputes between providers and consumers that require previous data.

Simple SLOs which can be specified in the SLA are used to define the QoS metrics measured in the services. This is done using the publishing service that also allows to temporary start and stop SLA monitoring. The listeners then register with our distributed monitoring engine and record the results in our distributed database. Every listener is valid only within a given period of time after which it expires.

The monitoring engine builds a time series data set for currently executing services at the edge of the network, this allows us to make use of time-series forecasting algorithms to detect if a service is about to fail or degrade in quality [11]. The ability to detect when a service is about to fail or degrade in quality is useful to help avoid SLA violations and to recompose the service before the user knows there is a problem. When a service is forecast to degrade in quality, the first step is to conduct an SLA negotiation to identify whether we can keep the current service but negotiate better QoS. If we are already using the best possible QoS values or the service provider is not negotiating then we send a message to the service composition engine to perform a dynamic recomposition using a suitable alternative service if one is available. If no suitable alternative service is available, then the service composition engine sends a message to the service discovery engine to discover new services in the environment.

We use an LSTM based neural network as our forecasting approach for hard QoS based on our previous work [11]. LSTMs have also developed from RNNs but improved on the initial architecture by storing and retrieving information over a long period of time with explicit gating mechanisms and a built-in error carousel [89]. Unlike the traditional RNN, which simply computes a weighted sum of the input and applies a nonlinear function, each j-th LSTM unit maintains a memory c_t^j at time t. The output h_t^j which is the activation of the LSTM unit is then:

$$h_t^j = o_t^j \tanh(c_t^j) \tag{9.9}$$

where o_t^j is an output gate that modulates the amount of memory content exposure. The output gate is computed by:

$$o_t^j = \sigma(W_o x_t + U_o h_{t-1} + V_o c_t)^j \tag{9.10}$$

where σ is a logistic sigmoid function and V_o is a diagonal matrix. This procedure of taking a linear sum between the existing state and the newly computed state is similar to the gated recurrent unit (GRU) unit, however LSTMs can control the degree to which its state is exposed. The memory cell c_t^j is updated by partially forgetting the existing memory and adding a new memory content $\tilde{c}_t^j$:

$$c_t^j = f_t^j c_{t-1}^j + i_t^j \tilde{c}_{t-1}^j \tag{9.11}$$

The introduction of gates allows the LSTM unit to decide whether to keep the existing memory or to overwrite it. This improves on the traditional recurrent unit, which overwrites its content at each time-step. Intuitively, if the LSTM unit detects an important feature from an input sequence at the early stage, it easily carries this information (the existence of the feature) over a long distance, hence, capturing potential long-distance dependencies. This process is illustrated graphically in Figure 9.7. We also used increased monitoring frequency for hard and soft QoS applications to ensure the level of QoS provides suitable reliability. For applications with less strict QoS guarantees, we use a simple persistence algorithm with less frequent monitoring to provide a basic level of forecasting.

Due to the large amount of services in the environment, it can be difficult to know which services to perform the SLA negotiation on to provide the best application for the user. We can use IoTPredict to create a shortlist of possible services to perform SLA negotiation with based on their previous invocations from previous users in the environment. We demonstrate the problem graphically in Figure 9.8. The QoS value of IoT service s observed by user u can be predicted by exploring the QoS experiences from a user similar to u. A user is

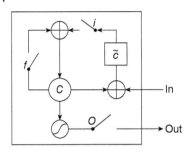

Figure 9.7 Long short-term memory network.

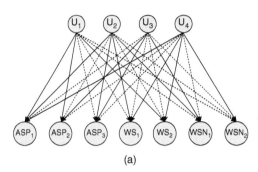

(a)

(b)

	ASP₁	ASP₂	ASP₃	WS₁	WS₂	WSN₁	WSN₂
U₁	0.9	0.23			0.12		
U₂	0.43		1.22	0.23		0.33	
U₃		0.30	1.39		0.37		
U₄	0.23			0.34	0.34		0.24

(c)

	ASP₁	ASP₂	ASP₃	WS₁	WS₂	WSN₁	WSN₂
U₁	0.9	0.23	0.91	0.25	0.12	0.22	0.30
U₂	0.43	0.23	1.22	0.23	0.25	0.33	0.36
U₃	0.80	0.30	1.39	0.38	0.36	0.37	0.15
U₄	0.23	0.23	1.20	0.34	0.34	0.22	0.24

Figure 9.8 Demonstration of QoS prediction in IoT. (a) invocation graph, (b) User-service matrix, and (c) predicted matrix.

similar to u if they share similar characteristics, which can be extracted from their QoS experiences with different services. By sharing local QoS experience among users, these approaches can predict the QoS value of a range of IoT services including autonomous service providers (ASPs), Web services, and WSNs even if the user u has never invoked the service s before. This can be modeled as a bipartite graph $G = (U \cup S, E)$, such that each edge in E connects a vertex in U to S. Let $U = \{u_1, u_2, \ldots, u_4\}$ be the set of component users and $S = \{ASP_1, ASP_2, \ldots, WSN_2\}$ denote the set of IoT services and E (solid lines) represent the set of invocations between U and S as shown in Figure 9.8a. Given a pair (i, j), $u_i \in U$ and $c_j \in S$, edge e_{ij} corresponds to the QoS value of that invocation. Given the set E, the task is to predict the weight of potential invocations (broken lines).

The task can be framed as a matrix completion problem, where we want to fill in the remaining values to have a fully completed matrix as shown in Figure 9.8c. In our framework, we use the IoTPredict algorithm to make predictions for the missing values in the matrix. This reduces the search space for possible SLA negotiations considerably reduces the time spent on negotiations that could not fulfill the QoS requirements, this speeds up the time to create these applications and also allows for more dynamic service re-compositions.

9.3.4   Fog-to-Fog Service Level Renegotiation

Renegotiation is defined as a negotiation to alter the terms of an existing agreement, allowing consumers and service providers to initiate updates in an established SLA, such as the modification on the availability when some of the sensors are damaged or running out of power at a certain period of time. In this work, two types of renegotiation are considered: proactive renegotiation and reactive renegotiation. In the proactive case, monitoring engine reports an alarm indicating that a service is about to fail or degrade in quality. Then the renegotiation is conducted to inform the provider about the violation possibility, and asking for a better QoS or a lower price. The proactive renegotiation result would not affect the trust value of a service; however, it provides instructions during the service composition stage which indicates that the proactive renegotiation should be performed in a timely manner to guarantee a short response time of service composition.

The reactive renegotiation is triggered when any SLA violation is detected during the service validation process, which can be used for amending the SLOs based on the monitored data and enhance the trustworthiness of the IoT services. The new proposal contains a different service level to which the provider must comply since the validation result indicates that the service cannot fulfill the claimed service level. If the service provider agrees to change the SLOs that conforms to the proposed monitored service level, its trust value would not be affected. If the service insists on the claimed service level, the trust value would decrease by $\Delta\tau$:

$$\Delta\tau = \frac{n_{\mathrm{vio}}}{N_{\mathrm{SLO}}} \tag{9.12}$$

where n_{vio} is the number of violated SLOs after renegotiation and N_{SLO} is the total number of SLOs in the SLA. For example, if the SLA of service C2 has two guarantee terms stating that the availability would bigger than 98% and the response time would <10 ms, while the monitored data indicate the response time is violated, then the monitored response time (e.g. 18 ms) is set as reserved value for renegotiation. The reserved value is regarded as the bottom line of the renegotiation, if the value is not acceptable by the provider, the trust value would decrease by 0.5. In our scenario, the reactive renegotiation does not have a strict demand on the responsiveness, however, it aims to argue for the best actual QoS under the current circumstance.

Although research on service negotiation in the IoT environment is very limited, it is possible to apply Web service negotiation techniques to IoT services by extending current existed negotiation frameworks. As WS-Agreement is an open standard implemented by a wide range of systems, we follow the renegotiation protocol of WS-Agreement Negotiation, which includes revoking the previous SLA and updating service terms such as guarantees and SLOs.

The WS-Agreement renegotiation model is described as following: the renegotiation initiator creates a new negotiation instance, which implements the Negotiation port type, then invokes the *initiateNegotiation* method of the renegotiation responder. During the negotiation initiation stage, the roles, responsibilities of the negotiating parties, and the renegotiated agreement instance are specified in the negotiation context. The renegotiation responder sends back its negotiation instance endpoint and the bilateral renegotiation process starts. This proposal exchange will last for several rounds until the initiator decides to create the renegotiated agreement based on an accepted offer by invoking the *createAgreement* method of the responder, and the original agreement's state is changed to "Completed."

WS-Agreement negotiation defines the type and format of negotiation information, and specifies the rules to generate and exchange negotiation information. WS-Agreement Negotiation formalizes negotiation information as *negotiation offers* [47]. The elements of a negotiation offer include the offer id, name, negotiation offer context, terms and negotiation constraints, which specifies negotiating restrictions of negotiable service properties. WS-Agreement Negotiation states that the *NegotiationConstraint* has the type of *Required* or *Optional*, which will be used to validate subsequent counter offers. Considering the message payload in the fog environment, we regulate that only the negotiable properties and the *Required* terms will be presented in negotiation offers. The state model of negotiation offers controls the interactions between negotiation parties and indicates the rules to take action after receiving a new offer. For example, the state *Advisory* indicates multiple back-and-forth interactions while the state *Solicited* indicates a single request–reply interaction. The state transition model can be found in [47].

There are various negotiation strategies that cover decision support system techniques for SLA renegotiation, such as game theory [50], reinforcement learning [90], Bayesian learning [91], multi-criteria decision-making (MCDM) techniques [92], etc. Considering the time-constrained requirement, the machine learning technique may not be suitable for SLA renegotiation in a dynamic environment due to the low runtime performance in readapting and relearning any changing situation. A game theory based multiobjective bilateral negotiation strategy has been proposed for service negotiation with incomplete information in the cloud-based IoT environment [93], which chooses to play *trade-off* or *concession* tactic in each round based on a predefined probability to balance the success rate and QoS utility. This mixed approach uses an exponential utility function to weight and calculate preferences from the counterpart, which forms the basis of our negotiation strategy. Considering a negotiation with incomplete information, the provider may propose multiple counteroffers with various SLO levels to increase the chance of successful negotiation, we use the cost performance to evaluate each offer and select the most optimized one (i.e. the one with highest cost performance) for further negotiation, which is defined as:

$$\text{CP}(x^t_{a \to b}) = \frac{\sum_{j=1}^{n} w_j u(x^t_{a \to b}[j])}{\text{P}_{\text{rice}}} \tag{9.13}$$

where $x^t_{a \to b}$ is the negotiation offer that proposed by a to b at time t, $u(x^t_{a \to b}[j])$ is the utility value of a requested service property j, and w_j is the weight of each attribute j ($\sum_{j=1}^{n} w_j = 1$). The utility value is computed using Eq. (9.14) [50]:

$$u(q_{ij}) = \begin{cases} \frac{1}{e-1}(e^x - 1), \text{higher-is-better attribute} \\ \frac{1}{e-1}(e^{1-x} - 1), \text{lower-is-better attribute} \end{cases} \tag{9.14}$$

We assume that the price model of a service is private to the service provider, the reputation system has no prior knowledge about the business objectives of a service provider. In other words, we only modify SLOs based on the monitoring data in the initial offer while keeping the business value list such as price and penalty the same. However, the service providers can modify the price based on their business benefit model in the subsequent offers. The preferred value of each attribute in the initial offer ($t = t_0$) is calculated as following:

$$x^{t_0}_{a \to b}[j] = \ln \left\{ \left(\frac{0.9 \times u(j)}{k} + b \right) \right\} / a \tag{9.15}$$

where $u(j)$ is the utility value of the monitored data for attribute j (i.e. reserved value). k, b, and a are constants that for lower-is-better attributes $k = \frac{e}{e-1}$, $b = \frac{1}{e}, a = -1$, for higher-is-better attributes $k = \frac{1}{e-1}, b = 1, a = 1$.

The bilateral negotiation process starts with the submission of initial offer from a fog node to a service provider. When the node g receives a counter offer $x^t_{p \to g}$ from provider p, it evaluates $x^t_{p \to g}$ using the utility function and taking actions at time t' based on the offer state and the interpretation formula of node g:

$$I^g(t', x(\text{ad})^t_{p \to g}) = \begin{cases} \text{accept}, \forall j \text{ in } J, u(x^t_{p \to g}[j]) \ge u(x^{t'}_{g \to p}[j]) \\ \text{reject}, \exists j \text{ in } J, u(x^t_{p \to g}[j]) < u(j) \\ \qquad \text{and } t' \ge t_{\text{deadline}} \\ x^{t'}_{g \to p}, \text{otherwise} \end{cases} \tag{9.16}$$

$$I^g(t', x(\text{so})^t_{p \to g}) = \begin{cases} \text{accept, } \forall j \text{ in } J, u(x^t_{p \to g}[j]) \geq u(j) \\ \text{reject, } \exists j \text{ in } J, u(x^t_{p \to g}[j]) < u(j) \end{cases} \tag{9.17}$$

where $x(\text{ad})^t_{p \to g}$ is an offer in advisory state that proposed by p to g at time t, while $x(\text{so})^t_{p \to g}$ represents an offer in solicited state. The j represents the collection of negotiated terms. The $u(j)$ is the utility of reserved value for attribute j, $x^{t'}_{g \to p}$ is the counter offer that g would propose to p at the time of the interpretation. The algorithm of playing the mixed concession-trade-off tactic is described in [93].

Each time when a fog node receives a negotiation offer from the service provider, the negotiation instance validates the offer based on the negotiation constraints and the state of the originating offer before interpretation. An offer will be regarded as invalid and rejected in the following situations:

- The service properties in the negotiation offer conflict with creation constraints specified in the referenced template.
- The service properties in the negotiation offer violate the negotiation constraints specified in the originating offer.
- The state of the counter offer violates the state transition model specified by WS-Agreement Negotiation (e.g. transiting from a solicited state to an advisory state).
- The counter offer which is in an accepted state is different from the originating offer.

The renegotiation is terminated when the maximum timeout occurs, or negotiating parties reach a consensus on the service properties (i.e. successful negotiation). If an offer is accepted, the renegotiation initiator (i.e. renegotiating fog node) creates a new pending agreement by using the *agreementFactoryEPR* specified in the negotiation context of the offer.

Figure 9.9 illustrates the renegotiation process at runtime when initiated by a fog node (i.e. negotiator), which includes following phases:

Template retrieval phase: The monitor engine reports an alarm of SLA violation (or possible SLA violation) with the monitored data and the reference of originating SLA. The negotiator g looks up the corresponding SLA template based on the template identification in the SLA.

Renegotiation customization phase: Negotiator sends an initiation request to the service provider to customize the renegotiation context. The purpose of this phase is testing the network connection, verifying the availability of the current SLA template with the service provider, specifying the roles and responsibilities of the negotiating parties, and avoiding possible communication ambiguity by specifying the negotiation protocol and SLA schema. If the provider refuses the request, the renegotiation process is terminated immediately.

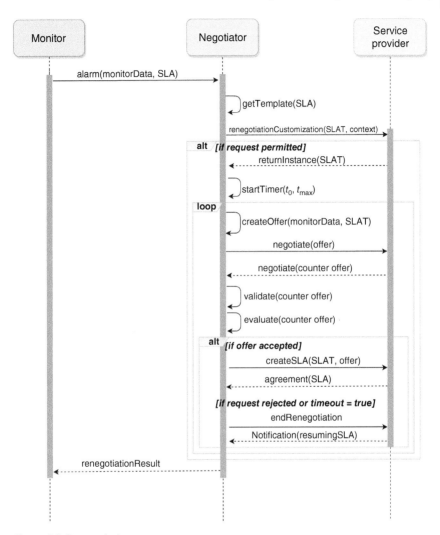

Figure 9.9 Renegotiation process.

Bilateral renegotiation phase: Once the renegotiation context is successfully initiated, negotiator *g* creates an initial offer based on the monitored data and the SLA template, and submits to the service provider. The service provider evaluates the offer based on its own business benefit model and chooses to accept the offer, reject the offer, or propose a counter offer. The offer exchanging process is repeated until an acceptable offer is reached, or the renegotiation process is terminated (i.e. the renegotiation allowable time is over, or a negotiation participant performs the terminate operation).

SLA creation phase: If an offer is accepted, the new pending SLA is created. Once the SLA is signed by both party, the state of originating SLA is updated to "Completed." Otherwise, the original SLA still holds as the new agreement has not been reached.

9.4 Conclusion and Open Issues

In this chapter, we have described the architecture and components for a service adaptation system in a fog environment. This chapter introduces a number of novel ideas for the performance validation in a fog environment and the use of a monitoring and negotiation component to enable the reputation system. This chapter provides an architecture for other researchers to develop an adaptation system in a fog environment. The components that we discuss have been designed specifically to handle the requirements of a fog environment. The service performance validation provides an effective mechanism to validate service performance. The runtime QoS monitor is designed to reduce the load on a fog network by making use of predictions where possible an only probing services when needed for applications with hard QoS requirements. The SLA renegotiation component provides an adaptation mechanism to the dynamic changing of fog environment, which helps to guarantee the service level and the continuity of composite service.

As part of our future work, we plan to implement a test implementation of the reputation system in our lab. We have a number of devices such as Raspberry Pi's and Intel Galileo's that can be used as a tested for the reputation system. We can simulate malicious users and services in the environment and evaluate how our reputation system handles these cases. We can evaluate how robust the system is as we increase the number of malicious users in the system. This initial demo can then be expanded to a university level, where we can deploy our devices across our university campus to evaluate how well the reputation system will scale. The reputation of services in a fog environment is a key research problem with a number of interesting challenges. We have introduced an architecture capable of providing an effective adaptation system and described in detail a number of components that could be used in this architecture.

References

1 Karnouskos, S., Savio, D., Spiess, P. et al. (2010). Real-world service interaction with enterprise systems in dynamic manufacturing environments. In: *Artificial Intelligence Techniques for Networked Manufacturing Enterprises Management*, 423–457. London: Springer.

2 Botta, A., De Donato, W., Persico, V., and Pescapé, A. (2016). Integration of cloud computing and internet of things: a survey. *Future Generation Computer Systems* 56: 684–700.

3 Bonomi, F., Milito, R., Natarajan, P., and Zhu, J. (2014). Fog computing: a platform for internet of things and analytics. In: *Big Data and Internet of Things: A Roadmap for Smart Environments*, 169–186. Springer.

4 Cisco (2018). Cisco global cloud index: forecast and methodology. 2016–2021 White Paper. https://www.cisco.com/c/en/us/solutions/collateral/service-provider/global-cloud-index-gci/white-paper-c11-738085.pdf.

5 Chen, N., Yang, Y., Zhang, T. et al. (2018). Fog as a service technology. *IEEE Communications Magazine* 56 (11): 95–101.

6 Razzaque, M.A., Milojevic-Jevric, M., Palade, A., and Clarke, S. (2016). Middleware for internet of things: a survey. *IEEE Internet of Things Journal* 3 (1): 70–95.

7 Chen, N., Cardozo, N., and Clarke, S. (2016). Goal-driven service composition in mobile and pervasive computing. *IEEE Transactions on Services Computing* 11 (1): 49–62.

8 White, G., Nallur, V., and Clarke, S. (2017). Quality of service approaches in IoT: a systematic mapping [Online]. *Journal of Systems and Software* 132: 186–203. http://www.sciencedirect.com/science/article/pii/S016412121730105X.

9 Ludwig, H., Keller, A., Dan, A. et al. (2003). *Web Service Level Agreement (WSLA) Language Specification*, 815–824. IBM Corporation.

10 White, G., Palade, A., Cabrera, C., and Clarke, S. (2018). IoTPredict: collaborative QoS prediction in IoT. 2018 IEEE International Conference on Pervasive Computing and Communications (PerCom 2018), Athens, Greece (March 2018).

11 White, G., Palade, A., and Clarke, S. (2018). Forecasting QoS attributes using LSTM networks. In: *International Joint Conference on Neural Networks (IJCNN)*, 1–8. IEEE.

12 Braubach, L., Murillo, J.M., Kaviani, N. et al. (eds.) (2018). Qos prediction for reliable service composition in IoT. In: *Service Oriented Computing – ICSOC 2017 Workshops*, 149–160. Cham, Switzerland: Springer International Publishing.

13 Mokhtar, S., Liu, J., Georgantas, N. et al. (2005). QoS-aware dynamic service composition in ambient intelligence environments. In: *Proceedings of the 20th IEEE/ACM International Conference on Automated Software Engineering*, 317–320. ACM.

14 de Medeiros, R.W.A., Rosa, N.S., and Pires, L.F. (2015). Predicting service composition costs with complex cost behavior. In: *IEEE International Conference on Services Computing (SCC)*, 419–426. IEEE.

15 Li, Y., Huai, J.P., Deng, T. et al. (2007). QoS-aware service composition in service overlay networks. In: *IEEE International Conference on Web Services (ICWS 2007)*, 703–710. IEEE.

16 Wang, Z., Xu, T., Qian, Z. et al. (2009). A parameter-based scheme for service composition in pervasive computing environment. In: *International Conference on Complex, Intelligent and Software Intensive Systems*, 543–548. IEEE.

17 Zhou, X., Ge, Y., Chen, X. et al. (2011). A distributed cache based reliable service execution and recovery approach in MANETs. In: *IEEE Asia-Pacific Services Computing Conference*, 298–305. IEEE.

18 Efstathiou, D., McBurney, P., Zschaler, S., and Bourcier, J. (2014). Efficient multi-objective optimisation of service compositions in mobile ad hoc networks using lightweight surrogate models. *Journal of Universal Computer Science* 20 (8): 1089–1108.

19 Aschoff, R.R. and Zisman, A. (2011). QoS-Driven proactive adaptation of service composition. In: *International Conference on Service-Oriented Computing*, 421–435. Berlin, Heidelberg: Springer.

20 Silas, S., Ezra, K., and Blessing Rajsingh, E. (2012). A novel fault tolerant service selection framework for pervasive computing. *Human-centric Computing and Information Sciences* 2 (1): 5.

21 Prinz, V., Fuchs, F., Ruppel, P. et al. (2008). Adaptive and fault-tolerant service composition in peer-to-peer systems. In: *Distributed Applications and Interoperable Systems. Lecture Notes in Computer Science*, vol. 5053 (eds. R. Meier and S. Terzis), 30–43. Berlin, Heidelberg: Springer.

22 Wang, H., Sun, H., and Yu, Q. (2013). Reliable service composition via automatic QoS prediction. In: *IEEE International Conference on Services Computing*, 200–207. IEEE.

23 Schuler, C., Weber, R., Schuldt, H., and Schek, H.-J. (2004). Scalable peer-topeer process management-the OSIRIS approach. *Web Service*.

24 Groba, C. and Clarke, S. (2011). Opportunistic composition of sequentially connected services in mobile computing environments. In: *IEEE International Conference on Web Services*, 17–24. IEEE.

25 Groba, C. and Clarke, S. (2012). Synchronising service compositions in dynamic Ad Hoc environments. In: *IEEE First International Conference on Mobile Services*, 56–63. IEEE.

26 Groba, C. and Clarke, S. (2013). Opportunistic service composition in dynamic ad hoc environments. *IEEE Transactions on Services Computing* 7 (4): 642–653.

27 Ferrer, A.J., HernáNdez, F., Tordsson, J. et al. (2012). OPTIMIS: a holistic approach to cloud service provisioning. *Future Generation Computer Systems* 28 (1): 66–77.

28 Koller, B., Frutos, H.M., and Laria, G. (2010). Service level agreements in brein. In: *Grids and Service-Oriented Architectures for Service Level Agreements*, 157–165. Springer.

29 Wieder, P., Butler, J.M., Theilmann, W., and Yahyapour, R. (2011). *Service Level Agreements for Cloud Computing*. Springer Science + Business Media.

30 Palade, A., Cabrera, C., Li, F. et al. (2018). Middleware for internet of things: an evaluation in a small-scale IoT environment. *Journal of Reliable Intelligent Environments*: 1–21.

31 Mubeen, S., Asadollah, S.A., Papadopoulos, A.V. et al. (2017). Management of service level agreements for cloud services in IoT: a systematic mapping study. *IEEE Access* 6: 30184–30207.

32 Gaillard, G., Barthel, D., Theoleyre, F., and Valois, F. (2014). Service level agreements for wireless sensor networks: a WSN operator's point of view. In: *2014 IEEE Network Operations and Management Symposium (NOMS)*, 1–8. IEEE.

33 Misura, K. and Zagar, M. (2014). Internet of things cloud mediator platform. In: *2014 37th International Convention on Information and Communication Technology, Electronics and Microelectronics (MIPRO)*, 1052–1056. IEEE.

34 Mingozzi, E., Tanganelli, G., and Vallati, C. (2014). A framework for qos negotiation in things-as-a-service oriented architectures. In: *2014 4th International Conference on Wireless Communications, Vehicular Technology, Information Theory and Aerospace 31 and Electronic Systems (VITAE)*, 1–5. IEEE.

35 Yao, Y. and Ma, L. (2008). Automated negotiation for web services. In: *11th IEEE Singapore International Conference on Communication Systems – ICCS 2008*, 1436–1440. IEEE.

36 Kim, J.-H., Kwon, H., Kim, D.-H. et al. (2008). Building a service-oriented ontology for wireless sensor networks. In: *Seventh IEEE/ACIS International Conference on Computer and Information Science – ICIS 08*, 649–654. IEEE.

37 Tsiatsis, V., Gluhak, A., Bauge, T. et al. (2010). The SENSEI Real World Internet Architecture. In: *Towards the Future Internet*, 247–256. Ios Pr Inc.

38 Villalonga, C., Bauer, M., Aguilar, F.L. et al. (2010). A resource model for the real world internet. In: *European Conference on Smart Sensing and Context*, 163–176. Springer.

39 Gaillard, G., Barthel, D., Theoleyre, F., and Valois, F. (2014). SLA specification for IoT operation-the WSN-SLA framework. Ph.D. dissertation. INRIA.

40 Chen, D. and Varshney, P.K. (2004). QoS support in wireless sensor networks: a survey. In: *Proceedings of the International Conference on Wireless Networks*, vol. 1, 227–233. ICWN'04.

41 Casola, V., De Benedictis, A., Rak, M. et al. (2013). An SLA-based approach to manage sensor networks as-a-service. In: *2013 IEEE 5th International Conference on Cloud Computing Technology and Science (CloudCom)*, vol. 1, 191–197. IEEE.

42 Swiatek, P. and Rucinski, A. (2013). IoT as a service system for eHealth. In: *2013 IEEE 15th International Conference on e-Health Networking, Applications and Services (Healthcom)*, 81–84. IEEE.

43 Czajkowski, K., Foster, I., Kesselman, C. et al. (2002). SNAP: a protocol for negotiating service level agreements and coordinating resource management in distributed systems. In: *Job scheduling strategies for parallel processing*, 153–183. Springer.

44 Hadidi, N., Dimopoulos, Y., and Moraitis, P. (2010). Argumentative alternating offers. In: *International Workshop on Argumentation in Multi-Agent Systems*, 105–122. Springer.

45 FIPA (2002). FIPA Contract Net Interaction Protocol Specification. Architecture, no. SC00029H, p. 9 [Online]. http://www.mit.bme.hu/projects/intcom99/9106vimm/fipa/XC00029E.pdf.

46 Zulkernine, F., Martin, P., Craddock, C., and Wilson, K. (2009). A policy-based middleware for web services sla negotiation. In: *IEEE International Conference on Web Services – ICWS 2009*, 1043–1050. IEEE.

47 Waeldrich, O., Battré, D., Brazier, F.F. et al. (2011). *WS-Agreement Negotiation Version 1.0*, vol. 35, 41. Open Grid Forum.

48 Mobach, D.G., Overeinder, B.J., and Brazier, F.M. (2006). A ws-agreement based resource negotiation framework for mobile agents. *Scalable Computing: Practice and Experience* 7 (1): 23–36.

49 Sim, K.M., Guo, Y., and Shi, B. (2009). BLGAN: Bayesian learning and genetic algorithm for supporting negotiation with incomplete information. *IEEE Transactions on Systems, Man, and Cybernetics, Part B (Cybernetics)* 39 (1): 198–211.

50 Zheng, X., Martin, P., Brohman, K., and Da Xu, L. (2014). Cloud service negotiation in internet of things environment: a mixed approach. *IEEE Transactions on Industrial Informatics* 10 (2): 1506–1515.

51 Miorandi, D., Sicari, S., Pellegrini, F.D., and Chlamtac, I. (2014 [Online]. http://www.sciencedirect.com/science/article/pii/S1570870512000674). Internet of things: vision, applications and research challenges. *Ad Hoc Networks* 10 (7): 1497–1516. (accessed May 2018).

52 Yachir, A. et al. (2016). Event-aware framework for dynamic services discovery and selection in the context of ambient intelligence and internet of things. *IEEE Transactions on Automation Science and Engineering* 13 (1): 85–102.

53 Su, G., Rosenblum, D.S., and Tamburrelli, G. (2016). Reliability of run-time quality-of-service evaluation using parametric model checking. In: *Proceedings of the 38th International Conference on Software Engineering*, 73–84. ACM.

54 Huang, Y., Huang, J., Cheng, B. et al. (2017). Time-aware service ranking prediction in the internet of things environment. *Sensors* 17 (5): 974.

55 Resnick, P. et al. (1994). GroupLens: an open architecture for collaborative filtering of netnews. In: *Proceedings of the 1994 ACM Conference on Computer Supported Cooperative Work*, 175–186. ACM.

56 Ma, H., King, I., and Lyu, M.R. (2007). Effective missing data prediction for collaborative filtering. In: *Proceedings of the 30th Annual International ACM SIGIR Conference on Research and Development in Information Retrieval*, 39–46. ACM.

57 Burke, R. (2002). Hybrid recommender systems: survey and experiments. *User Modeling and User-Adapted Interaction* 12 (4): 331–370.

58 Breese, J.S., Heckerman, D., and Kadie, C. (1998). Empirical analysis of predictive algorithms for collaborative filtering. In: *Proceedings of the Fourteenth Conference on Uncertainty in Artificial Intelligence*, 43–52. Morgan Kaufmann Publishers Inc.

59 Deshpande, M. and Karypis, G. (2004). Item-based top-N recommendation algorithms. *ACM Transactions on Information Systems* 22 (1): 143–177.

60 Xue, G.-R. et al. (2005). Scalable collaborative filtering using cluster-based smoothing. In: *Proceedings of the 28th Annual International ACM SIGIR Conference on Research and Development in Information Retrieval*, 114–121. ACM.

61 R. Salakhutdinov and A. Mnih (2008), Probabilistic matrix factorization. Advances in Neural Information Processing Systems, pp. 1257–1264.

62 Singla, P. and Richardson, M. (2008). Yes, there is a correlation: from social networks to personal behavior on the web. In: *Proceedings of the 17th International Conference on World Wide Web*, 655–664. ACM.

63 White, G., Palade, A., Cabrera, C., and Clarke, S. (2017). Quantitative evaluation of QoS prediction in IoT. In: *47th Annual IEEE/IFIP International Conference on Dependable Systems and Networks Workshops (DSN-W)*, 61–66. IEEE.

64 Zhang, Y., Zheng, Z., and Lyu, M.R. (2011). Exploring latent features for memory-based qos prediction in cloud computing. In: *IEEE 30th International Symposium on Reliable Distributed Systems*, 1–10. IEEE.

65 Lo, W., Yin, J., Deng, S. et al. (2012). An extended matrix factorization approach for qos prediction in service selection. In: *IEEE Ninth International Conference on Services Computing*, 162–169. IEEE.

66 Geebelen, D. et al. (2014). Qos prediction for web service compositions using kernel-based quantile estimation with online adaptation of the constant offset. *Information Sciences* 268: 397–424.

67 Zheng, Z. and Lyu, M.R. (2013). Personalized reliability prediction of web services. *ACM Transactions on Software Engineering and Methodology* 22 (2): 12:1–12:25.

68 Zheng, Z. et al. (2013). Collaborative web service qos prediction via neighborhood integrated matrix factorization. *IEEE Transactions on Services Computing* 6 (3): 289–299.

69 Xu, J., Zheng, Z., and Lyu, M.R. (2016). Web service personalized quality of service prediction via reputation-based matrix factorization. *IEEE Transactions on Reliability* 65 (1): 28–37.

70 Syu, Y., Kuo, J.-Y., and Fanjiang, Y.-Y. (2017). Time series forecasting for dynamic quality of web services: an empirical study. *Journal of Systems and Software* 134: 279–303.

71 Wagner, N., Michalewicz, Z., Khouja, M., and McGregor, R.R. (2007). Time series forecasting for dynamic environments: the DyFor genetic program model. *IEEE Transactions on Evolutionary Computation* 11 (4): 433–452.

72 Amin, A. and Colman, A. (2012). An approach to forecasting QoS attributes of web services based on ARIMA and GARCH models. In: *IEEE 19th ICWS*, 74–81. IEEE.

73 Godse, M., Bellur, U., and Sonar, R. (2010). Automating QoS based service selection. In: *2010 IEEE International Conference on Web Services*, 534–541. IEEE.

74 Xia, Y., Ding, J., Luo, X., and Zhu, Q. (2013). Dependability prediction of WS-BPEL service compositions using petri net and time series models. In: *2013 IEEE Seventh International Symposium on Service-Oriented System Engineering*, 192–202. IEEE.

75 Senivongse, T. and Wongsawangpanich, N. (2011). Composing services of different granularity and varying QoS using genetic algorithm. In: *Proceedings of the World Congress on Engineering and Computer Science*, vol. 1, 1–6. San Francisco, USA: WCECS.

76 Syu, Y., Fanjiang, Y.-Y., Kuo, J.-Y., and Ma, S.-P. (2015). Applying genetic programming for time-aware dynamic QoS prediction. In: *2015 IEEE International Conference on Mobile Services (MS)*, 217–224. IEEE.

77 Zadeh, M.H. and Seyyedi, M.A. (2010). QoS monitoring for web services by time series forecasting. In: *2010 3rd IEEE International Conference on Computer Science and Information Technology (ICCSIT)*, vol. 5, 659–663. IEEE.

78 Laptev, N., Yosinski, J., Li, L.E. et al. (2017). Time-series extreme event forecasting with neural networks at uber. In: *Proceedings of International Conference on Machine Learning*, vol. 34, 1–5.

79 Sak, H., Senior, A., and Beaufays, F. (2014). Long short-term memory recurrent neural network architectures for large scale acoustic modeling. In: *Fifteenth Annual Conference of the International Speech Communication Association, Singapore*, 338–342.

80 Huang, Z., Xu, W., and Yu, K. (2015). Bidirectional LSTM-CRF models for sequence tagging. *Computer Science*.

81 White, G. and Clarke, S. (2018). Smart Cities with Deep Edges. ECML PKDD 2018 Workshops, Dublin, Ireland, Springer, Cham, pp. 53–64.

82 White, G., Palade, A., Cabrera, C. et al. (2019). Autoencoders for QoS prediction at the edge. In: *IEEE International Conference on Pervasive Computing and Communications (PerCom)*, 1–9. Kyoto, Japan: IEEE.

83 White, G., Cabrera, C., Palade, A., and Clarke, S. (2018). Augmented reality in IoT. 8th International Workshop on Context-Aware and IoT Services (CIoTS 2018).

84 White, G., Liang, Z., and Clarke, S. (2019). A quantified-self framework for exploring and enhancing personal productivity. In: *Proceedings of CBMI*, vol. 19.

85 Ameller, D. and Franch, X. (2008). Service level agreement monitor (SALMon). In: *Seventh International Conference on Composition-Based Software Systems (ICCBSS 2008)*, 224–227. IEEE.

86 Abbaspour Asadollah, S. and Kian Chiew, T. (2011). Web service response time monitoring: architecture and validation. In: *Theoretical and Mathematical Foundations of Computer Science* (ed. Q. Zhou), 276–282. Berlin, Heidelberg: Springer.

87 Chen, N. and Clarke, S. (2014). A dynamic service composition model for adaptive systems in mobile computing environments. In: *International Conference on Service-Oriented Computing-ICSOC*, 93–107. Berlin, Heidelberg: Springer.

88 Serguei, P. (2017). *Tangle*. IOTA.

89 J. Chung, C. Gulcehre, K. Cho, and Y. Bengio (2014), Empirical evaluation of gated recurrent neural networks on sequence modelling, NIPS 2014 Workshop on Deep Learning, Canada .

90 Jiane, L. (2008). An agent bilateral multi-issue alternate bidding negotiation protocol based on reinforcement learning and its application in ecommerce. In: *2008 International Symposium on Electronic Commerce and Security*, 217–220. IEEE.

91 Silaghi, G.C., Şerban, L.D., and Litan, C.M. (2010). A framework for building intelligent SLA negotiation strategies under time constraints. In: *International Workshop on Grid Economics and Business Models*, 48–61. Springer.

92 Amato, A., Di Martino, B., and Venticinque, S. (2014). Agents based multicriteria decision-aid. *Journal of Ambient Intelligence and Humanized Computing* 5 (5): 747–758.

93 Zheng, X. (2014). *QoS Representation, Negotiation and Assurance in Cloud Services*. Canada: Queen's University.

10

Efficient Task Scheduling for Performance Optimization

Yang Yang[1], Shuang Zhao[2], Kunlun Wang[1], and Zening Liu[1]

[1] *Shanghai Institute of Fog Computing Technology (SHIFT), ShanghaiTech University, Shanghai, China*
[2] *Shanghai Institute of Microsystem and Information Technology (SIMIT), Chinese Academy of Sciences, Shanghai, China*

10.1 Introduction

With the advent of mobile Internet, more and more smart mobile terminals have been extended to multifunctional high-performance computing and communication platform for future fifth-generation (5G) networking and Internet of things (IoT) [1–3]. Examples include vehicles in vehicular networks, the mobile phones in device-to-device (D2D) networks, and robots in intelligent manufacturing. All these applications can be modeled as homogeneous fog networks, which are defined as a group of peer nodes with sharable computing, communication, and storage resources for node-to-node/device-to-device (N2N/D2D) communications and task scheduling [4, 5]. Compounded with these, more and more new mobile applications such as online gaming and movie, augmented reality (AR), and intelligent drive services come true and attract great attention. Since these novel applications require low latency, they are typically demanding intensive computation and communication energy for real time and high data exchange rate. However, due to physical size constraint, the mobile devices have finite resources. As a result, the finite battery lifetimes and limited computation capacities of mobile devices pose significant challenges for designing these new applications. Although the energy-efficient communication has been considered in many aspects, the energy-efficient task scheduling and executions have been rarely considered so far to the best of our knowledge, which is essential in fog computing.

On the other hand, many applications in fog computing have rigorous service delay and energy consumption requirements. However, the degree and urgency of these requirements vary in specific scenarios. For example, the primary challenge in designing vehicular networks is to ensure good delay performance.

Fog and Fogonomics: Challenges and Practices of Fog Computing, Communication, Networking, Strategy, and Economics, First Edition. Edited by Yang Yang, Jianwei Huang, Tao Zhang, and Joe Weinman.
© 2020 John Wiley & Sons, Inc. Published 2020 by John Wiley & Sons, Inc.

Thus, the requirements on energy consumption are relatively low [6]. In contrast with vehicular networks, IoT applications typically have rigid energy consumption requirements [7]. Online gaming on mobile terminals, such as AR and 3D modeling, always require low latency and high energy efficiency (EE) [8]. In intelligent manufacturing and tactile Internet, the average delay and delay jitter are more critical parameters [9, 10]. How to balance the trade-off performance between service delay and energy consumption in homogeneous fog networks is a very challenging problem, which directly affects users' quality of experience (QoE) and therefore is the focus of this research.

The trade-off between service delay and energy consumption in wireless networks has been extensively studied. Two energy-efficient resource management algorithms are developed in [11, 12] to achieve the optimal power-delay trade-off over various wireless networks. Yu et al. [13] focused on energy-aware network selection and resource allocation in cellular and Wi-Fi networks, while taking into account traffic delay. Sheng et al. [14] developed an EE and delay trade-off power control policy for D2D communications underlying cellular networks. Deng et al. [15] investigated the power-delay trade-off in fog–cloud computing system and developed an optimal workload allocation between fog and cloud toward the minimal power consumption with constrained service delay. Mao et al. [16] designed an online joint radio and computational resource management algorithm for multiuser MEC systems. They demonstrated that the algorithm can find a balance between energy consumption and service-delay performance. Most of the aforementioned and related work has focused on energy-delay trade-off analysis and algorithm design in the context of power control, scheduling, and admission control under specific wireless scenarios. To the best of our knowledge, delay energy balanced task scheduling (DEBTS) and execution in general homogeneous fog networks including the D2D and vehicle-to-vehicle (V2V) scenarios have not yet been thoroughly considered.

In this chapter, we first propose a typical fog network consisting of multiple fog nodes (FNs), wherein some task nodes (TNs) have heavy computation tasks, while some helper nodes (HNs) have spare resources for sharing with their neighboring nodes. To minimize the delay of every task in such a fog network, we formulate a noncooperative game called paired offloading of multiple tasks (POMT) to model the competition among TNs for the communication resources and computation capabilities of HNs. Further, we develop the corresponding distributed POMT task scheduling algorithm. Then, we propose a novel energy-efficient fog computing framework, which exploits collaboration among the user device nodes, but each node can have heterogeneous computing capacity and network quality. We develop a low complexity maximal energy-efficient task scheduling (MEETS) algorithm for achieving energy-efficient task scheduling among FNs, which can efficiently derive the modulation selection strategy and the offloading time slots alloca-tion for task scheduling utilizing cognitive spectrum access from the primary network. Further, in Section 10.4, we extend the jointly optimal management of computation and communication resources for computation offloading proposed in, e.g. [15, 16], where computation and communication resources are dynamically and beneficially shared among FNs via the assistance of a local fog controller. With a guarantee in service delay performance, this section develops a DEBTS algorithm. We conclude in Section 10.6.

10.2 Individual Delay-minimization Task Scheduling

10.2.1 System Model

Consider a typical fog network consisting of many FNs with different com-putation resources and capabilities. As shown in Figure 10.1, at any specific time slot, an FN is further classified as (i) a TN if it has a task to compute locally or to offload to a neighboring HN, (ii) a HN if it has spare resources to share with neighboring TNs, or (iii) a busy node (BN) if it is not available due to limited communication, computation, and storage capabilities. For example, the fog network in Figure 10.1 has $N = 4$ TNs, $K = 4$ HNs and 3 BNs at this particular time slot. It is worth noting that TNs, HNs, and BNs are not fixed. To be specific, TNs and BNs can be HNs when they are idle in the following slots, and vice versa.

For the sake of analysis, let $\mathcal{N} = \{1, 2, \dots, N\}$ and $\mathcal{K} = \{1, 2, \dots, K\}$ denote the set of TNs (i.e. tasks) and the set of HNs, respectively. Assume that a computation task can be either executed by its owner, i.e. the local TN or entirely offloaded to one neighboring HN. While, a HN can accommodate multiple tasks from different TNs, if its spare resources and capabilities permit. For example, as shown in Figure 10.1, TN-1 executes its task on local device, which may be due to (i) HN-2 does not have sufficient computation/storage resources for accommodating the task from TN-1, or (ii) the communication channel between TN-1 and TN-2 is not good at this slot. TN-2 offloads its task to HN-1 because HN-1 has more computation resource and better communication channel than HN-2. HN-4 accommodates multiple tasks from TN-3 and TN-4 simultaneously as it has sufficient resources and capabilities. In this section, tasks are assumed to be nonsplitable or indivisible, and the task scheduling problem is actually the pairing strategy between N TNs and K HNs for achieving the minimum processing delay for every task.

10.2.2 Problem Formulation

Every TN wants to minimize the time required to process its own task. How-ever, such a problem is difficult because the pairing strategies between TNs and

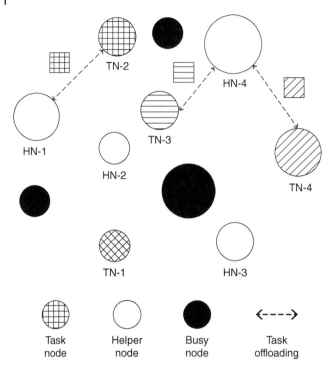

Figure 10.1 A fog network with four task nodes, four helper nodes, and three busy nodes. They are with different computation resources and capabilities, which are distinguished by the size of cycles.

HNs are coupled. To be specific, the time required to process task n depends on not only the pairing strategy of TN n but also other TNs' pairing strategies for the reason that there exists competition among TNs for the communication resources and computation resources of HNs. For example, TNs may compete for communication resources, e.g. time frames or radio resource blocks (RB), to transmit tasks to HNs. And since the computation capability of single HN is usually limited, tasks offloaded to the same HN may also compete for the computation capability of the HN.

In this work, we assume that HNs occupy orthogonal wireless channels, i.e. no interference between different HNs, and TNs transmit tasks to HNs utilizing the time division multiple access (TDMA) scheme at each time slot.[1] Further, we assume that every HN assigns its spare computation resources to all the existing tasks in proportion to their computation workloads.[2] Similar to many

---

1 This scenario can be seen as an upper bound for the transmission time of tasks being transmitted to HNs.
2 This scenario can be seen as an upper bound for the computation time of tasks being processed on HNs.

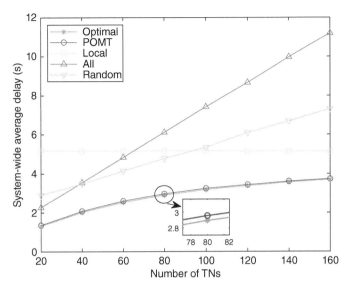

Figure 10.2 System-wide average delay with different number of TNs.

other works [17, 18], the communication phase of transmitting results back to TNs is ignored since the size of results is negligible compared to that of tasks.

10.2.3 POMT Algorithm

We formulate the interactions among TNs as a noncooperative game, i.e. POMT, and prove the existence of Nash Equilibrium (NE) utilizing the potential game theory [19]. Further, we design a distributed algorithm called POMT to achieve an NE [20]. Analytical and simulation results show that our POMT algorithm can offer the near-optimal performance in system average delay and obtain more number of beneficial TNs, at much lower complexity and higher efficiency, comparing to a centralized optimal algorithm for computation offloading.

Figures 10.2 and 10.3 compare the proposed POMT algorithm with the following baseline solutions in terms of both the system average delay and the number of beneficial TNs:

- local computing (local): every TN chooses to process its task on local device.
- all offloading (all): every TN chooses to offload its task to the HN with the highest data rate. It imitates the myopic behavior of individuals.
- random offloading (random): every TN randomly chooses to process its task on local device or offload it to a randomly selected HN.
- optimal offloading (optimal): the centralized near-optimal solution in terms of system-wide average delay is obtained utilizing the cross entropy method [21].

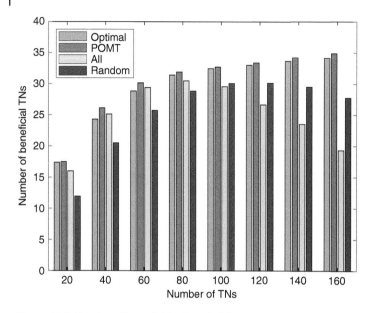

Figure 10.3 Number of beneficial TNs with different number of TNs.

The beneficial TNs are those who can reduce its delay by offloading tasks to HNs, compared with local computing. All numerical results of Figures 10.2 and 10.3 are averaged over 500 simulation rounds.

Figure 10.2 illustrates the system-wide average delay with different number of TNs. As illustrated in Figure 10.2, the system-wide average delay increases as the number of TNs increases, except for the local computing, and the proposed POMT algorithm can always achieve the near-optimal performance. Besides, it can be observed that our POMT algorithm can reduce 27–74%, 40–67%, and 49–53% system-wide delay than the local computing, all offloading and random offloading, respectively. Furthermore, it should be noted that the computation offloading is not the best choice for all TNs. As we can see, the average delay of all offloading exceeds that of local computing and random offloading as the number of TNs increases. The reason is that the network is too congested, i.e. the resources are scarce, for every TN to offload task. This phenomenon can also be observed from Figure 10.3 in the following context, where it is not beneficial for all TNs to offload tasks to HNs.

Figure 10.3 shows how the number of beneficial TNs changes with the number of TNs. As shown in Figure 10.3, the number of beneficial TNs achieved by the POMT algorithm increases with the number of TNs increasing and levels off. Moreover, the POMT algorithm can always achieve more number of beneficial TNs than other three schemes, especially when the number of TNs is large. This is because that, in the case of large number of TNs, the HNs' resources

Table 10.1 Average run time.

N	20	40	60	80	100	120	140	160
POMT ($\times 10^3$ ms)	0.02	0.05	0.12	0.21	0.30	0.41	0.55	0.68
Optimal ($\times 10^5$ ms)	0.02	0.10	0.26	0.49	0.79	1.18	1.69	2.29

are scarce, and there exists serious contentions among TNs. Whereas all TNs' offloading tasks to HNs will further worsen such contentions and finally damage each other's interests, and thus the number of beneficial TNs reduces. In contrast, the proposed algorithm can coordinate the contentions among TNs and always achieve more beneficial TNs until the network is saturated, i.e. the number of beneficial TNs levels off.

Table 10.1 compares the run time of the POMT algorithm and the optimal solution. We run our algorithms on a desktop with 2 GHz Intel Xeon CPU for 500 simulation rounds. As can be seen, the average run time of POMT algorithm is two orders of magnitude smaller than the optimal solution.

10.3 Energy-efficient Task Scheduling

10.3.1 Fog Computing Network

Consider a fog computing system shown in Figure 10.4 with one single-antenna task node S offloaded its computation task to a set of K nearby single-antenna nodes called the helper nodes, denoted by $\mathcal{K} = \{1, 2, \dots, K\}$. These helper nodes without computing tasks in their own task queues are required to compute the computation task from the same task node. There are also existing N nearby nodes with available computation resources, called idle nodes, which cannot help to compute the task due to long distance, deep fading, or unavailable spectrum resource. Assume that the established task-scheduling pairs opportunistically access the spectrum from the primary user (PU) having task computation, i.e. busy node, of the fog network, and the central controller, e.g. base station (BS), has perfect knowledge of channel gains, local computing energy per bit at all FNs, which can be obtained by feedback. Using the information, the center controller selects different offloading helper nodes, determines the offloaded task size, and allocates time resource to each task scheduling pair with the criterion of maximum EE.

For the mathematical tractability of task scheduling, we consider the TDMA system for an arbitrary time slot allocation to each established task-scheduling pair. For the TDMA system, time frame T is divided into slots each with a duration of τ_i seconds where τ_i is scheduled to the ith established task-scheduling

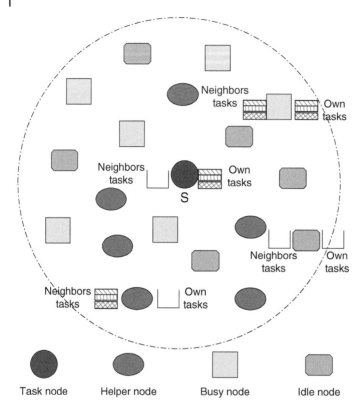

Task node   Helper node   Busy node   Idle node

Figure 10.4 A fog network with a task node, *K* helper nodes with sharable computing resource, *M* busy nodes with sharable spectrum resource, and *N* idle nodes. Each node maintains two first-in-first-out (FIFO) queues serving its own task and neighbors' task.

pair. In general, each allocated time slot comprises two sequential phases: (i) task offloading or local computing and (ii) task computing from neighbor helper nodes and downloading of computation results from the helper nodes to task node. Note that the size of the computation outcome for many applications is in general much smaller than the size of computation input data. Then, we specifically neglect the energy overhead for the neighbor helper nodes to send the computation outcomes back to the task node, which is similar to many existing works such as [22–26]. Moreover, the downloading rate is much faster than that of offloading due to relative smaller sizes of computation results. Under these conditions, we assume that the second phase has a negligible duration compared to the first phase and not considered in resource scheduling.

According to the arbitrary slots in TDMA, the central controller can efficiently schedule the time resources for the helper nodes achieving the energy-efficient task scheduling. If the task node has partial or no offloading tasks, the remaining tasks or all the tasks are computed by itself, respectively, using a local CPU.

10.3.2 Medium Access Protocol

In spectrum domain, we utilize the opportunistic spectrum access model. It is assumed that task-scheduling transmissions are allowed to opportunistically utilize those unused spectrum left by the busy nodes. The opportunistic spectrum access can be done by the task node sensing the environment and then cognitively adapting their spectrum parameters accordingly [27]. In general, the spectrum occupancy of the node can be modeled as a continuous-time Markov chain with available (the link is idle) and unavailable (the link is busy) states [28]. In this work, we assume that the transmissions of the busy nodes and task-scheduling system are both continuous. Thus, the spectrum occupied times from the busy nodes are independent and exponentially distributed with aggregated parameter $q_{i,j}^{-1}$ for the available state and $u_{i,j}^{-1}$ for the unavailable state on the link of established task-scheduling pair i utilizing jth resource block (RB) ($i \in \{1, \dots, K\}$). Under this model, the stationary distribution of the available and unavailable probabilities for the spectrum are respectively given by [28] as $a_{i,j}$ and $\tilde{a}_{i,j}$, respectively.

10.3.3 Energy Efficiency

If a mobile task node does not have enough computation capacity to execute the task, the task can be offloaded to execute on nearby homogeneous helper nodes, these nodes can establish task-scheduling pairs. In this case, the task transmission time slot allocation for each task-scheduling pair i is given by $\tau_i, \forall i \in \mathcal{K}$. As shown in Figure 10.4, mobile task node S is required to compute R_S-bit input data within the time slot, out of which l-bit is offloaded to the nearby helper nodes and $(R_S - l)$-bit can be computed locally by its own CPU.

Denoted by $\mathcal{M} = \{1, 2, \dots, N_m\}$ the set of the RBs. Assume that one RB includes one time-slot and 12 subcarriers and can be reused by only one secondary task node. With RB reuse, the jth RB allocated to rth busy node is reused by all the established task-scheduling links from the same task node. In order to offload the task from the task node to the helper node i on the jth RB, the transmit power is $\mathcal{P}_{i,j}$. Let σ^2 denote the noise power. In order to achieve the energy-efficient offloadings, the number of offloading bits per joule total energy consumption is considered as an important metric. To evaluate the EE, the circuit power has an important effect, and it cannot be ignored. Based on [29], we can get the circuit power consumption of all the circuit blocks. In this work, we assume that the total circuit energy consumption of the task node at idle state in a time slot T is E_0. Similar to traditional metric of EE, the offloading transmission energy E_{off} and the computation energy E_{comp} from all the task-scheduling pairs are also considered. Thus, the circuit power, the offloading power, and the computation power are the three parts of the power consumption.

Considering that the EE focus on the total energy consumption and the quantity of offloading task, which is often regarded as the traditional definition of EE. We propose to maximize the expected quantity of offloading task per energy consumption. Consequently, there is a fundamental trade-off between the energy consumption and quantity of offloading task. The concept of EE plays an important role in the task scheduling of fog networks, which can exploit the available nearby helper nodes to help computing the redundant tasks reducing the energy consumption of the task node. On the other hand, from the node's perspective, the energy-efficient task offloading ensures the efficient task computation, i.e. the nearby helper node would get the optimum quantity of offloaded task by adopting the energy-efficient approach.

The motivation of using the distributed task offloading is the limitation of the computation capacity and to fully enable the unused computation resources from the nearby homogeneous helper nodes, prior to the computation task execution. The key idea of the algorithm design is to optimize the time slot allocation for the multiple nearby helper nodes in a TDMA time slot T and the modulation scheme allocation for the task-scheduling transmissions, which can utilize the resources efficiently. Then, to make the task offloading and computing jointly efficiently, the average EE optimization problem under the constraints can be mathematically formulated. By this, we can ensure that as long as the task node S on its task offloading achieves the optimal EE, for any nearby helper node i, the energy-efficient quantity of offloading task l_i for each helper node i will depend on the optimal offloading rate and offloading time slot allocation τ_i.

10.3.4 Problem Properties

Generally speaking, a key challenge of solving the EE optimizing problem is that it requires the system information such as available neighbor nodes' resources, generated task description, and links connectivity. To facilitate the implementation, the system information can be obtained by the central controller, which determines the schedules for task offloading. Then, to enable practical deployment, the problem needs to be solved with a low-complexity algorithm. Then, we develop the MEETS algorithm [17] to solve the problem of energy-efficient fog computing in homogeneous fog networks. At first, we propose an optimization framework to determine an energy-efficient task scheduling decision that maximizes the computation and transmission EE, which is based on the fractional programming and can be solved with low computation complexity.

10.3.5 Optimal Task Scheduling Strategy

It can be observed that the energy-efficient offloading rate, defined as $b_{i,j}^*$ for helper node i, is determined by the corresponding spectrum access probability, the channel path loss, and the computing energy per bit $C_{re,i}P_{re,i}$, respectively. The main result of optimal modulation scheme allocation solving EE optimization problem for task offloading can be summarized as follows. For each helper node i, there exists a threshold ξ_i^* of spectrum access probability. When the channel of the task-scheduling link has deep fading, then it is obvious that ξ_i^* is small, which means that the probability of scheduling the minimum modulation scheme is large. If the established task-scheduling link N_i has larger spectrum access probability than ξ_i^*, the minimum available modulation scheme should be selected, which can achieve the optimal EE for the task scheduling in homogeneous fog networks. At the same time, the scheduling rate is a decreasing function of the computation energy consumption, and the probability of scheduling minimum modulation scheme for each task-scheduling link is larger with larger computation energy consumption. Then, the energy-efficient modulation scheme allocation policy has a threshold-based structure when task scheduling achieves EE. For the integer-relaxation task offloading problem, applying Karush-Kuhn-Tucker (KKT) conditions directly can lead to its optimal solution. It can be observed that for each task-scheduling link, link or node with deeper fading or higher computation energy should be allocated with less quantity of offloading task. Therefore, the initial resource and offloaded task size allocation is firstly determined by defining average channel gain and computation energy.

Figure 10.5 shows the EE versus different number of task-scheduling helper nodes, we compare the MEETS, equal-time offloading (ETO) and energy-efficient-time offloading (EETO) algorithms under different conditions. ETO and EETO allocate equal time slot and energy-efficient time slot for each helper node, respectively. We can observe from the figure that the MEETS can always offer better EE than that of other strategies with different number of task-scheduling helper nodes. Furthermore, we can observe that the optimal modulation scheme allocation strategy can also offer better EE than that of adaptive modulation of coding (AMC) strategy with the same time slot allocation strategy, which coincides with the analytic results. At the same time, another interesting remark is that with increasing number of helper nodes, the EE is increasing. This phenomenon indicates that with the number of helper nodes increasing, the quantity of offloading task will be larger, meaning that it is more beneficial to the task offloading. In addition, it can be also observed from the figure that the EE of the equal time slot allocation is larger than that of the energy-efficient time slot allocation while the number of helper nodes is less than 7, which is due to the dominance of the offloading power consumption in comparison to the computation power consumption when the

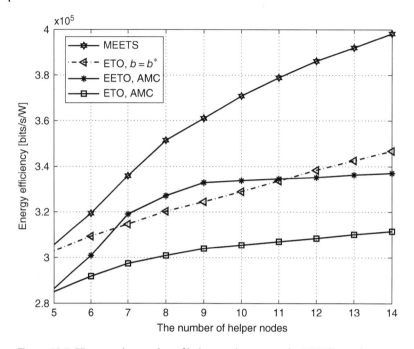

Figure 10.5 EE versus the number of helper nodes among the MEETS, equal-time offloading (ETO) and energy-efficient-time offloading (EETO), where AMC is the adaptive modulation and coding.

number of helper nodes is small. Still, the EE of the equal time slot allocation is larger than that of the energy-efficient time slot allocation while the number of helper nodes is larger than 11. This reflects that the energy-efficient time slot allocation strategy has the same influence with the equal time slot allocation strategy at the large region of the number of helper nodes, then the optimal modulation scheme allocation strategy offers better EE than the AMC.

10.4 Delay Energy Balanced Task Scheduling

10.4.1 Overview of Homogeneous Fog Network Model

We consider a fog network with a set of $\mathcal{U} = \{1, \ldots, |\mathcal{U}|\}$ FNs sharing spectrum and computation resources, as shown in Figure 10.6. Under the assistance of a local fog controller, each FN is running computationally intensive tasks that are generated locally or offloaded from different neighbors. The local fog controller could be a small data center installed at a base station deployed by the telecom operator. Thus, it can be accessed by the FNs through wireless channels and determine computation offloading management and resource

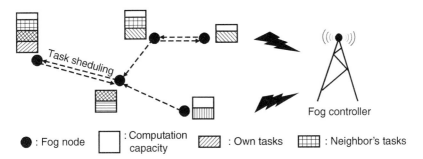

Figure 10.6 A fog network with five FNs and one local fog controller. Boxes with slants and boxes with grids denote a fog node's own computation tasks and its neighbor's tasks, respectively.

scheduling with readily available wireless channel state information at the fog controller. We assume that each FN is equipped with a frequency-tunable CPU. Computation and communication resources are shared among those FNs. By establishing communication links with nearby nodes, each node can opportunistically leverage the under-utilized computation resources to execute computation tasks. By offloading part of the computation tasks to nearby FNs with spare computation resources, the mobile FNs can enjoy a higher QoE and reduce battery energy consumption [16].

For convenience, we assume that the network operates in a time-slotted structure, indexed by $t \in \{0, 1, 2, \dots\}$, and the time slot length is τ. Motivated by the fact that the local fog controller has the global network information, including the location of FNs, computation resources, and task arrivals, we consider a network-assisted architecture, where the fog controller conducts resource discovery, node connections, and task scheduling at each time slot for each FN. We should emphasize here that in our model, the computation offloading from the FNs to the base station is not considered, and offloading can only occur between peer mobile nodes. Under these conditions, we can clearly see the benefits of task offloading and effectively relieve the burden on the network.

To encourage mobile FNs to participate in collaborations in the long run and prevent the FNs from overly aggressive offloading, we introduce an incentive constraint which ensures that a mobile FN exploits more resources from other nodes only if it contributes more resources for others [30]. The incentive constraint model reflects the idea that the resources exploited by one FN is proportional to that contributed by it.

10.4.2 Problem Formulation and Analytical Framework

We consider the energy consumed during task execution, as well as the power consumed during task scheduling. The energy consumed at the fog controller is ignored for simplicity. The time-averaged energy consumption of different mobile FNs is adopted as one performance metric, the average traffic delay is adopted as another performance metric. According to Little's law [31], the average traffic delay experienced by each mobile FN is proportional to the averaged number of unexecuted computation tasks, which is the average sum queue length of the task buffers.

It is very desirable, but quite challenging, to simultaneously reduce service delay and energy consumption in homogeneous fog networks for delay-sensitive and energy-constraint applications. To study the balance between service delay and energy consumption, we formulate an average energy consumption minimization problem, with a guarantee in service delay performance. It is a stochastic optimization problem, where the CPU-cycle frequency scheduling for each mobile FN, binary task offloading, and the transmission power allocation need to be determined in each time slot. This is a highly challenging problem with a large amount of stochastic information to be handled. Besides, the problem requires that future information about the network, such as task traffic, available computation resources, and FN connectivity, be known in advance. This information is difficult to predict due to the mobility of FNs.

10.4.3 Delay Energy Balanced Task Offloading

We use Lyapunov optimization to solve the average energy consumption minimization problem mentioned previously, the key idea behind which is to turn time-average constraints into a pure queue stability problem.

In [32], we proposed a DEBTS mechanism for such a homogeneous fog network, which mainly consists of three stage operations in each time slot: CPU-cycle frequency scheduling, transmission power allocation, and binary task offloading. DEBTS is an online algorithm, which optimizes the task scheduling in each time slot. It only utilizes the current slot network state information, including locally generated computation task, and wireless channel fading information. In each time slot, the DEBTS algorithm first computes the optimal CPU-cycle frequency, and then alternately updates the transmission power and binary task offloading indicator in an iterative way. It converges to the optimal transmission power allocation and suboptimal task scheduling at each iteration. DEBTS is an online algorithm, which optimizes the task scheduling in each time slot. It only utilizes the current slot network state information, including locally generated computation task, and wireless channel fading information.

10.4.4 Performance Analysis

The average energy consumption of DEBTS algorithm is inversely proportional to inversely proportional to control parameter V, which is

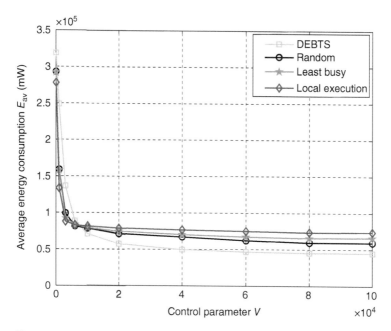

Figure 10.7 Energy consumption performance.

predetermined according to different performance requirements. This is depicted in Figure 10.7, where we compare the energy consumption performance for the proposed DEBTS algorithm with other three scheduling algorithms: Random, Least Busy, and Local Execution algorithm. The stable average energy consumption for the proposed DEBTS algorithm is much smaller than the other three scheduling algorithms. Specifically, the DEBTS algorithm can save approximately 26%, 29%, and 35% energy than random scheduling, least busy scheduling, and local execution algorithms for a typical V value of 4×10^4.

The service delay performance for DEBTS algorithm is demonstrated in Figure 10.8. It can be observed that the average delay increases linearly with V and becomes unbounded when V goes to infinity. The DEBTS algorithm under any V value outperforms the other three scheduling algorithms regarding to the service delay performance. In addition, the delay reduction brought by DEBTS becomes even more obvious with increased V. For example, the DEBTS algorithm can reduce service delay roughly 29%, 32%, and 36% over random scheduling, least busy scheduling, and local execution algorithms for the V value of 4×10^4.

The delay jitter performance under the DEBTS algorithm and the other three scheduling algorithms is evaluated in Figure 10.9. The delay jitter is estimated by the mean-square deviation of the time series of task queue length. In general, the delay jitter performance of DEBTS is still much better than

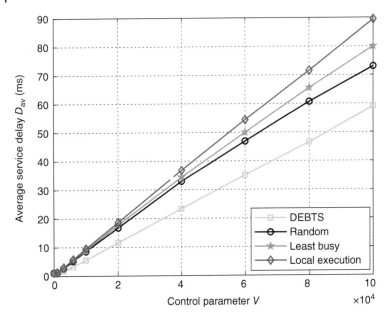

Figure 10.8 Service delay performance.

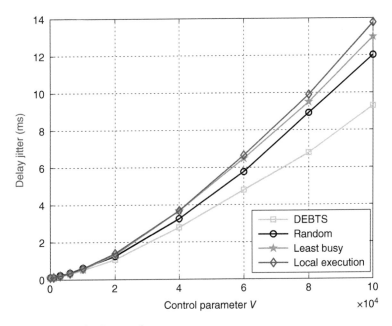

Figure 10.9 Delay jitter performance.

the others, i.e. about 14%, 24%, and 24% performance gain for the V value of 4×10^4 compared with random, least busy, and local execution algorithm, respectively. These comparisons demonstrate the advantage of the proposed DEBTS algorithm. Besides, the delay jitter under all the scheduling algorithms aggravates with the increase of V. Along with Figure 10.7, those observations imply that there is a trade-off between average energy consumption and delay jitter. Large control parameter V under DEBTS algorithm helps to save energy consumption but will lead to large service delay and serious delay jitter. In practical applications, the average delay and delay jitter can be controlled by DEBTS with a small V.

10.5 Open Challenges in Task Scheduling

Although many efforts for task scheduling in fog computing systems have been made, there are still some open issues need to be addressed.

10.5.1 Heterogeneity of Mobile Nodes

With the rapid realization of IoT, there are tens of billions of heterogeneous devices around us, which can connect to the networks. Under this circumstance, mobile helper nodes are heterogeneous in their buffer, battery level, computing capability, etc. At the same time, tasks are heterogeneous in their size, deadline, requirement for computing resources, etc. Then, it is challenging to decide how to schedule heterogeneous tasks to heterogeneous mobile nodes. However, most of previous works put their emphasis either on the same sizes of all tasks or executing all the tasks in parallel.

10.5.2 Mobility of Mobile Nodes

In task scheduling, node mobility is exploited to create communication opportunities to schedule task among mobile nodes. Unfortunately, node's mobility in fog computing system is a double-edged sword, which not only brings advantages that traditional cloud computation system or cellular network do not have but also creates the instability problem. Thus, the end-to-end path between mobile nodes is variable, and the scheduling of task totally depends on instable contact between mobile nodes. For this problem, the critical challenge is how to effectively manage and utilize the mobility of mobile nodes.

10.5.3 Joint Task and Traffic Scheduling

Generally speaking, mobile helper nodes with sufficient computing resources may help other mobile nodes with insufficient computing resources to perform task computation, while mobile nodes with sufficient traffic usage may help other mobile nodes with insufficient traffic usage to download/upload data. Although many outstanding works have been dedicated to studying task or traffic scheduling, most of the existing researches either focus on traffic scheduling alone or study computation task scheduling separately. Few works consider both traffic scheduling and task scheduling simultaneously. Clearly, in order to efficiently utilize computation, communication, and storage resource and achieve remarkable performance, joint task and traffic scheduling design in fog computing system is beneficial.

10.6 Conclusion

In this chapter, we have investigated different efficient task-scheduling problems in fog computing networks. We first proposed a fog computing framework, wherein some TNs had heavy computation tasks, while some HNs had spare computation resources for sharing with their neighboring nodes and developed a distributed algorithm called POMT to offer the near-optimal performance in system average delay. Then, we proposed an energy-efficient fog computing framework with multiple neighbor helper nodes sharing their computation resource and presented a low complexity MEETS algorithm. Further, we developed the online DEBTS algorithm to minimize the overall energy consumption while reducing average service delay and delay jitter among collaborative FNs. For each of them, performance analyses are conducted, and both exhibit significant performance gains under different conditions and service scenarios.

References

1 Wang, C.-X., Haider, F., Gao, X. et al. (2014). Cellular architecture and key technologies for 5G wireless communication networks. *IEEE Communications Magazine* 52 (2): 122–130.
2 Yang, Y., Xu, J., Shi, G., and Wang, C.-X. (2017). *5G Wireless Systems: Simulation and Evaluation Techniques*. Springer.
3 Chiang, M. and Zhang, T. (2016). Fog and IoT: an overview of research opportunities. *IEEE Internet of Things Journal* 3 (6): 854–864>.
4 Chen, X. and Zhang, J. (2017). When D2D meets cloud: hybrid mobile task offloadings in fog computing. 2017 IEEE International Conference on Communications (ICC), May 2017, pp. 16.
5 Hou, X., Li, Y., Chen, M. et al. (2016). Vehicular fog computing: a viewpoint of vehicles as the infrastructures. *IEEE Transactions on Vehicular Technology* 65 (6): 38603873.

6 Karagiannis, G., Altintas, O., Ekici, E. et al. (2011). Vehicular networking: a survey and tutorial on requirements, architectures, challenges, standards and solutions. *IEEE Communications Surveys & Tutorials* 13 (4): 584616.

7 Mainetti, L., Patrono, L., and Vilei, A. (2011). Evolution of wireless sensor networks towards the internet of things: a survey. 2011 19th International Conference on Software, Telecommunications and Computer Networks (SoftCOM), pp. 16.

8 Tachi, S., Inami, M., and Uema, Y. (2014). Augmented reality helps drivers see around blind spots. *IEEE Spectrum* 31.

9 Wang, L., Chen, X., and Liu, Q. (2017). A lightweight intelligent manufacturing system based on cloud computing for plate production. *Mobile Networks and Applications* 22 (6): 1170–1181.

10 Fettweis, G.P. (2014). The tactile internet: applications and challenges. *IEEE Vehicular Technology Magazine* 9 (1): 6470.

11 Zhang, X. and Tang, J. (2013). Power-delay tradeoff over wireless networks. *IEEE Transactions on Communications* 61 (9): 36733684.

12 Berry, R.A. (2013). Optimal power-delay tradeoffs in fading channels–small delay asymptotics. *IEEE Transactions on Information Theory* 59 (6): 39393952.

13 Yu, H., Cheung, M.H., Huang, L., and Huang, J. (2016). Power-delay tradeoff with predictive scheduling in integrated cellular and Wi-Fi networks. *IEEE Journal on Selected Areas in Communications* 34 (4): 735742.

14 Sheng, M., Li, Y., Wang, X. et al. (2016). Energy efficiency and delay tradeoff in device-to-device communications underlaying cellular networks. *IEEE Journal on Selected Areas in Communications* 34 (1): 92106.

15 Deng, R., Lu, R., Lai, C. et al. (2016). Optimal workload allocation in fog-cloud computing toward balanced delay and power consumption. *IEEE Internet of Things Journal* 3 (6): 11711181.

16 Mao, Y., Zhang, J., Song, S.H., and Letaief, K.B. (2017). Stochastic joint radio and computational resource management for multi-user mobile-edge computing systems. *IEEE Transactions on Wireless Communications* 16 (9): 59946009.

17 Yang, Y., Wang, K., Zhang, G. et al. (2018). MEETS: maximal energy efficient task scheduling in homogeneous fog networks. *IEEE Internet of Things Journal* 5 (5): 4076–4087.

18 Zhao, P., Tian, H., Qin, C., and Nie, G. (2017). Energy-saving offloading by jointly allocating radio and computational resources for mobile edge computing. *IEEE Access* 5: 1125511268.

19 Monderer, D. and Shapley, L.S. (1996). Potential games. *Games and Economic Behavior* 14 (1): 124143.

20 Yang, Y., Liu, Z., Yang, X. et al. (2019). POMT: paired offloading of multiple tasks in heterogeneous fog networks. *IEEE Internet of Things Journal*, Early Access.

21 Rubinstein, R.Y. and Kroese, D.P. (2004). *The Cross Entropy Method: A Unified Approach To Combinatorial Optimization, Monte-Carlo Simulation.* New York: Springer-Verlag.

22 Chen, X., Jiao, L., Li, W., and Fu, X. (2016). Efficient multi-user computation offloading for mobile-edge cloud computing. *IEEE/ACM Transactions on Networking* 24 (5): 27952808.

23 Rudenko, A., Reiher, P., Popek, G.J., and Kuenning, G.H. (1998). Saving portable computer battery power through remote process execution. *ACM SIGMOBILE Mobile Computing and Communications Review* 2 (1): 1926.

24 Huertacanepa, G. and Lee, D. (2008). An adaptable application offloading scheme based on application behavior. Proceedings of the 22nd International Conference on Advanced Information Networking and Applications Workshops, pp. 387392.

25 Xian, C., Lu, Y., and Li, Z. (2007). Adaptive computation offloading for energy conservation on battery-powered systems. Proceedings of IEEE ICDCS, vol. 2, pp. 18.

26 Huang, D., Wang, P., and Niyato, D. (2012). A dynamic offloading algorithm for mobile computing. *IEEE Transactions on Wireless Communications* 11 (6): 19911995.

27 Sakr, A.H., Tabassum, H., Hossain, E., and Kim, D.I. (2015). Cognitive spectrum access in device-to-device-enabled cellular networks. *IEEE Communications Magazine* 53 (7): 126133.

28 Zhao, Q., Geirhofer, S., Tong, L., and Sadler, B.M. (2008). Opportunistic spectrum access via periodic channel sensing. *IEEE Transactions on Signal Processing* 56 (2): 785796.

29 Cui, S., Goldsmith, A.J., and Bahai, A. (2004). Energy-efficiency of MIMO and cooperative MIMO techniques in sensor networks. *IEEE Journal on Selected Areas in Communications* 22: 10891098.

30 Pu, L., Chen, X., Xu, J., and Fu, X. (2016). D2D fogging: an energy-efficient and incentive-aware task offloading framework via network-assisted D2D collaboration. *IEEE Journal on Selected Areas in Communications* 34 (12): 38873901.

31 Ross, S.M. (2014). *Introduction to Probability Models.* Academic Press.

32 Yang, Y., Zhao, S., Zhang, W. et al. (2018). Debts: delay energy balanced task scheduling in homogeneous fog networks. *IEEE Internet of Things Journal* 5 (3): 20942106.

11

Noncooperative and Cooperative Computation Offloading

Xu Chen and Zhi Zhou

School of Data and Computer Science, Sun Yat-sen University, Guangzhou, China

11.1 Introduction

As smart phones are gaining enormous popularity, more and more new mobile applications such as face recognition, natural language processing, interactive gaming, and augmented reality are emerging and attract great attention [1]. This kind of mobile applications are typically resource-hungry, demanding intensive computation and high energy consumption. Due to the physical size constraint, however, mobile devices are in general resource-constrained, having limited computation resources and limited battery life [1]. The tension between resource-hungry applications and resource-constrained mobile devices hence poses a significant challenge for the future mobile platform development.

As an interesting and promising solution to relieving such tension, task offloading has been emerged as a key focus in both academia and industry. In the last decade, many researchers focus on mobile cloud computing (see [2] and the references therein) where mobile users can offload their computation-intensive tasks to the resource-rich remote clouds via wireless access. Although this paradigm is already utilized as a form of commercial cloud services such as Windows Azure, it often suffers from labile wireless connections (e.g. weak cellular signal) and wide area network (WAN) delay between mobile devices and clouds [2]. As an extension, fog computing (particularly mobile edge computing) is an emerging paradigm that leverages a multitude of collaborative end-user and/or near-user devices to carry out a substantial amount of computation tasks they can each benefit from [3]. As illustrated in Figure 11.1, fog computing is implemented at the network edge, it can provide low-latency as well as agile computation augmenting services for device users [3].

Fog and Fogonomics: Challenges and Practices of Fog Computing, Communication, Networking, Strategy, and Economics, First Edition. Edited by Yang Yang, Jianwei Huang, Tao Zhang, and Joe Weinman.
© 2020 John Wiley & Sons, Inc. Published 2020 by John Wiley & Sons, Inc.

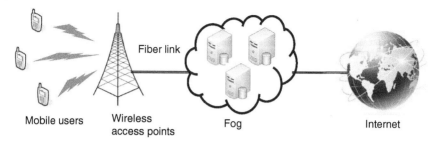

Mobile users   Wireless   Fog   Internet

access points

Figure 11.1 An illustration of fog computing.

Although the fog-based approach can significantly augment computation capability of mobile device users in a low-latency manner, the task of developing a comprehensive and reliable fog computing system remains challenging. A key challenge is how to achieve an efficient computation offloading coordination among mobile device users. One critical factor of affecting the performance of fog computing is the wireless access efficiency [4]. If too many mobile device users choose to offload the computation to the fog via wireless access simultaneously, they may generate severe interference to each other, which would reduce the data rates for computation data transmission. This hence can lead to low energy efficiency for computation offloading and long data transmission time. In this case, it would not be beneficial for the mobile device users to offload computation to the fog.

To address the aforementioned challenge, in this chapter, we study the problem of efficient multiuser computation offloading for fog computing. Specifically, we consider two distinctive scenarios. The first is the case that multiple users are selfish and thus offloading computation in a noncooperative manner. While in another case, multiple users can be cooperated to perform offloading in a cooperative manner. For the noncooperative offloading problem, we propose a game theoretic approach for achieving efficient computation offloading for fog computing. By formulating the decentralized computation offloading decision-making problem among mobile device users as a decentralized computation offloading game, we analyze the structural property of the game and show that the game always admits a Nash equilibrium. Finally, we design a decentralized computation offloading mechanism that can achieve a Nash equilibrium of the game and quantify its efficiency ratio over the centralized optimal solution.

For the cooperative offloading problem, we propose HyFog, a device-to-device (D2D)-assisted task offloading framework, which coordinates a massive crowd of devices at the network edge to support hybrid and cooperative computation offloading. A key objective of this framework is to achieve energy-efficient collaborative task executions at network-edge for mobile users. To this end, we develop a novel three-layer graph matching algorithm for efficient hybrid task offloading among the devices. Specifically, we first construct a three-layer graph to capture the choice space enabled by these three execution approaches, and then, the problem of minimizing the total task execution cost is recast as a minimum weight matching problem over the constructed three-layer graph, which can be efficiently solved using the Edmonds' Blossom algorithm.

11.2 Related Works

Fog computing has attracted significant attention in recent years. Most previous work has investigated the efficient computation offloading mechanism design from the perspective of a single mobile device user. For example Im et al. in [5] proposed AMUSE, which is a cost-aware Wi-Fi offloading framework that takes into account a user's throughput-delay trade-offs and her cellular budget constraint. Mao et al. in [6] studied a single-device dynamic computation offloading problem for fog computing with energy harvesting capability. Wen et al. in [7] presented an efficient offloading policy by jointly configuring the clock frequency in the mobile device and scheduling the data transmission to minimize the energy consumption. Barbera et al. in [4] showed by realistic measurements that the wireless access plays a key role in affecting the performance of mobile cloud computing.

Recently, the problem of efficient multiuser computation offloading for fog computing has begun to receive great attention. In particular, Yang et al. in [8] studied the scenario that multiple users share the wireless network bandwidth and solved the problem of maximizing the fog computing performance by a centralized heuristic genetic algorithm. Chen et al. in [9] studied the multiuser computation offloading problem for fog computing in a multichannel wireless environment and proposed a distributed game-theoretical computation offloading algorithm that could achieve a Nash equilibrium. Tang et al. in [10] proposed a general computation, communication, and caching framework that enable mobile users to share all three types of resources through D2D connections, in a cooperative manner. To incentivize cooperation for computation resource sharing among multiple users, Prasad et al. in [11] proposed an auction-based pricing mechanism to incentivize cooperation for fog computation resource sharing.

Considering different application scenarios, in this chapter, we study the multiuser computation offloading problem from both competitive and cooperative perspectives. In particular, in scenarios where the devices belong to different users that are selfish, competitive offloading captures the profit-seeking nature of those selfish users. While in some scenarios, an operator can control multiple devices, e.g. considering a sensing service operator may host multiples sensors and a surveillance service operator may control multiple cameras. Besides, with the wide penetration of online mobile social medias such as Facebook and WeChat, social ties among them can be exploited to achieve trustworthy cooperation [12]. Note that in such scenarios, multiple users or devices can be coordinated to offload in a cooperative manner.

11.3 Noncooperative Computation Offloading

11.3.1 System Model

In this section, we introduce the system model of fog computing illustrated in Figure 11.1. We consider a set of $\mathcal{N} = \{1, 2, \ldots, N\}$ collocated mobile device users and each of which has a computationally intensive and delay sensitive task to be completed. There exists a wireless access base-station s, through which the mobile device users can offload the computation to the fog. Similar to many previous studies in fog computing [9, 10] and mobile networking [13, 14], to enable tractable analysis and get useful insights, we consider a quasi-static scenario where the set of mobile device users $\mathcal{N}$ remains unchanged during a computation offloading period (e.g. within several seconds), while may change across different periods.[1] The general case that mobile users may depart and leave dynamically within a computation offloading period will be considered in a future work. Since both the communication and computation aspects play a key role in fog computing, we next introduce the communication and computation models in details.

11.3.1.1 Communication Model

We first introduce the communication model for wireless access. The wireless access base-station s can be either a Wi-Fi access point or a Femtocell network access point [15], or a macrocell base-station in cellular networks that manages the uplink/downlink communications of mobile device users. We denote $a_n \in \{0, 1\}$ as the computation offloading decision of mobile device user n. Specifically, we have $a_n = 1$ if user n chooses to offload the computation to the fog via wireless access. We have $a_n = 0$ if user n decides to compute its task locally on the mobile device. Given the decision profile $\mathbf{a} = (a_1, a_2, \ldots, a_N)$ of all the mobile device users, we can compute the uplink data rate for computation offloading of mobile device user n as [16]

$$R_n(\mathbf{a}) = W \log_2 \left(1 + \frac{P_n H_{n,s}}{\omega_n + \sum_{m \in \mathcal{N} \setminus \{n\}: a_m = 1} P_m H_{m,s}} \right) \tag{11.1}$$

---

1 This assumption holds for many applications such as face recognition and natural language processing, in which the size of computation input data is not large, and hence, the computation offloading can be finished in a smaller time scale (e.g. within several seconds) than the time scale of users' mobility.

Here, W is the channel bandwidth and P_n is user n's transmission power which is determined by the wireless access base-station according to some power control algorithms such as [17, 18]. Furthermore, $H_{n,s}$ denotes the channel gain between the mobile device user n and the base-station, and $\omega_n = \omega_n^0 + \omega_n^1$ denotes the background interference power including the noise power ω_n^0 and the interference power ω_n^1 from other mobile device users who carry out wireless transmission but do not involve in the fog computing.

From the communication model in Eq. (11.1), we see that if too many mobile device users choose to offload the computation via wireless access simultaneously, they may incur severe interference, leading to low data rates. As we discuss latter, this would negatively affect the performance of fog computing.

11.3.1.2 Computation Model

We then introduce the computation model. We consider that each mobile device user n has a computation task $\mathcal{I}_n \triangleq (B_n, D_n)$ that can be computed either locally on the mobile device or remotely on the fog via computation offloading. Here, B_n denotes the size of computation input data (e.g. the program codes and input parameters) involving in the computation task $\mathcal{I}_n$ and D_n denotes the total number of CPU cycles required to accomplish the computation task $\mathcal{I}_n$. A mobile device user n can apply the methods in [1, 8] to obtain the information of B_n and D_n. We next discuss the computation overhead in terms of both energy consumption and processing time for both local and fog computing approaches.

For the local computing approach, a mobile device user n executes its computation task $\mathcal{I}_n$ locally on the mobile device. Let F_n^l be the computation capability (i.e. CPU cycles per second) of mobile device user n. Here, we allow that different mobile devices may have different computation capability. The computation execution time of the task $\mathcal{I}_n$ by local computing is then given as

$$T_n^l = \frac{D_n}{F_n^l} \tag{11.2}$$

For the computational energy, we have that

$$E_n^l = v_n D_n \tag{11.3}$$

where v_n is the coefficient denoting the consumed energy per CPU cycle. According to the realistic measurements in [7, 19], we can set $v_n = 10^{-11}(F_n^l)^2$.

According to Eqs. (11.2) and (11.3), we can then compute the overhead of the local computing approach in terms of computational time and energy as

$$Z_n^l = \gamma_n^T T_n^l + \gamma_n^E E_n^l \tag{11.4}$$

where $0 \le \gamma_n^T, \gamma_n^E \le 1$ denote the weights of computational time and energy for mobile device user n's decision-making, respectively. To provide rich modeling flexibility and meet user-specific demands, we allow that different users can choose different weighting parameters in the decision-making. For example when a user is at a low battery state, the user would like to put more weight on energy consumption (i.e. a larger γ_n^E) in the decision-making, in order to save more energy. When a user is running some application that is sensitive to the delay (e.g. video streaming), then the user can put more weight on the processing time (i.e. a larger γ_n^T), in order to reduce the delay. Note that the weights could be dynamic if a user runs different applications or has different policies/demands at different computation offloading periods. For ease of exposition, in this paper, we assume that the weights of a user are fixed within one computation offloading period, which can be changed in different periods.

For the fog computing approach, a mobile device user n will offload its computation task $\mathcal{I}_n$ to the fog, and the fog will execute the computation task on behalf of the mobile device user.

For the computation offloading, a mobile device user n would incur the extra overhead in terms of time and energy for transmitting the computation input data to the fog via wireless access. According to the communication model in Section 11.3.1.1, we can compute the transmission time and energy of mobile device user n for offloading the input data of size B_n as, respectively,

$$T_{n,\text{off}}^c(\mathbf{a}) = \frac{B_n}{R_n(\mathbf{a})} \tag{11.5}$$

and

$$E_n^c(\mathbf{a}) = \frac{P_n B_n}{R_n(\mathbf{a})} \tag{11.6}$$

After the offloading, the fog will execute the computation task $\mathcal{I}_n$. Let F_n^c be the computation capability (i.e. CPU cycles per second) assigned to user n by the fog. The execution time of the task $\mathcal{I}_n$ of mobile device user n on the fog can be then given as

$$T_{n,\text{exe}}^c = \frac{D_n}{F_n^c} \tag{11.7}$$

According to Eqs. (11.5)–(11.7), we can compute the overhead of the fog computing approach in terms of processing time and energy as

$$Z_n^c(\mathbf{a}) = \gamma_n^T (T_{n,\text{off}}^c(\mathbf{a}) + T_{n,\text{exe}}^c) + \gamma_n^E E_n^c(\mathbf{a}) \tag{11.8}$$

Similar to many studies such as [20–24], we neglect the time overhead for the fog to send the computation outcome back to the mobile device user, due to the fact that for many applications (e.g. face recognition), the size of the computation outcome in general is much smaller than the size of computation input data including the mobile system settings, program codes, and input parameters.

According to the communication and computation models aforementioned, we see that the computation offloading decisions **a** among the mobile device users are coupled. If too many mobile device users simultaneously choose to offload the computation task to the fog via wireless access, they may incur severe interference, and this would lead to a low data rate. When the data rate $R_n(\mathbf{a})$ of a mobile device user n is low, it would consume high energy in the wireless access for offloading the computation input data to fog and incur long transmission time as well. In this case, it would be more beneficial for the user to compute the task locally on the mobile device to avoid the long processing time and high energy consumption by the fog computing approach. In the following Section 11.3.2, we will adopt a game theoretic approach to address the issue of how to achieve efficient computation offloading decision-makings among the mobile device users.

11.3.2 Decentralized Computation Offloading Game

In this section, we develop a game theoretic approach for achieving efficient computation offloading decision-makings among the mobile device users. The primary rationale of adopting the game theoretic approach is that the mobile devices are owned by different individuals, and they may pursue different interests. Game theory is a powerful framework to analyze the interactions among multiple mobile device users who act in their own interests and devise incentive-compatible computation offloading mechanisms such that no user has the incentive to deviate unilaterally. Moreover, by leveraging the intelligence of each individual mobile device user, game theory is a useful tool for devising decentralized mechanisms with low complexity, such that the users can self-organize into a mutually satisfactory solution. This can help to ease the heavy burden of complex centralized management by the fog and reduce the controlling and signaling overhead between the fog and mobile device users.

11.3.2.1 Game Formulation
We consider the decentralized computation offloading decision-making problem among the mobile device users within a computation offloading period. Let $a_{-n} = (a_1, ..., a_{n-1}, a_{n+1}, ..., a_N)$ be computation offloading decisions by all other users except user n. Given other users' decisions a_{-n}, user n would like to select a proper decision $a_n \in \{0, 1\}$ (i.e. local computing or fog computing) to minimize its computation overhead in terms of energy consumption and processing time, i.e.

$$\min_{a_n \in \{0,1\}} V_n(a_n, a_{-n}), \quad \forall n \in \mathcal{N}$$

According to Eqs. (11.4) and (11.8), we can obtain the overhead function of mobile device user n as

$$
V_n(a_n, a_{-n}) = \begin{cases} Z_n^l, & \text{if } a_n = 0 \\ Z_n^c(\mathbf{a}), & \text{if } a_n = 1 \end{cases}
\tag{11.9}
$$

We then formulate the aforementioned problem as a strategic game $\Gamma = (\mathcal{N}, \{A_n\}_{n \in \mathcal{N}}, \{V_n\}_{n \in \mathcal{N}})$, where the set of mobile device users $\mathcal{N}$ is the set of players, $A_n \triangleq \{0, 1\}$ is the set of strategies for user n, and the overhead function $V_n(a_n, a_{-n})$ of each user n is the cost function to be minimized by player n. In the sequel, we call the game Γ as the decentralized computation offloading game. We now introduce the concept of Nash equilibrium [25].

Definition 11.1 A strategy profile $\mathbf{a}^* = (a_1^*, \ldots, a_N^*)$ is a Nash equilibrium of the decentralized computation offloading game if at the equilibrium $\mathbf{a}^*$, no player can further reduce its overhead by unilaterally changing its strategy, i.e.

$$
V_n(a_n^*, a_{-n}^*) \leq V_n(a_n, a_{-n}^*), \quad \forall a_n \in A_n, \ n \in \mathcal{N}
\tag{11.10}
$$

The Nash equilibrium has the nice self-stability property such that the users at the equilibrium can achieve a mutually satisfactory solution, and no user has the incentive to deviate. This property is very important to the decentralized computation offloading problem, since the mobile devices are owned by different individuals and they may act in their own interests.

11.3.2.2 Game Property

We then study the existence of Nash equilibrium of the decentralized computation offloading game. To proceed, we first introduce an important concept of best response [25].

Definition 11.2 Given the strategies a_{-n} of the other players, player n's strategy $a_n^* \in A_n$ is a best response if

$$
V_n(a_n^*, a_{-n}) \leq V_n(a_n, a_{-n}), \quad \forall a_n \in A_n
\tag{11.11}
$$

According to Eqs. (11.10) and (11.11), we see that at the Nash equilibrium, all the users play the best response strategies toward each other. Based on the concept of best response, we have the following observation for the decentralized computation offloading game.

Lemma 11.1 *Given the strategies a_{-n} of other mobile device users in the decentralized computation offloading game, the best response of a user n is given as the following threshold strategy:*

$$a_n^* = \begin{cases} 1, & \text{if } \sum_{m \in \mathcal{N} \setminus \{n\}: a_m=1} P_m H_{m,s} \le L_n \\ 0, & \text{otherwise} \end{cases} \tag{11.12}$$

where the threshold

$$L_n = \frac{P_n H_{n,s}}{2^{\frac{(\gamma_n^T + \gamma_n^E P_n) B_n}{W (\gamma_n^T T_n^l + \gamma_n^E E_n^l - \gamma_n^T T_{n,\exp}^c)}} - 1} - \omega_n \tag{11.13}$$

According to Lemma 11.1, we see that when the received interference $\sum_{m \in \mathcal{N} \setminus \{n\}: a_m=1} P_m H_{m,s}$ is lower enough, it is beneficial for user n to offload the computation to the fog. Otherwise, the user n should compute the task on the mobile device locally. Since the wireless access plays a critical role in fog computing, we next discuss the existence of Nash equilibrium of the decentralized computation offloading game in both homogeneous and heterogeneous wireless access cases.

Homogeneous Wireless Access Case We first consider the case that users' wireless access is homogenous, i.e. $P_m H_{m,s} = P_n H_{n,s} = K$, for any $n, m \in \mathcal{N}$. This can correspond to the scenario that all the mobile device users experience the similar channel condition and are assigned with the same transmission power by the base-station. However, different users may have different thresholds L_n, i.e. they are heterogeneous in terms of computation capabilities and tasks.

For the homogenous wireless access case, without loss of generality, we can order the set $\mathcal{N}$ of mobile device users so that $\frac{L_1}{K} \ge \frac{L_2}{K} \ge \cdots \ge \frac{L_N}{K}$. Based on this, we have the following useful observation.

Lemma 11.2 *For the decentralized computation offloading game with homogenous wireless access, if there exists a nonempty beneficial fog computing group of mobile device users $S \subseteq \mathcal{N}$ such that*

$$|S| \le \frac{L_i}{K} + 1, \quad \forall i \in S \tag{11.14}$$

and further if $S \subset \mathcal{N}$,

$$|S| > \frac{L_j}{K}, \quad \forall j \in \mathcal{N} \setminus S \tag{11.15}$$

then the strategy profile wherein users $i \in S$ play the strategy $a_i = 1$, and the other users $j \in \mathcal{N} \setminus S$ play the strategy $a_j = 0$ is a Nash equilibrium.

Algorithm 11.1   Algorithm for Finding Beneficial Fog Computing Group

---

1:   **Input: the set of ordered mobile device users with** $\frac{L_1}{K} \geq \frac{L_2}{K} \geq \dots \geq \frac{L_N}{K}$ **and** $\frac{L_1}{K} \geq 0$.

2:   **Output**: a beneficial fog computing group S.

3:   **set** $S = \{1\}$.

4:   **for** $t = 2$ to N **do**

5:         **set** $\widetilde{S} = S \cup \{t\}$

6:         **if** $|\widetilde{S}| > \frac{L_t}{K} + 1$ **then**

7:               **stop** and go to **return**.

8:         **else set** $S = \widetilde{S}$.

9:         **end if**

10:   **end for**

11:   **return** S

For example for a set of four users with

$$\left(\frac{L_1}{K}, \frac{L_2}{K}, \frac{L_3}{K}, \frac{L_4}{K} \right) = (5, 4, 3, 2)$$

the beneficial fog computing group is $S = \{1, 2, 3\}$. In general, when $\frac{L_1}{K} \geq 0$, we can construct the beneficial fog computing group by using Algorithm 11.1. Thus, we have the following result.

Theorem 11.1   *The decentralized computation offloading game with homogenous wireless access always has a Nash equilibrium. More specifically, when $\frac{L_1}{K} < 0$, all users $n \in \mathcal{N}$ playing the strategy $a_n = 0$ is a Nash equilibrium. When $\frac{L_1}{K} \geq 0$, we can construct a beneficial fog computing group $S \neq \emptyset$ by Algorithm 11.1 such that the strategy profile wherein users $i \in S$ play the strategy $a_i = 1$ and the other users $j \in \mathcal{N} \backslash S$ play the strategy $a_j = 0$ is a Nash equilibrium.*

Since the computational complexity of ordering operation (e.g. quicksort algorithm) is typically $\mathcal{O}(N \log N)$ and the construction procedure in Algorithm 11.1 involves at most N operations (with each operation of the complexity of $\mathcal{O}(1)$), the beneficial fog computing group construction algorithm has a low computational complexity of $\mathcal{O}(N \log N)$. This implies that we can compute the Nash equilibrium of the decentralized computation offloading game in the homogenous wireless access case in a fast manner.

General Wireless Access Case   We next consider the general case including the case that users' wireless access can be heterogeneous, i.e. $P_m H_{m,s} \neq P_n H_{n,s}$. Since mobile device users may have different transmission power P_n, channel gain $H_{n,s}$, and thresholds L_n, the analysis based on the beneficial fog computing group in the homogenous case cannot apply here. We hence resort to a power tool of potential game [26].

Definition 11.3 A game is called a potential game if it admits a potential function $\Phi(\mathbf{a})$ such that for every $n \in \mathcal{N}$, $a_{-n} \in \prod_{i \neq n} \mathcal{A}_i$, and $a'_n, a_n \in \mathcal{A}_n$, if

$$V_n(a'_n, a_{-n}) < V_n(a_n a_{-n}) \tag{11.16}$$

we have

$$\Phi(a'_n, a_{-n}) < \Phi(a_n, a_{-n}) \tag{11.17}$$

Definition 11.4 The event where a player n changes to an action a'_n from the action a_n is a better response update if and only if its cost function is decreased, i.e.

$$V_n(a'_n, a_{-n}) < V_n(a_n, a_{-n}) \tag{11.18}$$

An appealing property of the potential game is that it admits the finite improvement property, such that any asynchronous better response update process (i.e. no more than one player updates the strategy at any given time) must be finite and leads to a Nash equilibrium [26]. Here, the potential function to a game has the same spirit as the Lyapunov function to a dynamical system. If a dynamic system is shown to have a Lyapunov function, then the system has a stable point. Similarly, if a game admits a potential function, the game must have a Nash equilibrium.

We now prove the existence of Nash equilibrium of the general decentralized computation offloading game by showing that the game is a potential game. Specifically, we define the potential function as

$$\Phi(\mathbf{a}) = \frac{1}{2} \sum_{n=1}^{N} \sum_{m \neq n} P_n H_{n,s} P_m H_{m,s} I_{\{a_n=1\}} I_{\{a_m=1\}} + \sum_{n=1}^{N} P_n H_{n,s} L_n I_{\{a_n=0\}} \tag{11.19}$$

where $I_{\{A\}}$ is the indicator function such as $I_{\{A\}} = 1$ if the event A is true and $I_{\{A\}} = 0$ otherwise.

Theorem 11.2 *The general decentralized computation offloading game is a potential game with the potential function as given in* Eq. (11.19), *and hence always has a Nash equilibrium and the finite improvement property.*

Theorem 11.2 implies that any asynchronous better response update process is guaranteed to reach a Nash equilibrium within a finite number of iterations. This motivates the algorithm design in Section 11.3.3.

11.3.3 Decentralized Computation Offloading Mechanism

In this section, we propose a decentralized computation offloading mechanism in Algorithm 11.2 for achieving the Nash equilibrium of the decentralized computation offloading game.

11.3.3.1 Mechanism Design

The motivation of using the decentralized computation offloading mechanism is to coordinate mobile device users to achieve a mutually satisfactory decision-making, prior to the computation task execution. The key idea of the mechanism design is to utilize the finite improvement property of the decentralized computation offloading game and let one mobile device user improve its computation offloading decision at a time. Specifically, by using the clock signal from the wireless access base-station for synchronization, we consider a slotted time structure for the computation offloading decision update. Each decision slot t consists the following two parts:

- *Interference measurement*: Each mobile device user n locally measures the received interference $\mu_n(t) = \sum_{m \in \mathcal{N} \setminus \{n\}: a_m(t)=1} P_m H_{m,s}$ generated by other users who currently choose the decisions of offloading the computation tasks to the fog via wireless access. To facilitate the interference measurement, for example the users m who choose decisions $a_m(t) = 1$ at the current slot will transmit some pilot signals to the base-station. And each mobile device user can then enquire its received interference $\mu_n(t)$ from the base-station.

- *Decision update contention*: We exploit the finite improvement property of the game by having one mobile device user carry out a decision update at each decision slot. We let users who can improve their computation performance compete for the decision update opportunity in a decentralized manner. More specifically, according to Lemma 11.1, each mobile device user n first computes its set of best response update based on the measured interference $\mu_n(t)$ as

$$\Delta_n(t) \triangleq \{a_n^* : V_n(a_n^*, a_{-n}(t)) < V_n(a_n(t), a_{-n}(t))\}$$

$$= \begin{cases} \{1\}, & \text{if } a_n(t) = 0 \text{ and } \mu_n(t) \leq L_n \\ \{0\}, & \text{if } a_n(t) = 1 \text{ and } \mu_n(t) > L_n \\ \emptyset, & \text{otherwise} \end{cases} \quad (11.20)$$

The best response here is similar to the steepest descent direction selection to reduce user's overhead. Then, if $\Delta_n(t) \neq \emptyset$ (i.e. user n can improve), user n will contend for the decision-update opportunity. Otherwise, user n will not contend and adhere to the current decision at next decision slot, i.e. $a_n(t+1) = a_n(t)$. For the decision-update contention, for example we can adopt the random backoff-based mechanism by setting the time length of decision-update contention as τ^*. Each contending user n first generates a backoff time value τ_n according to the uniform distribution over $[0, \tau^*]$ and countdown until the backoff timer expires. When the timer expires, if the user has not received any request-to-update (RTU) message from other mobile device users yet, the user will update its decision for the next slot as $a_n(t+1) \in \Delta_n(t)$ and then broadcast a RTU message to all users to indicate that it wins the decision-update contention. For other users, on hearing the RTU message, they will not update their decisions and will choose the same decisions at next slot, i.e. $a_n(t+1) = a_n(t)$.

According to the finite improvement property in Theorem 11.2, the mechanism will converge to a Nash equilibrium of the decentralized computation offloading game within finite number of decision slots. In practice, we can implement that the computation offloading decision update process terminates when no RTU messages are broadcasted for multiple consecutive decision slots (i.e. no decision update can be further carried out by any users). Then, each mobile device user n executes the computation task according to the decision a_n obtained at the last decision slot by the mechanism. Due to the property of Nash equilibrium, no user has the incentive to deviate from the achieved decisions. This is very important to the decentralized computation offloading problem, since the mobile devices are owned by different individuals and they may act in their own interests. By following the decentralized computation offloading mechanism, the users adopt the best response to improve their decision-makings and eventually self-organize into a mutually satisfactory solution (i.e. Nash equilibrium).

Algorithm 11.2   Decentralized Computation Offloading Mechanism

---

1: initialization:

2:    each mobile device user n **chooses** the computation decision $a_n(0) = 1$.

3: **end initialization**

4: **repeat** for each user n and each decision slot t in parallel:

5:    **measure** the interference $\mu_n(t)$.

6:    **compute** the best response set $\Delta_n(t)$.

7:    **if** $\Delta_n(t) \neq \emptyset$ **then**

8:      **contend** for the decision update opportunity.

9:      **if win** the decision update contention **then**

10:        **choose** the decision $a_n(t+1) \in \Delta_n(t)$ for next slot.

11:        **broadcast** the RTU message to other users.

12:      **else choose** the original decision $a_n(t+1) = a_n(t)$ for next slot.

13:      **end if**

14:    **else choose** the original decision $a_n(t+1) = a_n(t)$ for next slot.

15:    **end if**

16: **until** no RTU messages are broadcasted for M consecutive slots.

We then analyze the computational complexity of the algorithm. In each iteration, N mobile users will execute the operations in Lines 5–15. Since the operations in Lines 5–15 only involve some basic arithmetical calculations, the computational complexity in each iteration is $\mathcal{O}(N)$. Suppose that it takes C iterations for the algorithm to converge. Then the total computational complexity of the algorithm is $\mathcal{O}(CN)$.

11.3.3.2 Performance Analysis

We then discuss the efficiency of Nash equilibrium by the decentralized computation offloading mechanism. Note that the decentralized computation offloading game may have multiple Nash equilibria, and the proposed decentralized computation offloading mechanism will randomly select one Nash equilibrium (since a random user is chosen for decision update). Following the definition of price of anarchy (PoA) in game theory [27], we will quantify the efficiency ratio of the worst-case Nash equilibrium over the centralized optimal solution. Let Υ be the set of Nash equilibria of the decentralized computation offloading game. Then, the PoA is defined as

$$\text{PoA} = \frac{\max_{\mathbf{a} \in \Upsilon} \sum_{n \in \mathcal{N}} V_n(\mathbf{a})}{\min_{\mathbf{a} \in \prod_{n=1}^{N} A_n} \sum_{n \in \mathcal{N}} V_n(\mathbf{a})} \tag{11.21}$$

which is lower bounded by 1. A larger PoA implies that the set of Nash equilibrium is less efficient (in the worst-case sense) using the centralized optimum as a benchmark. Let

$$\overline{Z_n^c} = \frac{(\gamma_n^T + \gamma_n^E P_n) B_n}{W \log_2 \left(1 + \frac{P_n H_{n,s}}{\omega_n} \right)} + \gamma_n^T T_{n,\text{exe}}^c \tag{11.22}$$

We can show the following result.

Theorem 11.3 *The PoA of the decentralized computation offloading game is at most* $\dfrac{\sum_{n=1}^{N} Z_n^l}{\sum_{n=1}^{N} \min\{Z_n^l, \overline{Z_n^c}\}}$.

Theorem 11.3 indicates that when users have lower cost of local computing (i.e. Z_n^l is smaller), the Nash equilibrium is closer to the centralized optimum, and hence, the PoA is lower. Moreover, when the communication efficiency is higher (i.e. $P_n H_{n,s}$ is larger and hence $\overline{Z_n^c}$ is larger), the performance of Nash equilibrium can be improved.

11.4 Cooperative Computation Offloading

11.4.1 HyFog Framework Model

As illustrated in Figure 11.2, we consider that the HyFog framework involves a set $\mathcal{N} = \{1, 2, \dots, N\}$ of user devices. A device can establish the cellular connection with its associated base station (e.g. via LTE-Cat and LTE-M for smartphones and Internet of Things [IoT] devices, respectively) as well as the D2D connection with a device in proximity (e.g. using cellular D2D or Wi-Fi-direct).

11.4.1.1 Resource Model

We first introduce the model to describe the device resources of a user device for mobile task computing as follows:

- *Computation capacity*: For a device i, let c_i be its CPU working frequency, and hence the total computation capacity is c_i (in CPU cycles per unit time).
- *Cellular link*: Each device i can establish a cellular link with its associated base station. The cellular transmission and receiving power of device i are denoted as P_{it}^c and P_{ir}^c, respectively, and the corresponding upload and download cellular data rates are D_i^t and D_i^r, respectively.
- *D2D link*: Each device i can also establish a D2D link with another device in proximity. We denote the D2D data rate from a user i to another user j as D_{ij}, and P_{it}^d and P_{ir}^d are the D2D transmission power and receiving power of

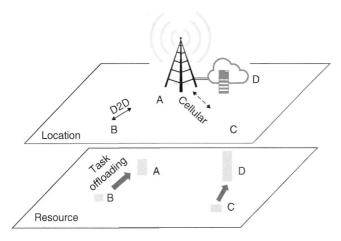

Figure 11.2 An illustration of D2D and fog offloaded task executions.

a device i, respectively. From the network operator's perspective,[2] we introduce the D2D connectivity graph $G = \{\mathcal{N}, \mathcal{E}\}$, where the set of devices $\mathcal{N}$ is the vertex set and $\mathcal{E} = \{(i,j) : e_{ij} = 1, \ \forall i,j \in \mathcal{N}\}$ is the edge set where $e_{ij} = 1$ if it is feasible for devices i and j to establish a D2D link between them. Similar to FlashLinQ [28], due to the time and resource constraint, we impose a practically relevant constraint that a device is allowed to establish and maintain at most one D2D link during a task offloading slot.

To further facilitate the task execution, we consider that a fog server is deployed at the network edge by the operator. In particular, the fog server consists of a set of virtual machines $\mathcal{M} = \{1, 2, ..., M\}$, where a virtual machine $m \in \mathcal{M}$ has a computing capacity (in terms of CPU working frequency) of K_m and can host a task from a single device [29].

11.4.1.2 Task Execution Model

In the HyFog framework, we adopt a parameter tuple $\langle I_i, R_i, O_i \rangle$ to characterize the task of a device i, where I_i is the input data size of the task, R_i is the amount of computing resource required for the task (i.e. the number of CPU cycles), O_i is the output data size of the task. Note that our model can be also easily extended to account for other types of resources (e.g. storage) by introducing more parameters in the tuple.

We next introduce the task execution model such that a device has three different options for processing its task as follows:

- *Local mobile execution*: A device i can locally execute its own task. According to the task parameter tuple, the execution time for computation is given by $T_i^l = R_i/c_i$, and the energy consumption is given by $E_i^l = \rho_i^c R_i$, where ρ_i^c is the energy cost per CPU cycle for computation, which depends on the energy efficiency of the processor model and can be measured in practice [30]. Accordingly, we can then obtain the cost of the local mobile execution approach by device i in terms of both time and energy overhead as

$$\theta_i^l = \lambda_i^t T_i^l + \lambda_i^e E_i^l \tag{11.23}$$

where $\lambda_i^t, \lambda_i^e \in \{0, 1\}$ denote the weighting parameters of computational time and energy for device i's decision-making, respectively. For example when a device user is at a low battery state and cares about the energy consumption, the user can set $\lambda_i^e = 1$ and $\lambda_i^t = 0$ in the decision-making. When a user is running some application that is sensitive to the delay (e.g. interactive gaming) and hence concerns about the processing time, then the user can set $\lambda_i^e = 0$ and $\lambda_i^t = 1$ in the decision-making. To provide rich modeling flexibility, our model can also apply to the generalized case where $\lambda_i^t, \lambda_i^e \in [0, 1]$

---

2 Through network-assisted device discovery and local information report by the devices, the network operator can gather sufficient D2D connectivity information in practice.

such that a device user can take both computational time and energy into the decision-making at the same time. In practice, the proper weights that capture a user's valuations on computational energy and time can be determined by applying the multiattribute utility approach in the multiple criteria decision-making theory [31].

- *D2D offloaded execution*: A device i can offload its own task to a nearby device j via D2D link. In this case, the transmission time and energy consumption for task input and output data transfer through D2D transmission between these two devices are given by $T_{ij}^{d1} = I_i/D_{ij} + O_i/D_{ji}$ and $E_{ij}^{d1} = (P_{it}^d + P_{jr}^d)I_i/D_{ij} + (P_{jt}^d + P_{ir}^d)O_i/D_{ji}$, respectively. In addition, the computing time and energy overhead for executing the offloaded task in device j are given by $T_{ij}^{d2} = R_i/c_j$ and $E_{ij}^{d2} = \rho_j^c R_i$, respectively. Accordingly, we can then obtain the cost of the D2D offloaded execution approach in terms of both time and energy overhead as

$$\theta_{ij}^d = \lambda_i^t(T_{ij}^{d1} + T_{ij}^{d2}) + \lambda_i^e(E_{ij}^{d1} + E_{ij}^{d2}) \tag{11.24}$$

Due to the constraint of physical size, mobile devices typically have limited resource capacity, and hence, we will assume that a device can execute at most one task.

- *Fog offloaded execution*: A device i can offload its own task to the fog server at the network edge via the cellular link. In this case, the transmission time and energy consumption for task input and output data transfer through D2D transmission between these two devices are given by $T_i^{c1} = I_i/D_i^t + O_i/D_i^r$ and $E_i^c = P_{it}^c I_i/D_i^t + P_{ir}^c O_i/D_i^r$, respectively. Suppose the virtual machine $m \in \mathcal{M}$ is assigned to process the task of device i at the fog server. Then, the computing time of the virtual machine m is $T_i^{c2} = R_i/K_m$. Accordingly, we can obtain the cost of the fog offloaded execution approach by device i in terms of time and energy as

$$\theta_{im}^c = \lambda_i^t(T_i^{c1} + T_i^{c2}) + \lambda_i^e E_i^c \tag{11.25}$$

As discussed in Section 11.1, different approaches have heterogeneous requirements for achieving efficient task execution, and the decisions on choosing proper task execution approaches among the task devices are coupled, making the problem of minimizing the total task execution cost very challenging.

Inspired by the key observation that a task should be assigned to a single task execution spot (i.e. its local device, a nearby device, or a virtual machine), we are going to devise a graph matching based task offloading algorithm, in order to minimize the total cost of the task executions by all the devices.

11.4.2 Inadequacy of Bipartite Matching–Based Task Offloading

As a common approach, one might consider to adopt the minimum weight bipartite matching solution for solving the task offloading problem. That is, as

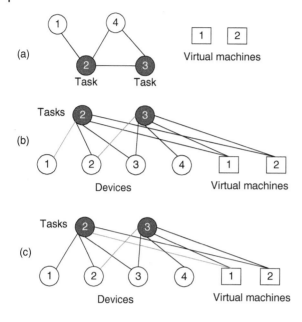

Figure 11.3 An illustration of bipartite matching–based task offloading, with (a) the D2D connectivity graph and virtual machines and (b) and (c) the transformed bipartite matching with two illustrative matching solutions (lines labeled in light gray).

illustrated in Figure 11.3, a weighted bipartite graph can be constructed, such that on one side, a node i represents a task of a device i, and on the other side, the nodes represent the set of devices as well as the set of virtual machines for task executions. We then define the edge set for the bipartite graph as follows: (i) There exists an edge between a task node i and its corresponding local device node i, and the edge weight is the cost of local mobile execution θ_i^l; (ii) there exists an edge between a task node i and a device node j if it is feasible to establish a D2D link between devices i and j, and the edge weight denotes the cost of D2D offloaded execution θ_{ij}^d; and (iii) there exists an edge between a task node i and each virtual machine node m and the edge weight indicates the cost of fog offloaded execution θ_{im}^c. Then, the minimum task execution cost solution is equivalent to the minimum weight bipartite matching solution over the constructed bipartite graph.

However, the aforementioned approach would fail to work in this setting. Taking Figure 11.3a for instance, suppose the minimum weight bipartite matching solution is that device 2's task is offloaded to device 1 and device 3's task is offloaded to device 2. In this case, device 2 needs to simultaneously establish and maintain two D2D links with devices 1 and 3 for task offloading, which would demand excessive resource and overhead. Otherwise, if device 3's task is offloaded to device 2, it can happen only after device 2 has finished its task offloading, disconnected the D2D link with device 1 and then established a new D2D link with device 3, which would also incur significant latency. As another counter example in Figure 11.3b, where the minimum weight bipartite matching solution is that device 2's task is offloaded to the virtual machine via cellular link and device 3's task is offloaded to device 2 via D2D link. In this case, a significant cost is charged to device 2 since device 2 has to simultaneously take up the cellular communication (for finishing its task offloading to the fog) and D2D communication as well as the task execution (for finishing device 3's D2D-offloaded task). This would lead to the issue of overloading for a single device.

11.4.3   Three-Layer Graph Matching Based Task Offloading

To tackle the disadvantages of the bipartite matching–based approach, we next propose a novel graph matching algorithm for the hybrid task offloading. A key idea is to first construct a three-layer graph such that the graph components in these three layers represent the task execution spots for the local mobile execution, D2D offloaded execution, and fog offloaded execution, respectively. Then, we carry out the minimum weight graph matching over the constructed three-layer graph, in order to minimize the total cost of task executions by the mobile devices.

In general, the three-layer graph (as illustrated in Figure 11.4) is initially constructed on the basis of D2D connectivity graph *G* in the D2D layer, and then, it is expanded to the local and fog layers. The structures of the graph are outlined as follows:

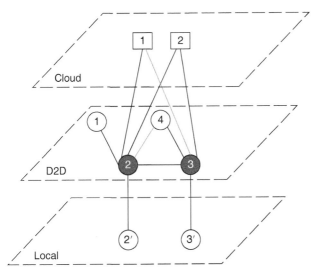

Figure 11.4 An illustration of the constructed three-layer graph matching–based task offloading, with an illustrative matching solution (lines labeled in light gray).

- *D2D layer*: We adopt the D2D connectivity graph G as the subgraph in this layer to ensure the offloading feasibility for D2D offloaded execution approach. If device node has a task to be executed, then the task node is labeled with gray color in Figure 11.4 (e.g. nodes 2 and 3). Moreover, for a device k that does not have a task, if all its neighboring devices also do not have any tasks, then we will remove the device node k in the graph, since the device k will not be assigned with any task. This also helps to reduce the graph size for matching.
- *Local layer*: Then for a device node i in the D2D layer that has a task to be executed, we add a new replicated node i' in this layer and then add an edge connecting node i' and node i in the D2D layer (e.g. nodes 2 and 3 have tasks and hence two replicated nodes are added in the local layer in Figure 11.4). This represents the option of choosing the local mobile execution approach.
- *Fog layer*: Next, we add the set of virtual machine nodes at the fog layer. For each virtual machine node m, we add an edge connecting it with every device node i having a task in the D2D layer. This indicates the option of choosing the fog offloaded execution approach.

By doing so, for the matching over the constructed graph, we have the following representations for different cases:

- *Case 1*: If node i in the D2D layer is matched with its replicated node i', then it means that the device i will execute its task locally.
- *Case 2*: If node i in the D2D layer is matched with another node j in the D2D layer without a replicated node, then it represents that the task of device i will be offloaded to device j via the D2D link.
- *Case 3*: If node i (having the replicated node i') in the D2D layer is matched with another node j in the D2D layer having a replicated node j', then it indicates that devices i and j exchange their tasks for the mutually D2D offloaded executions. For example one device with strong computing power but a lightweight task, whereas the other device with week computing power but a heavy task. Such mutual task offloading can be beneficial.
- *Case 4*: If node i in the D2D layer is matched with a virtual machine node m, then it means that the device i will offload its task to the virtual machine m in the fog server at the network edge.

Moreover, as illustrated in Figure 11.4, in the constructed graph, each edge is associated with a device node with a task. This ensures that each device that has a task will choose one execution approach to compute its task in the matching solution. On the other hand, due to the property of graph matching, each node will be matched with at most one node in the constructed graph. This prohibits the situations that a device has to simultaneously establish multiple D2D links with multiple devices and take up unbalanced workloads with excessive overhead. For example the illustrative matching solutions in Figure 11.3a,b are not a valid matching solution for the constructed three-layer graph in Figure 11.4.

Based on the previous discussions, we next define the weights for the edges of the constructed three-layer graph as follows:

- *Case 1*: For an edge that connects node i in the D2D layer with its replicated node i' in the local layer, we set the weight $w_{ii'} = \theta_i^l$, i.e. the cost of local mobile execution.
- *Case 2*: For an edge that connects node i in the D2D layer with another node j without a replicated node in the D2D layer, we set the weight $w_{ij} = \theta_{ij}^d$, i.e. the cost of D2D offloaded execution from device i to device j.
- *Case 3*: For an edge that connects node i (having the replicated node i') in the D2D layer with another node j in the D2D layer that has a replicated node j', we set the weight $w_{ij} = \theta_{ij}^d + \theta_{ji}^d$, i.e. the total cost of the mutual D2D offloaded executions.
- *Case 4*: For an edge that connects node i in the D2D layer with a virtual machine node m in the fog layer, we set the weight $w_{im} = \theta_{im}^c$, i.e. the cost of the fog offloaded execution.

Based on the aforementioned constructed three-layer weighted graph, we can obtain the minimum cost task offloading solution by finding the minimum weight graph matching solution. As illustrative example, we illustrate a feasible matching solution (labeled with gray circles) in Figure 11.4 where node 2 in the D2D layer is matched with node 4 in the D2D layer (i.e. device 2 offloads its task to device 4 via D2D link) and node 3 in the D2D layer is matched with node 1 in the fog layer (i.e. device 3 offloads its task to the virtual machine 1 in the fog server via cellular link). In general, we can compute the minimum weight graph matching solution using the well-known Edmonds' Blossom algorithm [32], which possesses a polynomial time complexity and can compute the solution in a fast manner as shown in the performance evaluation section later.

To fully explore the optimization space for the hybrid task offloading in HyFog, in this paper we adopt a global point of view to develop the three-layer graph matching algorithm, which works in a centralized manner. This is highly relevant to the emerging network-assisted architecture such as 5G D2D overlaid cellular networks, cellular IoT, and software-defined mobile networks, where the operator serves as the controller for system management and scheduling as well as provisioning device assistance functionality. The extension of the proposed algorithm for the distributed implementation is challenging and will be considered in a future work.

11.5 Discussions

In the previously mentioned Sections 11.3 and 11.4, we study the problem of multiuser fog computation offloading and focus on addressing the key issue of energy-efficient task offloading from the competitive and cooperative perspectives. In this part, we will further discuss the several important directions for extending the proposed solutions into a full-fledged framework by accounting for variety of application factors.

11.5.1 Incentive Mechanisms for Collaboration

For practical implementation, the D2D-assisted task offloading framework HyFog strongly relies on device's collaboration, and hence, a good incentive mechanism that can prevent the overexploiting and free-riding behaviors that harm device user's motivation for collaboration is highly desirable. As an initial attempt, we next discuss a network-assisted incentive mechanism tailored to the D2D-assisted task offloading framework as follows.

The incentive mechanism is motivated by the resource tit-for-tat scheme in peer-to-peer systems. The key idea is to ensure that a device is allowable to exploit more resources from other devices only if it has contributed sufficient resources to the others. We can regard the resource contribution as user credit, which will be maintained by the base station. The resource tit-for-tat constraint reflects that the resource amount that a device exploits from others is in proportion to that it contributes to the others. If a device wants more resources, it needs to share more in return. To promote the collaboration and resource contribution, when a device does not satisfy resource tit-for-tat constraints above, the network operator will not assign device i's offloaded task.

11.5.2 Coping with System Dynamics

In order to gain useful insights, in the aforementioned Sections 11.3 and 11.4, we mainly study multiuser fog computation offloading in the static setting and consider the task assignment issue during each task offloading round. To implement the proposed solutions in practical systems, we need to consider its generalization to deal with the system dynamics. In a dynamic setting, many system factors are time varying, e.g. devices' D2D connections can dynamically change due to device users' mobility, and the cellular link quality can vary from time to time due to fading effect. Moreover, different number of new tasks can be generated at a device across different time frames, and hence, we need to carefully address the task queueing issues, to prevent the queue explosion from a long-run perspective.

To cope with the system dynamics, we can consider to generalize the proposed multiuser fog computation offloading frameworks by resorting to the tool of Lyapunov optimization [15]. Generally speaking, two key salient features enable Lyapunov optimization suitable for addressing dynamic task assignment problem: (i) Lyapunov optimization deals with online stochastic optimization problems with time-average objective, and in general, it only utilizes the system information at the current time period; (ii) it also enables the feature of stabilizing the queues by providing a drift-plus-penalty function for joint queue stability and time-average objective optimization. Thus, in a future work, we can explore to leverage the Lyapunov optimization approach to design efficient online task assignment that can be adaptive to the system dynamics, and meanwhile can ensure the stability of the task queues.

11.5.3 Hybrid Centralized–Decentralized Implementation

A key focus of the cooperative offloading framework HyFog is to embrace the benefit of network-assisted D2D collaboration for energy-efficient fog computing. The network-assisted architecture enables an efficient centralized management paradigm and hence is advocated in many future networking systems, e.g. 5G D2D overlaid cellular networks, cellular IoT, and software-defined mobile networks. Another important direction of further extending the proposed D2D crowd framework is to consider its hybrid centralized–decentralized implementation. This can be highly relevant to the application scenario that we would like to achieve a synergetic scheduling across multiple heterogeneous networks such as cellular network and Wi-Fi network. The decentralized nature of CSMA access in Wi-Fi requires a hybrid centralized–decentralized design. Also, in some cases, some gateway device can have better communications links with its peripheral devices and hence can be selected as a leader device for achieving efficient local coordination.

One possible approach for implementing the hybrid centralized–decentralized paradigm is that we can first decompose the D2D connectivity graph into multiple communities. Within a community, one leader device can be selected or elected to manage the local task assignment (e.g. using the proposed graph matching based scheme locally) for D2D collaboration-based fog computing. The leader devices will also negotiate among themselves for task assignment synchronization and conflict resolution. How to design the lightweight and efficient protocol for such a hybrid centralized–decentralized implementation can be very interesting and challenging.

11.6 Conclusion

In this chapter, we study the problem of energy of efficient multiuser computation offloading for fog computing. By considering different application scenarios, we study the problem from competitive and cooperative perspectives. Specifically, for competitive offloading decision-making problem among mobile device users, we propose as a decentralized computation offloading game formulation. We show that the game always admits a Nash equilibrium for both cases of homogenous and heterogeneous wireless access. We also design a decentralized computation offloading mechanism that can achieve a

Nash equilibrium of the game and further quantify its PoA. For the cooperative offloading problem, we propose a D2D-assisted task offloading framework that coordinates a massive crowd of devices at the network edge to support hybrid and cooperative computation offloading. We also developed a three-layer graph matching algorithm for efficient hybrid task offloading and solved the problem of minimizing the total task execution cost by transferring it into the minimum weight matching problem over the constructed three-layer graph.

For the future work, we are going to study the incentive mechanism design for device collaboration, the online computation offloading that addresses the systems' dynamics, as well as the hybrid centralized–decentralized implementation of multiuser fog computation offloading.

References

1 Cuervo, E., Balasubramanian, A., Cho, D.-k. et al. (2010). MAUI: making smartphones last longer with code offload. In: *MobiSys'10 Proceedings of the 8th International Conference on Mobile Systems, Applications, and Services*, 49–62. New York, NY: ACM.

2 Dinh, H.T., Lee, C., Niyato, D., and Wang, P. (2013). A survey of mobile cloud computing: architecture, applications, and approaches. *Wireless Communications and Mobile Computing* 13: 1587–1611.

3 Bonomi, F., Milito, R., Zhu, J., and Addepalli, S. (2012). Fog computing and its role in the internet of things. In: *ACM SIGCOMM Workshop on Mobile Cloud Computing*, 13–16. New York, NY: ACM.

4 Barbera, M. V., Kosta, S., Mei, A., and Stefa, J. (2013). To offload or not to offload? The bandwidth and energy costs of mobile cloud computing. *2013 Proceedings IEEE INFOCOM*, (April 14–19, 2013) , Turin, Italy.

5 Im, Y., Joe-Wong, C., Ha, S. et al. (2016). AMUSE: empowering users for cost-aware offloading with throughput-delay tradeoffs. *IEEE Transactions on Mobile Computing* 15 (5): 1062–1076.

6 Mao, Y., Zhang, J., and Letaief, K.B. (2016). Dynamic computation offloading for mobile-edge computing with energy harvesting devices. *IEEE Journal on Selected Areas in Communications* 34 (12): 3590–3605.

7 Wen, Y., Zhang, W., and Luo, H. (2012). Energy-optimal mobile application execution: taming resource-poor mobile devices with cloud clones. In: *2012 Proceedings IEEE INFOCOM (March 25–30, 2012)*, 2716–2720. Orlando, Florida, USA: IEEE.

8 Yang, L., Cao, J., Yuan, Y. et al. (2013). A framework for partitioning and execution of data stream applications in mobile cloud computing. *ACM SIGMETRICS Performance Evaluation Review* 40 (4): 23–32.

9 Chen, X., Jiao, L., Li, W., and Fu, X. (2016). Efficient multi-user computation offloading for mobile-edge cloud computing. *IEEE/ACM Transactions on Networking* 24 (5): 2795–2808.

10 Tang, M., Gao, L., and Huang, J. (2017). A general framework for crowdsourcing mobile communication, computation, and caching. *GLOBECOM 2017 – 2017 IEEE Global Communications Conference*, Singapore (4–8 December 2017).

11 Prasad, A.S., Arumaithurai, M., Koll, D., and Fu, X. (2017). RAERA: a robust auctioning approach for edge resource allocation. In: *MECOMM'17 Proceedings of the Workshop on Mobile Edge Communications*, 49–54. New York, NY: ACM.

12 Chen, X., Zhou, Z., Wu, W. et al. (2018). Socially-motivated cooperative mobile edge computing. *IEEE Network* 32: 177–183.

13 Iosifidis, G., Gao, L., Huang, J., and Tassiulas, L. (2013). An iterative double auction mechanism for mobile data offloading. *2013 11th International Symposium on Modeling and Optimization in Mobile, Ad Hoc, and Wireless Networks (WiOpt)* (May 13–17, 2013), Tsukuba Science City, Japan.

14 Wu, Y., Chou, P.A., and Kung, S. (2005). Minimum-energy multicast in mobile ad hoc networks using network coding. *IEEE Transactions on Communications* 53 (11): 1906–1918.

15 Lopez-Perez, D., Chu, X., Vasilakos, A.V., and Claussen, H. (2013). On distributed and coordinated resource allocation for interference mitigation in self-organizing LTE networks. *IEEE/ACM Transactions on Networking (TON)* 21 (4): 1145–1158.

16 Rappaport, T.S. (1996). *Wireless Communications: Principles and Practice*, vol. 2. Upper Saddle River, NJ: Prentice Hall.

17 Xiao, M., Shroff, N.B., and Chong, E.K. (2003). A utility-based power-control scheme in wireless cellular systems. *IEEE/ACM Transactions on Networking* 11 (2): 210–221.

18 Saraydar, C.U., Mandayam, N.B., and Goodman, D.J. (2002). Efficient power control via pricing in wireless data networks. *IEEE Transactions on Communications* 50 (2): 291–303.

19 Miettinen, A.P. and Nurminen, J.K. (2010). Energy efficiency of mobile clients in cloud computing. In: *HotCloud'10 Proceedings of the 2nd USENIX Conference on Hot Topics in Cloud Computing*, 1–7. Berkeley, CA: USENIX Association.

20 Kumar, K. and Lu, Y. (2010). Cloud computing for mobile users: can offloading computation save energy? *Computer* 43 (4): 51–56.

21 Huang, D., Wang, P., and Niyato, D. (2012). A dynamic offloading algorithm for mobile computing. *IEEE Transactions on Wireless Communications* 11 (6): 1991–1995.

22 Rudenko, A., Reiher, P., Popek, G.J., and Kuenning, G.H. (1998). Saving portable computer battery power through remote process execution. *ACM SIGMOBILE Mobile Computing and Communications Review* 2 (1): 19–26.

23 Huerta-Canepa, G. and Lee, D. (2008). An adaptable application offloading scheme based on application behavior. *AINAW'08 Proceedings of the 22nd International Conference on Advanced Information Networking and Applications – Workshops* (March 25–28, 2008), GinoWan, Okinawa, Japan.

24 Xian, C., Lu, Y., and Li, Z. (2007). Adaptive computation offloading for energy conservation on battery-powered systems. In: *ICPADS'07 Proceedings of the 13th International Conference on Parallel and Distributed Systems*, vol. 2, 1–8. Washington, DC: IEEE Computer Society.

25 Osborne, M.J. (1994). *A Course in Game Theory*. Cambridge, MA: MIT Press.

26 Monderer, D. and Shapley, L.S. (1996). Potential games. *Games and Economic Behavior* 14 (1): 124–143.

27 Roughgarden, T. (2005). *Selfish Routing and the Price of Anarchy*. Cambridge, MA, USA: MIT Press.

28 Wu, X., Tavildar, S., Shakkottai, S. et al. (2013). FlashLinQ: a synchronous distributed scheduler for peer to-peer ad hoc networks. *IEEE/ACM Transactions on Networking (TON)* 21: 1215–1228.

29 Shi, C., Lakafosis, V., Ammar, M.H., and Zegura, E.W. (2012). Serendipity: enabling remote computing among intermittently connected mobile devices. In: *MobiHoc'12 Proceedings of the 13th ACM International Symposium on Mobile Ad Hoc Networking and Computing*, 145–154. New York, NY: ACM.

30 Kwak, J., Choi, O., Chong, S., and Mohapatra, P. (2015). Processor-network speed scaling for energy–delay tradeoff in smartphone applications. *IEEE/ACM Transactions on Networking (TON)* 24: 1647–1660.

31 Triantaphyllou, E. (2013). *Multi-criteria Decision Making Methods: A Comparative Study*, vol. 44. Springer Science & Business Media.

32 Even, S. (2011). *Graph Algorithms*. Cambridge, UK: Cambridge University Press.

12

A Highly Available Storage System for Elastic Fog

Jaeyoon Chung[1], Carlee Joe-Wong[2], and Sangtae Ha[3]

[1] *Myota Inc., Malvern, PA, USA*
[2] *Department of Electrical and Computer Engineering, Carnegie Mellon University (CMU), Pittsburgh, PA, USA*
[3] *Department of Computer Science, University of Colorado Boulder, Boulder, CO, USA*

12.1   Introduction

Fog computing and dispersed computing are emerging technologies that leverage the proliferation of Internet-connected "edge" devices to enable next-generation applications [1]. These devices, such as smartphones, tablets, or set-top boxes, are not only increasing in number but also gaining more CPU and storage capabilities, allowing them to viably host parts of a truly distributed application. For instance, they can store data files or preprocess sensor data before sending it to a central application server. In this work, we consider "edge" resources, defined as small-scale devices like desktop computers available over a fixed local network, as well as "thing" resources, which can be smaller personal or mobile devices like smartphones.

Today's applications are generally not fully distributed: they often rely on cloud services, which allow applications to consume virtualized resources on remote datacenter servers that are shared with other applications. Utilizing cloud resources allows the applications to quickly scale up or down their resources depending on their needs. Fog devices, however, can lie on a "cloud-to-things continuum" and provide cloud-like services on devices near the end user. Leveraging these devices for some application functionalities can provide a variety of economic and technological benefits, e.g. eliminating the cost of renting cloud services and reducing the latency of communicating with the user device. In order to coordinate between multiple devices running an application, however, a fog-edge system should provide intermediate services, such as proxy, data aggregation, and service delegation. One of the main services is the ability to store application data and make it available to other devices as needed. In this work, we develop a platform for elastic fog storage

Fog and Fogonomics: Challenges and Practices of Fog Computing, Communication, Networking, Strategy, and Economics, First Edition. Edited by Yang Yang, Jianwei Huang, Tao Zhang, and Joe Weinman.

that seamlessly integrates distributed storage resources. Our work enables the development of other fog-based applications that can use our system for their underlying data storage.

12.1.1   Fog Versus Cloud Services

Cloud services offer applications the powerful advantage of elastic resource allocation to multiple users, leading to reduced server maintenance costs for users and better resource utilization for cloud providers. Users can delegate complicated resource management tasks to the cloud provider, simplifying their application hosting and operations, but they can experience a "vendor lock-in" problem: it is difficult for users to switch between cloud providers without significantly interrupting their applications, for instance if they have already stored a large amount of application data on a given cloud provider's servers. The cloud providers can then manage their physical and logical resources with complete ownership and controllability, possibly ignoring users' individual preferences for application performance. For instance, some users may desire more stringent privacy protection for their application data than that offered by a generic cloud service.

Users can avoid the vendor lock-in problem and customize their services by leveraging more "edge" resources owned by the user or his/her company. These resources have the additional advantage of being cheaper or even free to use. Fog systems provide a natural way to integrate these resources with cloud services. To take full advantage of fog architectures, users should then decompose their application tasks and select the proper resources for each task, considering resource heterogeneity, proximity, capacity, functionality, etc. In doing so, they can combine advantages offered by different types of devices on the cloud-to-things continuum (Figure 12.1). Furthermore, they also need to determine the resource allocation among different types of fog nodes, giving them the freedom to configure their own custom services. From a storage service perspective, fog allows users to store different parts of their data at locations near client devices, such as enterprise storage nodes and personal storage devices.

Developing an elastic fog storage system, however, comes with challenges. Most significantly, edge storage resources may not always be available to all users, and users may not have direct control over the management of these resources. For example, some enterprise storage locations can only be accessed when users are connected to company networks, and users may not have enough knowledge to configure them. Users may not even be able to fully trust these services or devices to keep their data private and not expose it to unauthorized parties, given their lack of full control. As a result, we cannot expect that these services or devices will automatically coordinate with each other to reliably store and provide access to users' files.

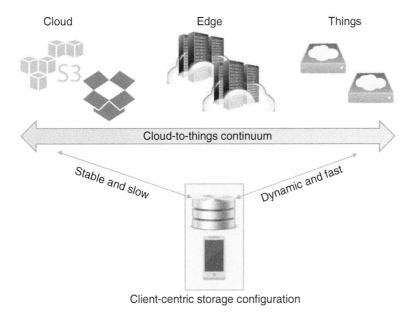

Figure 12.1 The cloud-to-things continuum, from a data storage perspective.

Existing distributed storage solutions such as Hadoop distributed file system (HDFS) cannot handle these scenarios: they are designed for data center environments where the network topology is well designed and failures are strictly managed. Latency in the data center is also low and stable enough to optimize the locations of file replicas, while heterogeneous fog devices can be offered by different vendors and are connected through the Internet, which can have widely variable latency. The assumptions of existing distributed storage solutions are therefore not valid in a fog environment, requiring us to develop a fundamentally new storage architecture.

12.1.2 A Fog Storage Service

Elastic fog storage enables us to extend local storage devices such as SSD and hard disk drive (HDD) into edge/cloud storage resources. Unlike the cloud, it is reasonable to expect that the fog/edge would consist of heterogeneous equipment with different capacities, functionalities, and availability in geographically different locations. This resource heterogeneity and unreliability raises performance challenges, which we aim to address in this work.

In the remainder of this chapter, we use the term "client" interchangeably with the user device. We define five desirable properties of elastic fog storage as follows:

- *Availability*: Edge locations may not always be accessible to users due to connection failures or storage nodes leaving the system. We use redundancy to ensure reliable connectivity to multiple distributed storage locations.
- *Scalability*: The major performance bottleneck of a distributed storage system is metadata lookup, which reduces read/write performance. We therefore minimize the metadata handling by decoupling it from the files stored. We also scale to multiple storage nodes by making the file management operations transparent to the client.
- *Flexibility*: Our elastic fog storage allows storage nodes to join and leave the system by incorporating resource discovery and automated configuration. It is also customizable according to users' performance requirements.
- *Efficiency*: It is necessary to minimize data transmission between client and storage nodes in order not to overload the network. Our elastic fog storage will also reuse existing resources and implementations instead of redesigning new components.
- *Security*: Instead of redesigning a security layer, our elastic fog storage will leverage security functions that are already implemented at each storage node. In addition, information dispersion algorithms such as Shamir's Secret Sharing Scheme and Erasure Coding will be used to ensure that access to the data is allowed only if a predefined number of data shares is collected from distributed storage nodes.

In providing the abovementioned five properties, we make several research contributions.

We design a fog storage system with full stack functions including resource discovery, automated configuration, data encoding, optimal data placement, and a robust request protocol. We first discuss our main design considerations, system architecture, and file operations. We then present our algorithms for uploading and downloading files.

We implemented a prototype system that uses devices along a cloud-to-things continuum, including local servers as "edge" devices, Raspberry Pis as "thing" devices and the cloud. Our prototype implements all POSIX file interfaces related to file operations on the user device, which provides a seamless extension of the local storage space to the fog setting. It incorporates practical considerations such as metadata design, reusing the native file system, and bidirectional communication beyond network address translation (NAT). After presenting our implementation, we evaluate our prototype in a variety of realistic settings. We show that our file download algorithms outperform simple heuristics and that file download times are robust to failures at edge storage locations.

We finally discuss some future extensions of this work before giving an overview of related works and concluding.

12.2 Design

We first outline our design considerations before discussing our system architecture and file operations.

12.2.1 Design Considerations

Our elastic fog storage system aims to provide an elastic storage layer that extends clients' local storage space by attaching remote storage resources, such as edge devices and the cloud storages, with location transparency. We thus focus on building a seamless and practical extension of the client's storage with minimum I/O performance degradation. At the same time, however, we cannot expect to have any control over the edge storage nodes, and these nodes may independently join or leave the system, without necessarily communicating with each other. Thus, our storage system should be client-driven, with dynamic configuration and data reallocation among storages fully available at the client. Given these challenges, we make the following design choices to achieve the five elastic fog storage properties of availability, scalability, flexibility, efficiency, and security introduced above:

Client-centric architecture: We introduce a client-centric architecture that directly and actively controls multiple storage nodes using a storage layer at the client. This architecture gives users more freedom and control over how their files are stored (i.e. flexibility), while also allowing them to adapt to possible storage failures (availability). By contrast, traditional distributed file systems have focused on shared storage space and storage extensions; they are designed to provide a server-centric architecture that gives clients an interface for a large remote storage pool.

Decoupling metadata: To ensure scalability and flexibility of fog storage, we decouple the file metadata (i.e. the addresses of the file content) from file content. By doing so, we can leverage most backend storages, regardless of edge or cloud, if they are able to perform basic operations such as CREATE, READ, UPDATE, and DELETE. It allows our elastic fog storage to extend backend storages while avoiding vendor lock-in or any other dependencies. In addition, the file content in the backend storages, which is not readable itself, is decoded only if an application obtains corresponding metadata that contains the references of required data shares.

Client-side data deduplication: We ensure efficiency by employing data deduplication, which is a well-known technique to eliminate redundant data by splitting data into smaller pieces (called chunks). One challenge of data deduplication, however, is that it needs to lookup a table of chunks in order to check if the chunk is already stored or not; it becomes inefficient if we store duplicate chunks or chunks that are no longer referred to by any files. We solve this challenge by using a reference count that indicates how many files refer to each chunk. We can then delete chunks with a zero-reference count. If version control is enabled, deleting chunks is unnecessary because those chunks may be used to recover data.

Fault tolerance: Some edge storage nodes may not be available due to system and network failures. Distributed edge storage systems should gracefully handle these failures. Since we use a client-centric architecture, we employ erasure coding at the client to improve system availability even if some storage nodes fail. Unlike other approaches such as Byzantine fault tolerance systems, erasure coding can also guarantee security by preventing each edge storage from reconstructing any meaningful user data.

Elasticity and configurability: Our elastic storage layer allows a client to flexibly store data in multiple storage nodes using the same interface as with local storage. We automatically add and remove storage nodes by using services like storage discovery and attachments, which are separated from file operations to ensure easy scalability to multiple remote storage nodes.

Service discovery: Service discovery is a process for discovering available resources in the cloud-to-things continuum, so that the client can utilize them. We assume a three-tier fog infrastructure service that includes things, the edge, and the cloud. Fog devices in the things layer advertise themselves to users by registering with one or more fog servers in the edge layer, which form a hierarchy of management domains. Clients or fog devices query the fog servers for the list of available resources; if the client cannot find enough fog resources in its own domain, the request is forwarded to the next highest tier, and finally to the cloud resource tier. Clients issue these requests using a resource directory service that is a DNS-like naming service for the relaxed hierarchy of fog resources. We have implemented a simple resource discovery protocol in the fog server that keeps track of fog resources.

12.2.2 Architecture

The continuous resource pool from the cloud to user/"thing" devices provides more elastic resource allocation for diverse application requirements, but it can also require more coordination effort between resource nodes and user devices. Our architecture is designed to integrate these devices so that we can combine their various advantages – for instance, using edge resources to reduce storage latency and cloud resources to provide security.

Our elastic fog storage framework maps data into distributed fog storage nodes, including local physical volumes and (remote) commodity storage nodes. We assume that the commodity storage nodes are object storage services that store data using simple key-value pairs, where the key is the ID of the stored data and the value is the data content. The commodity storage can be cloud (e.g. Amazon S3) or edge devices (e.g. network-attached storage and smartphones). Applications read/write data to these storage locations through an elastic storage layer that provides a POSIX interface:

Data encoding: Files are divided into smaller pieces called chunks for data deduplication and efficient transmission. We slice each file into 4MB chunks, following the client applications of other cloud storage services like Dropbox [2]. Even though the typical chunk size of data deduplication ranges from 8kB to 64kB [3] to maximize deduplication rates, employing small chunks degrades data transmission performance over wide area networks, as in our fog scenario. The chunk is also encoded into shares using erasure coding (e.g. Reed–Solomon codes [4]) for redundancy of data and efficiency of storage space. Clients thus require any t from a total of n created shares to decode their original data, where t and n are configurable parameters.

Metadata: Metadata information indicates how to reconstruct a file, including information on its constituent chunks and their share addresses. Thus, only clients with access to the metadata can properly decode and reconstruct the given file. The metadata also needs to store the chunk reference counts (the number of files referencing each chunk). Reference counters are necessary for deleting chunks that have no referring files, since common chunks can be shared with multiple files. The metadata is stored in a directory (e.g. ~/.meta) as a file for each data file, with the same directory hierarchy. Clients can share the metadata files with each other in order to share access to data files.

Mapping files to objects: Since edge storage nodes may not be trustworthy, we consider the client device to be a trusted storage resource. We therefore store the file metadata in the client storage and assume that the storage is secure and large enough to store the metadata. Users still need to back up the metadata regularly, but the metadata is much smaller than the actual file content. Read/write operations access the file content in the form of chunks. We also leverage local storage space for caching chunks. Frequently required chunks are expected to be stored in local storage. If some chunks are rarely used, we encode these chunks into shares and store them to remote storage locations so that they can be removed from the local storage space.

12.2.3 File Operations

In this section, we explain selected file operations that are supported by our system. We design these operations as an extension of native file operations on metadata files, which allows us to reuse file stat information already implemented in the kernel of the user device.

Readdir: Since we store metadata in the same directory hierarchy with the files, the readdir operation that shows the list of files in the directory is implemented as a hook to the metadata directory.

Getattr: Keeping the same directory hierarchy between metadata files and the data files allows us to reuse the file stat information of metadata files for the actual files. The required file stat data is automatically generated and stored when the metadata file is updated.

Chmod and chown: Like getattr, access control operations are also executed on metadata files.

Open: The file open operation opens the metadata file containing the desired file's chunk information. We reuse the native file system to handle flags and modes on metadata files. We call the native open for metadata files and wrap its file handler with these metadata files, but the actual data content is stored at the places referred to by the metadata.

Read: We implemented a lazy file handler that does not load the file content stored in chunk files until read is called at a corresponding offset. If read is called at the client, we handle the following three cases: first, if the chunk data is already loaded in local memory, we directly read the data from the memory buffer. Second, if the chunk data is not loaded in memory but the chunk is stored in local storage, we load the content of the chunk on memory and then read the data. Third, if the chunk data is not loaded in memory and the chunk is not in local storage, we lookup the metadata file and collect the required shares from the appropriate remote storage locations. We then reconstruct and store the chunk on the client's local device, load the chunk on memory, and continue the read operation.

Write: The write operation is similar to the read operation until the chunk is loaded on memory. If write is called, the input data is written to the chunk data loaded on memory, and we check the dirty flag on this chunk to indicate that it has new data. We do not write the dirty data to storage until fsync is called to synchronize the file.

Fsync: When fsync is called, we write all dirty chunks back to local storage and update the metadata files. Updating the chunk information in the metadata includes updating the reference counters of old and new chunks. To reuse the file stat information, we store the reference counter in st_nlink, which can be controlled by using the link command. Another way to control the st_nlink value is to use the directory's st_nlink. Linux implements the st_nlink of a given directory as the number of subdirectories and files, while Mac OS implements it as the number of sub-directories only. Thus, for either operating system, we can create an empty directory with the name of the referring file path under the corresponding chunk directory; this indicates how many files refer to the chunk. To update the reference counter, we delete the file path directory from the old chunk directory and create path directories to the new chunk directories. If the reference counter is zero, we can delete the chunk, unless version control has been enabled.

Delete: This operation is the most difficult one in a distributed file system, due to the difficulty of synchronizing metadata and maintaining its consistency.

Since we strictly check the chunk reference counters using the native local file system, we easily design a delete operation instead of keeping unused chunks. As with fsync, we visit the chunk directory and delete the corresponding file path if the st_nlink of the chunk directory is zero. We then delete the corresponding metadata file.

12.3 Fault Tolerant Data Access and Share Placement

In this section, we first discuss our data encoding and share placement requirements before considering our handling of share requests. We then present our algorithms to choose the remote storage locations from which a given chunk's shares are downloaded and where they are stored.

12.3.1 Data Encoding and Placement Scheme

A fog file system aims at extending the local storage space, while maintaining the same functionality as a local file system. Data sharing and version control could be achieved by accumulating metadata changes and data content in a distributed cloud environment [5], but that is not the goal of our work here. Distributing storage to fog nodes introduces a different set of challenges: for instance, file deletion is more challenging than version control, since we keep strongly consistent metadata and data content. Our system has a baseline assumption that guarantees at least the same robustness of a single local storage device and a cloud service: we suppose that local storage and at least one cloud storage are always available. We store file metadata in local storage devices, which gives us the required information on how to reconstruct user data. By using the local metadata, the directory and file stat operations do not need to access remote servers, providing protection against their unavailability.

Fog storage systems should consider both data availability and storage efficiency: user data should always be accessible whenever a user wants to access a file. We encode the file data using erasure coding (e.g. Reed–Solomon codes), which spreads the data into n shares and requires any t of them to reconstruct the data. Each share is stored at a separate storage location. Figure 12.2 illustrates how to disperse data by leveraging edge/thing resources and the cloud. To protect user privacy, edge/thing resources should not be able to access user data even if they collude, so we store at most $t - 1$ encoded shares in edge/thing storage locations. Thus, these devices cannot reconstruct the user data from their combined shares. We store t shares in the cloud, where they are always accessible so that users can always use cloud shares for the data reconstruction if edge or thing devices are unavailable. To reconstruct the data, the user device should collect as many shares as required from the edge/things and then collect the rest of the required shares from the cloud. Optionally, a client device

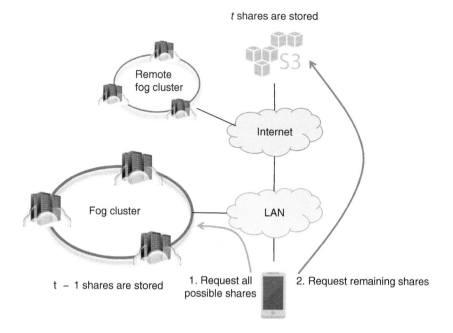

Figure 12.2 Robust yet efficient share requests.

can keep one additional share in local storage, which allows the client device to reconstruct data using shares in edge/things only. We discuss the specific choice of different edge/things below.

12.3.2 Robust and Exact Share Requests

Since edge/thing devices are not always available, depending on the locations of users, resource status, network policy, etc., we now consider how to send requests for the required shares. The baseline solution in previous studies is based on a Byzantine quorum system, which sends requests to all available receivers and takes only correct responses when some receivers' behavior is not trustable. If we apply this solution to our storage scenario, the client should send requests to all storage nodes and take the fastest responses. However, this is not a scalable solution since it wastes network bandwidth from $n - t$ slow requests. Thus, it is necessary to develop an efficient request mechanism that sends the minimum number of requests after detecting request failures.

Figure 12.3 shows a snapshot of our achieved throughput with four edge nodes (i.e. devices) and three cloud nodes. We distributed the data shares evenly to the storage nodes simultaneously without modifying the TCP/IP layer, as used in the current Internet. If the edge nodes are in the same local

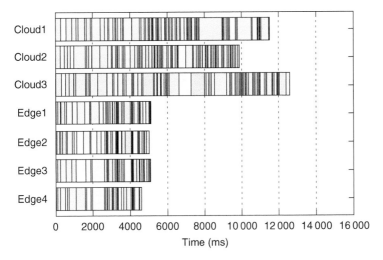

Cloud1

Cloud2

Cloud3

Edge1

Edge2

Edge3

Edge4

0   2000   4000   6000   8000   10 000   12 000   14 000   16 000

Time (ms)

Figure 12.3 Snapshot of throughput while transmitting data pieces evenly into four edges and three clouds simultaneously.

area network (LAN) with the client (the case of Figure 12.3), cloud nodes beyond the Internet junction link become the connection bottleneck. In other words, all of the edge nodes finish receiving the data shares before the client finishes transmitting its data to the cloud. This observation leads us to an *edge first* strategy that is a simple and effective way to decide the order of requests to collect required data shares from edge and clouds. To collect t shares, the client requests $t - d$ shares from edge devices and d remaining shares from the cloud.

12.3.3 Clustering Storage Nodes

Even though edge resources are connected to wide area networks, which have a random topology of end-to-end connections among clients and storage locations, it is necessary to identify the logical distance between storage locations in order to apply data replication schemes that guard against concurrent storage failures. In other words, we are able to achieve properties of existing distributed file storage systems if we extract and utilize datacenter-like storage hierarchy patterns from the user. For example, HDFS stores a replica at a node in the same rack and another replica at a server in a different rack in order to minimize failure dependency [6].

We propose a method to cluster storage locations using simple latency measures. Each user device can have a different set of clusters depending on its latency to the different storage nodes. Our challenge is to infer these clusters for each user without any direct measurements between storage nodes. The clustering task aims to find storage nodes with significant latency differences, using

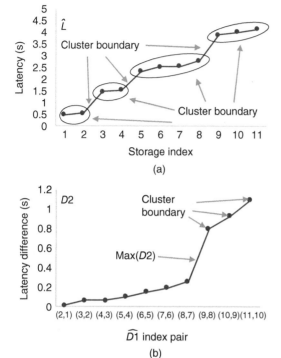

Figure 12.4 Clustering example of 11 locations of AWS S3 using measurements of single client. (a) Clustering using sorted latency. (b) Consecutive differences of the sorted latency.

these to define the node clusters. We first find the average latency of the user to each storage device and sort these latencies into a list $\hat{L}$ in ascending order as illustrated in Figure 12.4a. We then calculate the consecutive differences in latencies $D1$ using the sorted latency $\hat{L}$. We then sort $D1$, which is $\widehat{D1}$, and calculate the consecutive differences of this sorted list, denoting the resulting list of second differences as $D2$. We use the maximum $D2$ as a clustering threshold: the cluster boundaries of the sorted list of latencies $\hat{L}$ are then determined by all differences $D1$ whose consecutive differences $D2$ lie above the threshold (Figure 12.4b). In other words, whenever two storage nodes have a sufficiently large difference in latencies, we put them into different clusters.

12.3.4 Storage Selection

We now turn to the problem of selecting the remote storage locations from which clients should download shares, and the related problem of selecting the edge and thing devices at which shares should be stored. Our objective in developing these algorithms is to use these near-user storage locations to minimize the file download times, while also leveraging shares stored at the cloud to compensate for the edge/thing storage locations' possible unavailability. We must also consider that multiple users may store data at the same remote storage locations, forcing us to balance these users' shares among the available locations. Since the upload process is transparent to the user, we do not need to optimize over the upload latency. Thus, we first consider the file download times, given a set of remote share locations, before giving an algorithm to optimize the different share locations.

12.3.4.1 File Download Times

Given a file to download, users must download shares of each constituent chunk. We can therefore find the file download time by computing the download time of each constituent chunk; note that shares of different chunks are downloaded in parallel. Since the set of constituent chunks is fixed for a given file, minimizing the file download time is then equivalent to minimizing the chunk download times.

We suppose that each user u records the latency of downloading a 1kB file from the cloud, which we denote as l_u^c. Each user also records the current 1kB download latency to each edge/thing storage location, which we denote in a vector l_u; each storage location is indexed by $e = 1, 2, \ldots, E$. Note that since each user has a limited number of pipeline requests, the shares are generally downloaded in sequence in order to ensure request consistency; thus, we can assume that the total latency for downloading a share is the sum of the latencies to the chosen locations. A baseline algorithm would always select t shares from the cloud, for a total download latency of tl_u^c, multiplied by the share size in kB. Note that all shares of a given chunk are the same size.

To improve on this baseline algorithm, we use a simple greedy algorithm that takes advantage of the edge storage locations. We let f denote the set of remote edge/thing storage locations where shares of a given file f are located, and l_u^f the vector of current latencies to each of these locations; we also suppose that l_u^f is ordered, so that $l_u^f(1) \leq l_u^f(2) \leq \ldots \leq l_u^f(t-1)$. The user then chooses the t (or $t - 1$ if the file has a share in local storage) storage locations with lowest latencies, including the cloud latency l_u^c. Mathematically, let d denote the maximum index for which $l_u^c \leq l_u^f(d)$; if no such index d exists (i.e. all edge storage locations have higher latency than the cloud), then we take $d = 0$. The user then downloads d shares from the edge locations 1, 2, ..., d and $t - d$ shares from the cloud. Note that this algorithm guarantees lower download times than only downloading shares from the cloud.

12.3.4.2 Optimizing Share Locations

Given the aforementioned strategy for downloading file shares, the user wishes to choose the upload storage locations so as to (i) choose the locations with lowest network latency, and (ii) distribute shares of different chunks to different edge/thing storage locations, so that no single device is overwhelmed by requests from multiple users. Edge/thing devices are particularly likely to be overwhelmed by multiple requests, since they may have limited I/O capacity. Yet satisfying either of these objectives perfectly is difficult due to uncertainty in both the future network latencies and requests coming from other users. Both of these factors are hard to predict: both users and the edge/thing storage locations can be mobile, leading to high latency and thus user request variability, and the storage nodes themselves can enter or exit the system at any time, which makes prediction even more difficult. Thus, instead of pursuing online learning approaches [7, 8], which can asymptotically optimize the download time but make no guarantees on transient performance, we consider simpler heuristics in this chapter that are robust to user and storage variability.

We suppose that each user u knows the probability distribution π_u^e of its network latency l_u^e to each edge/thing storage location e, which can be computed from the time history of the latencies to this storage location. Even knowing the π_u^e, however, it is difficult to find the distribution of the file download times if $l_u^e > l_u^c$ with nonzero probability: then the number of locations that we use to download shares, d, is itself a random variable that depends on the realizations of all the edge/thing latencies. However, in many cases these locations (when available) exhibit lower latency than the cloud, as shown in Figure 12.6. Thus, we first suppose that users always download shares from all $t-1$ edge/thing locations and derive the resulting optimal share locations; this upload strategy can also be applied to cases where fewer than $t-1$ edge/thing locations are used, though it may not be optimal in these scenarios.

If service delays due to multiple user requests at a given storage location are not a concern, then a user need only contend with uncertain future latencies. Since the distributions of these latencies are known, however, users can simply choose the locations that minimize a function of these distributions. Some users, for instance, might want to minimize the expected file download time, i.e. choosing the locations λ_f that minimize $\mathbb{E}\left[\sum_{i=1}^{t-1} l_u^f(i) + l_u^c\right] = \mathbb{E}\left[\sum_{i=1}^{t-1} l_u^f(i)\right] + l_u^c$. The user would then choose the $t-1$ locations with the lowest expected latencies $\mathbb{E}[l_u^e]$. Other, more risk-averse users might care more about the worst-case latencies, choosing the $t-1$ locations with the lowest $\max l_u^e$. Either of these objectives can be achieved by ranking the edge/thing storage locations according to an individual user's preferences – in the above examples, by expected or worst-case latencies – and choosing the $t-1$ best locations according to this ranking. Note that the ranking algorithm can be applied regardless of the shape of the π_u^e distributions, and in particular regardless of whether edge storage locations can have larger latencies than cloud storage locations.

We now modify the ranking algorithm to ensure that the share locations are spread out among the users. We first note that different users will have different latency distributions $\pi_{u'}^e$. In fact, different users may have access to different subsets of the edge/thing storage locations. If the latencies are sufficiently distinct

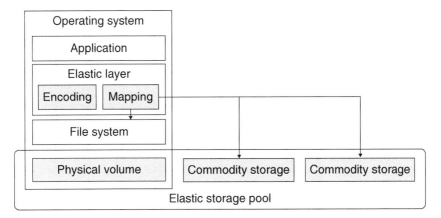

Figure 12.5 Implementation overview.

for different users, these users would employ different locations for their files, and the ranking algorithm above would itself ensure that shares for different users are sufficiently well distributed among the different locations. However, since an individual user would not know the distribution of latencies at other users, we cannot take this automatic distribution for granted. Instead, we propose to first cluster the edge storage locations as in Figure 12.2, and then rank the clusters according to their overall latency distributions. As mentioned earlier, users could choose to rank the clusters by expected latency, worst-case latency, etc. given their own preferences. They would then choose to store the shares in the $t - 1$ highest-ranked clusters, selecting the specific storage location uniformly at random within each cluster.

Clustering the storage locations by their latency measurements ensures that users do not only store their shares at the storage locations with the lowest latencies: these locations would fall into the same clusters, so only one of them could be used for each file. However, if different users have similar clusters, as in Figure 12.5a's experimental results, then the same storage locations would still receive more file shares than the rest. The locations with lowest latency would have a particularly large number of shares, since they would always be part of the clusters with the lowest latencies, and thus would often be chosen to store file shares. Yet these locations are exactly those that can handle a larger number of requests, since their lower network latency could compensate for larger queuing delays.

12.4 Implementation

We implemented a prototype, a standalone client application, for Linux and OSX using *LibFUSE*, which allows us to implement a file system in user space.

Our source code is written in Python with the *fusepy* package. Since we keep our prototype as simple and stable as possible, approximately 2700 lines of the client were written in Python, and 700 lines of the server were written in Golang. We implemented all FUSE operations except for readlink, mknod, link, symlink, and utime, which are not needed for our file operations. Our prototype works perfectly with advanced applications on an OSX GUI environment where unknown temporary files are frequently created, truncated, updated, renamed, and deleted for updating files.

Our prototype provides an *elastic layer* that encapsulates modules for file encoding and mapping shares to storage nodes (Figure 12.5). Our prototype mounts a virtual directory (e.g. /Fog) that extends the local storage space into multiple edge/cloud storages without disturbing users. When a user application places a file system call from the virtual directory, our prototype implements the actual logic of the corresponding event emitted by *LibFUSE*. To upload chunks to remote storage nodes, chunks are encoded to shares using the *Zfec* library [9], an implementation of a fast erasure codec. When a file is stored in the virtual directory, the actual file content is stored in edge/cloud storage nodes, while its metadata is stored locally at the device. The location where the file is stored is encapsulated from the user, but the interface to the user is the same as with the native file system.

Figure 12.6 shows an example of how to store three files: a "hello world" text file, an "image.jpg" image file, and a "video.mp4" video file. A user can list the files at /Fog, which can be directly obtainable using metadata in the user's local storage. When writing a file, the *elastic layer* splits the file into multiple chunks. Each chunk is encoded into n shares that have at least n candidate storage locations. To minimize the delay caused by blocking on storage interface threads until the previous upload request completes, we define one encoding thread that pushes share requests to a request queue of the corresponding storage interface thread. The storage interface thread keeps monitoring the request queue and executing requests in the queue.

12.4.1 Metadata

To avoid an exhaustive implementation effort for the file directory system, we reuse the operating system's native directory system to maintain the directory structure and file stat information. Since metadata and files in fog storage have a one-to-one map, we hook directory operations in fog storage to the metadata path, where metadata are stored with the same directory structure as the fog files. We wrapped a file handler whose file descriptor indicates the metadata file, but the actual data content is stored at the places referred to by the metadata. The file path, stat, permission control, and locking are automatically managed by the native file system for metadata file operations.

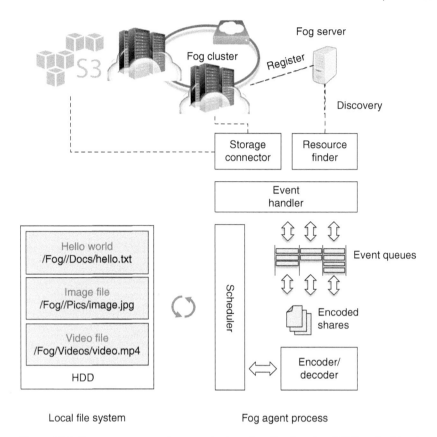

Figure 12.6 Fog agent process that encodes/transfers files from local HDD to fog/cloud and vice versa.

We thus implemented a completely transparent file directory system without database service.

12.4.2 Access Counting

Counting the number of accesses to each chunk is necessary to decide when to upload which chunks to remote storage locations. We also reuse st_mtime and st_atime of the local chunk files, where st_mtime indicates the last modification time and st_mtime − st_atime indicates how many times the chunk is read from the last check point. Since we override the st_atime value for access counting, we present st_mtime as st_atime for the file stat operation. The access count is reset to zero when the chunk is written or downloaded from remote storages. If the access count is lower than a predefined threshold and the chunk has not been assessed for a predefined period of time, we schedule the chunk to be uploaded into the remote storage locations.

12.4.3 NAT Traversal

In a fog environment, NAT is a common technique to configure LANs with user or enterprise fog resources. Fog servers maintain peer clustering to discover resources horizontally, and bidirectional session handling is necessary among fog servers as well as between the fog server and resource/client. Without a specific configuration, such as port forwarding, it is not possible to know the internal address of a given destination from outside the NAT equipment. We use the WebSocket protocol [10] to provide bidirectional communication among fog nodes. There are two ways to establish a communication session. First, a node listens for an advertisement message from the fog server containing address information. The advertisement message is broadcasted to the subnet of each network interface. Using the address information in the advertisement, the fog node initiates a WebSocket session establishment. Second, a fog node can actively join a fog server network using known address information. Regardless of whether NAT is present, the WebSocket session guarantees bidirectional communication through its light-weight protocol. WebSocket keeps the bidirectional communication session open until ping-pong messages are not successfully exchanged or one node explicitly closes the session. Fog nodes are also able to notice join and leave events immediately.

12.5 Evaluation

In this section, we evaluate the performance of our fog system using a cloud-only solution as the baseline. We consider both file download and upload times, as well as web page loading times when the web server uses different types of storage. We configure one Amazon Web Services (AWS) S3 storage, four (S)FTP servers in a campus network as edge resources, and one (S)FTP server on a Raspberry Pi 3 as a "thing" resource near our client device. The client device is a MacBook Pro from early 2015, equipped with a 2.9GHz Intel i5 core and 16GB DDR3 memory. We also access the storage resources from an off-campus residence with the same client device.

Figure 12.7a shows the clustering results of 11 AWS S3 storage nodes in the United States, Europe, and Asia. Clustering data center nodes is even more challenging than clustering edge nodes, because of their similar latencies when the nodes are located near backbone networks. We performed latency measurements from three different client locations without any prior information on the S3 locations. From Figure 12.7b's results, we found two cluster boundary points for location 1 and three each for locations 2 and 3. Thus, the client at

Figure 12.7 Clustering 11 nodes of AWS S3 from three different client locations. (a) Clustering AWS S3 from three locations. (b) Consecutive differences of (a).

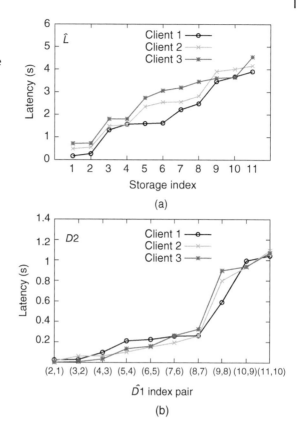

(a)

(b)

location 1 sees three clusters, and those at locations 2 and 3 divide the storage nodes into four clusters.

Figure 12.8 shows the quantiles and averages of the access latency to each storage resource, taken over 30 measurements. We define the latency of a given storage location as the completion time of writing 1kB data to the storage node and then deleting it. We observed that the edge storage latency is much more sensitive to the client location than that of the cloud: when on campus, the edge storage locations are connected to the client over the local campus network, leading to much lower latencies as they do not pass through a campus bottleneck link to the Internet. The residence area is approximately 7.5 km away from campus, which increases the edge storage latency more than 50 ms, while the average cloud latency barely changes. We also measured the latency between the edge resources and a remote AWS node, which is even higher than the cloud latency. Thus, accessing data from remote edge/things can be slower than accessing data from the cloud if the edge resources are far away from the client; in our approach, we only leverage edge/things storage devices in the closest

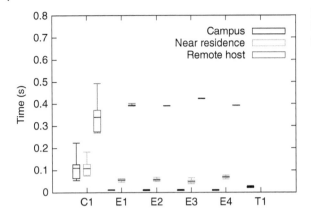

Figure 12.8 Storage latency measured for writing 1kB of data.

cluster to account for this possibility. The cloud's performance variation is much larger than the edge locations' due to the longer path to cloud servers through the Internet, which can be influenced by Internet traffic congestion and various queueing delays. The Raspberry Pi 3 also showed slower speeds than the edge servers because it is connected to the user client through a Wi-Fi network and its SD card has poor I/O performance. We do not report the Raspberry Pi latency from the residential location since off-campus clients cannot reach it, unless they have the mapping information between the device's public and private addresses.

Figure 12.9 shows the volume of transmitted data between the client and storage locations while changing the storage selection algorithm and the t and n parameters. We consider two different selection algorithms: edge-first, in which we prioritize downloading shares from edge locations; and random, in which we choose the locations uniformly at random. We store $n = 2t - 1$ shares (t shares at the cloud and $t - 1$ shares at edge nodes), so one share must be downloaded from the cloud in the edge-first approach. Thus, the ratio of data downloaded from the cloud to that downloaded from the edge is $1/(t - 1)$. For the random selection algorithm, the ratio of data downloaded follows the number of cloud and edge shares, and is $t/(t - 1)$. In our experiment, we download a 100MB file with a 4MB chunk size, and Figure 12.9 shows that the traffic volumes conform to these ratios.

Figure 12.10 shows the upload completion times from the campus locations with different t and n configurations. Note that we store t shares in the cloud, which together have the same size as the original data. Each of the t shares is stored in a different bucket but in the same data center. Regardless of t and n, the upload completion time is determined by the slowest cloud storage location. Since the bandwidth to this cloud storage does not saturate the client's link capacity, we can parallelize share uploads to the edge nodes with the upload to the slower cloud node. Previous studies using erasure coding for distributed

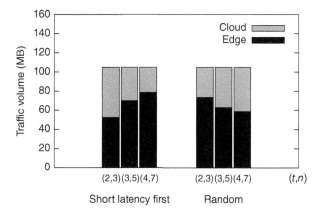

Figure 12.9 Traffic volume from cloud and edge storage locations.

Figure 12.10 Upload completion time.

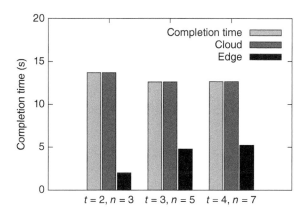

storage, such as DepSky [11] and CYRUS [5], showed that the upload completion time depends on the total size of data to upload: $S = B * n/t$ where B is
the original data size, n is the number of encoded shares, and t is the number
of shares required for decoding. Storing $t - 1$ shares at edge locations does not
interfere with data sessions to cloud storage locations since edge traffic is only
processed in the local network.

Figure 12.11 shows the download completion times for $(t, n) = (3, 5)$ for one
file while changing the file size and storage selection algorithm. We consider
both edge-first and random selection, as well as cloud-only downloads. Due to
the fixed 4MB chunk size, the download completion times for all three algorithms exhibit large increases at 4MB file size increments. We do not observe a
significant difference in completion times between the edge-first and random
selection algorithms. However, cloud-only selection consistently takes longer
than edge-first and random selection. This is consistent with the cloud latency

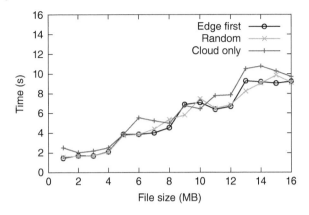

Figure 12.11 Download time while changing file size.

being much higher than edge latencies. Since we have four edge and one cloud storage locations, the random algorithm would tend to select edge storage locations over cloud ones, leading to similar download latencies as with edge-first selection.

Figure 12.12 shows more details of the download performance for different t and n values. We downloaded 100MB of data with a 4MB chunk size from campus, where edge resources are connected through LAN, and from the residential area, where the same campus edge resources are connected through a commercial Internet service provider's link from the client. Since the latency between the client and edge resources is short on the campus network (<20 ms in Figure 12.8), edge-first selection yields much shorter download times on campus. We observed that the campus's Internet junction link is the bottleneck link connecting to the cloud storage locations: with the cloud-only storage selection, establishing more cloud connections as t increases degrades the download performance. At the residential location, however, we observe the opposite result. The Internet junction link of the campus network now bottlenecks connections to the edge resources, so downloading more shares from the campus edge leads to slower download speeds. Simultaneous share download from the cloud lowers the total download times due to parallelizing the data downloaded from multiple cloud locations. The random selection algorithm, as we would expect, consistently performs in between the edge-first and cloud-only algorithms.

Figure 12.13a shows the download completion times when different numbers of edge nodes fail, again using a 100MB file. Since we change the number of failure nodes from 0 to 3, we used the encoding parameters $t = 4$ and $n = 7$. At both the campus and residence locations, the download times increase by <24%, even when $t - 1 = 3$ edge nodes fail. To emulate failures, we observe a thrown exception event when (S)FTP service is turned off while transmitting a file. After two successful data transmissions from a given edge node,

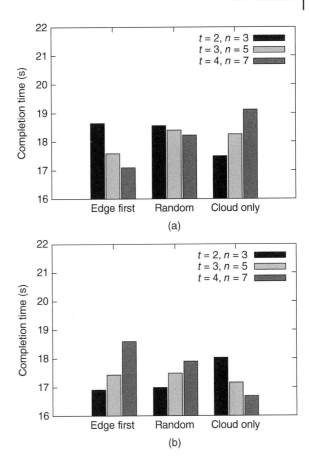

Figure 12.12 Download completion time while changing *t* and *n*. (a) Campus. (b) Residence.

we emulated the exception event, a Broken Pipe, from the connector module. Because there is no timeout until the Broken Pipe exception occurs (this exception is immediately thrown from the client when data transmission fails), the client sends a request message to the next target node if a failure is detected. If a hang request throws the Request Timeout exception, we mark fail2 on the corresponding storage connector to prune successive requests to the storage node. Most failures are immediately detectable at the client as client-side exceptions, so the performance degradation is made negligible by requesting the required data from the next target storage immediately after a failure detection. As with Figure 12.11's download completion times, we observe a step function-like trend at 4MB intervals as the file size increases (Figure 12.13b), but the completion times do not increase much with the number of node failures, especially when the file size is small.

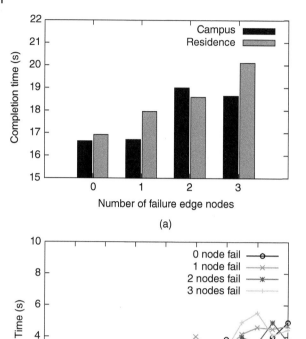

Figure 12.13 Download completion times for different (a) cloud selection algorithms and (b) numbers of failed edge nodes. (a) Download times with edge failures. (b) Download time with different file sizes.

12.6   Discussion and Open Questions

Two challenges of deploying an elastic fog storage system are further improving the system performance and handling device mobility. We discuss our proposed solutions here.

An ongoing trial using our system for data backup for university servers and equipment showed an approximate writing speed of 150MB/s, with a 10Gbps network link, four network-attached storage nodes, and 33% parity overhead, which is slightly slower than 7200 rpm HDD [12]. Storing shares in edge/thing devices with memory-based high-performance storage (e.g. Memcached [13, 14]) will realize true networked storage with fog environment. Thus, if we attach high-performance edge storage nodes through a high-speed network, our system I/O performance can be close to that of a local storage device. Overcoming bottleneck links at wireless hops, however, is still challenging.

Routing traffic for mobile devices is out-of-scope of this chapter, as we focus on end-to-end session management between clients and storage nodes. However, mobility support not only for clients but also for storage nodes (e.g. things with mobility and migration of storage) is an important requirement. One practical solution to maintain this communication with device mobility is to separate the device identifier and locator. Since the identifier stays fixed while the locator (or address) changes with mobile clients, end-to-end communication using device identifiers instead of addresses is an ideal solution. Related works in networking are making such a solution increasingly possible. For instance, Locator and Identifier Separation Protocol (LISP) allows data transmission using a host ID with an address whose map is managed by a DNS-like ID/address translation service [15]. Moving to IPv6 is also one possible solution for handling device mobility.

12.7 Related Work

Many traditional network file systems (NFSs) have been studied since the 1980s, and some of them are still being used. The NFS [16] is a distributed file system that allows clients to access files stored on remote servers as if files are located in the local file system. The Andrew file system (AFS) [17] is a distributed file system that uses a set of trusted servers across several networks to share a homogeneous, location-transparent file name space for the files stored on the servers.

Several related works have proposed integrating storage space at multiple cloud providers, which introduces a similar problem of coordinating file storage across multiple locations, though it does not raise as many dependability questions since edge nodes are significantly less reliable than cloud locations. Other works have built distributed file systems such as HDFS, Dynamo, and Ceph; however, these systems generally distribute files among multiple locations within a single cloud datacenter, and can, therefore, depend on reliable network connectivity and location availability to a much greater degree than is possible in fog settings [6, 18–22].

Byzantine fault-tolerance systems, among others, have been studied for reliable distributed storage [23–27]. These works proposed protocols that can read and write with unreliable cloud service providers. Realizing these protocols, however, requires running servers at the cloud providers, which is not generally possible at edge storage locations. Similarly, object-based distributed storage systems [22, 28] also require running code at the servers to handle multiple transactions and synchronization. In these systems, user data are split into smaller objects and distributed on multiple object storage nodes. To reconstruct files, the systems maintain metadata, including the required objects and their storage locations.

Another approach to integrating multiple cloud providers is to use proxy servers [29–32]. The proxy can scatter and gather user data to and from multiple providers, providing transparent access for users. Yet while the proxy server is a data sharing point among multiple clients, allowing greater deduplication efficiency, it is also a single point of failure. Cloud integration from the client has been proposed in [5, 11, 33–35]. While some of these works include user customization [5], they assume relatively stable latencies between the client and cloud servers, and infrequent cloud arrivals and departures. Edge nodes, however, may have highly variable latency and unpredictable availability, requiring more predictable strategies for handling file metadata and storing shares at these different locations.

12.8 Conclusion

In this work, we develop an elastic fog storage system that seamlessly extends local file storage to multiple remote storage locations. Unlike traditional cloud-based storage extensions, we integrate remote storage resources along the "cloud-to-things continuum," which allows us to take advantage of edge and thing resources' lower latency to the client, while leveraging cloud resources to maintain reliable access to users' data files. We implement a prototype of this system that makes minimal changes to existing file systems, e.g. reusing existing file directory structures to avoid excessive overhead from metadata management. Our prototype uses data deduplication and erasure coding to ensure high availability of file access and to handle possible unavailability at some edge storage resources. Compared to a cloud-only storage solution, our prototype achieves lower file download times and is robust to failures at edge nodes.

Acknowledgments

This work was supported by NSF grant CNS-1525435.

References

1 OpenFog Consortium Architecture Working Group. (2016). OpenFog architecture overview, White Paper. https://www.openfogconsortium.org/wp-content/uploads/OpenFog-Architecture-Overview-WP-2-2016.pdf (accessed 1 September 2018).

2 Drago, I., Mellia, M., Munafo, M.M. et al. (2012). Inside dropbox: understanding personal cloud storage services. In: *Proceedings of the 2012 ACM*

Conference on Internet Measurement Conference, ser. IMC'12, 481–494. New York, NY: ACM http://doi.acm.org/10.1145/2398776.2398827.

3 Meyer, D.T. and Bolosky, W.J. (2011). A study of practical deduplication. In: *Proceedings of the 9th USENIX Conference on File and Storage Technologies, ser. FAST'11*, 1. Berkeley, CA: USENIX Association http://dl.acm.org/citation.cfm?id=1960475.1960476.

4 Reed, I.S. and Solomon, G. (1960). Polynomial codes over certain finite fields. *Journal of the Society for Industrial and Applied Mathematics* 8 (2): 300–304. http://www.jstor.org/stable/2098968.

5 Chung, J.Y., Joe-Wong, C., Ha, S. et al. (2015). CYRUS: towards client-defined cloud storage. In: *Proceedings of the 10th European Conference on Computer Systems, ser. EuroSys'15*, 17:1–17:16. New York, NY: ACM http://doi.acm.org/10.1145/2741948.2741951.

6 Shvachko, K., Kuang, H., Radia, S., and Chansler, R. (2010). The hadoop distributed file system. In: *Proceedings of the 2010 IEEE 26th Symposium on Mass Storage Systems and Technologies (MSST), ser. MSST'10*, 1–10. Washington, DC: IEEE Computer Society http://dx.doi.org/10.1109/MSST.2010.5496972.

7 J. Vermorel and M. Mohri, "Multi-armed bandit algorithms and empirical evaluation," in *European Conference on Machine Learning*. Springer Porto, Portugal, 2005, pp. 437–448.

8 Le, T., Szepesvari, C., and Zheng, R. (2014). Sequential learning for multi-channel wireless network monitoring with channel switching costs. *IEEE Transactions on Signal Processing* 62 (22): 5919–5929.

9 Wilcox-O'Hearn, Z. (2012). zfec package 1.4.24. https://pypi.python.org/pypi/zfec (accessed 1 September 2019).

10 I. Fette and A. Melnikov. (2011). The WebSocket protocol. RFC 6455.

11 Bessani, A., Correia, M., Quaresma, B. et al. (2011). DepSky: dependable and secure storage in a cloud-of-clouds. In: *EuroSys'11 Proceedings of the Sixth Conference on Computer Systems*, 31–46. New York, NY: ACM http://doi.acm.org/10.1145/1966445.1966449.

12 UserBenchmark. (2016). December 2016 HDD rankings. http://hdd.userbenchmark.com.

13 Fitzpatrick, B. (2004). Distributed caching with memcached. *Linux Journal* 2004 (124): 5. http://dl.acm.org/citation.cfm?id=1012889.1012894.

14 Ousterhout, J., Agrawal, P., Erickson, D. et al. (2010). The case for RAM-clouds: scalable high-performance storage entirely in DRAM. *ACM SIGOPS Operating Systems Review* 43 (4): 92–105. http://doi.acm.org/10.1145/1713254.1713276.

15 Farinacci, D., Fuller, V., Meyer, D., and Lewis, D. (2013). The locator/ID separation protocol (LISP). RFC 6830.

16 Sandberg, R., Goldberg, D., Kleiman, S. et al. (1985). Design and implementation of the Sun network filesystem. In: *Proceedings of the Summer USENIX Conference*, 119–130. Berkeley, California: USENIX.

17 Howard, J.H., Kazar, M.L., Menees, S.G. et al. (1988). Scale and performance in a distributed file system. *ACM Transactions on Computer Systems (TOCS)* 6 (1): 51–81.

18 Rao, J., Shekita, E.J., and Tata, S. (2011). Using Paxos to build a scalable, consistent, and highly available datastore. *The Proceedings of the VLDB Endowment* 4 (4): 243–254. http://dx.doi.org/10.14778/1938545.1938549.

19 DeCandia, G., Hastorun, D., Jampani, M. et al. (2007). Dynamo: Amazon's highly available key-value store. In: *Proceedings of Twenty-First ACM SIGOPS Symposium on Operating Systems Principles, ser. SOSP'07*, 205–220. New York, NY: ACM http://doi.acm.org/10.1145/1294261.1294281.

20 Chang, F., Dean, J., Ghemawat, S. et al. (2008). Bigtable: a distributed storage system for structured data. *ACM Transactions on Computer Systems* 26 (2): 4:1–4:26. http://doi.acm.org/10.1145/1365815.1365816.

21 Apache Software Foundation (2014). HBase. http://hbase.apache.org (accessed 1 September 2019).

22 Weil, S.A., Brandt, S.A., Miller, E.L. et al. (2006). Ceph: a scalable, high-performance distributed file system. In: *OSDI'06 Proceedings of the 7th Symposium on Operating systems design and implementation*, 307–320. Berkeley, CA: USENIX Association http://dl.acm.org/citation.cfm?id=1298455.1298485.

23 Cachin, C. and Tessaro, S. (2006). Optimal resilience for erasure-coded Byzantine distributed storage. In: *Proceedings of International Conference on Dependable Systems and Networks (DSN'06)*, 115–124. Washington, DC: IEEE Computer Society http://dx.doi.org/10.1109/DSN.2006.56.

24 Goodson, G.R., Wylie, J.J., Ganger, G.R., and Reiter, M.K. (2004). Efficient Byzantine-tolerant erasure-coded storage. In: *DSN'04 Proceedings of the 2004 International Conference on Dependable Systems and Networks*, 135. Washington, DC: IEEE Computer Society http://dl.acm.org/citation.cfm?id=1009382.1009729.

25 Malkhi, D. and Reiter, M. (1997). Byzantine quorum systems. In: *STOC'97 Proceedings of the Twenty-Ninth Annual ACM Symposium on Theory of Computing*, 569–578. New York, NY: ACM http://doi.acm.org/10.1145/258533.258650.

26 Malkhi, D. and Reiter, M.K. (1998). Secure and scalable replication in Phalanx. In: *SRDS'98 Proceedings of the The 17th IEEE Symposium on Reliable Distributed Systems*, 51. Washington, DC: IEEE Computer Society http://dl.acm.org/citation.cfm?id=829523.831001.

27 Martin, J.-P., Alvisi, L., and Dahlin, M. (2002). Minimal Byzantine storage. In: *DISC'02 Proceedings of the 16th International Conference on Distributed*

Computing, 311–325. London, UK: Springer-Verlag http://dl.acm.org/citation.cfm?id=645959.676126.

28 Bowers, K.D., Juels, A., and Oprea, A. (2009). HAIL: a high-availability and integrity layer for cloud storage. In: *CCS'09 Proceedings of the 16th ACM Conference on Computer and Communications Security*, 187–198. New York, NY: ACM http://doi.acm.org/10.1145/1653662.1653686.

29 Abu-Libdeh, H., Princehouse, L., and Weatherspoon, H. (2010). RACS: a case for cloud storage diversity. In: *SoCC'10 Proceedings of the 1st ACM Symposium on Cloud Computing*, 229–240. New York, NY: ACM http://doi.acm.org/10.1145/1807128.1807165.

30 Hu, Y., Chen, H.C.H., Lee, P.P.C., and Tang, Y. (2012). NCCloud: applying network coding for the storage repair in a cloud-of-clouds. In: *FAST'12 Proceedings of the 10th USENIX Conference on File and Storage Technologies*, 21. Berkeley, CA: USENIX Association http://dl.acm.org/citation.cfm?id=2208461.2208482.

31 Papaioannou, T.G., Bonvin, N., and Aberer, K. (2012). Scalia: an adaptive scheme for efficient multi-cloud storage. In: *SC'12 Proceedings of the International Conference on High Performance Computing, Networking, Storage and Analysis*, 20:1–20:10. Los Alamitos, CA: IEEE Computer Society Press http://dl.acm.org/citation.cfm?id=2388996.2389024.

32 Resch, J.K. and Plank, J.S. (2011). AONT-RS: blending security and performance in dispersed storage systems. In: *FAST'11 Proceedings of the 9th USENIX Conference on File and Storage Technologies*, 14. Berkeley, CA: USENIX Association http://dl.acm.org/citation.cfm?id=1960475.1960489.

33 Ling, C.W. and Datta, A. (2014). InterCloud RAIDer: a do-it-yourself multi-cloud private data backup system. In: *Distributed Computing and Networking* (eds. M. Chatterjee, J. Cao, K. Kothapalli and S. Rajsbaum), 453–468. Berlin, Heidelberg: Springer.

34 Machado, G., Bocek, T., Ammann, M., and Stiller, B. (2013). A cloud storage overlay to aggregate heterogeneous cloud services. In: *38th Annual IEEE Conference on Local Computer Networks*, 597–605. New York City, NY: IEEE.

35 Mu, S., Chen, K., Gao, P. et al. (2012). μLibCloud: providing high available and uniform accessing to multiple cloud storages. In: *2012 ACM/IEEE 13th International Conference on Grid Computing*, 201–208. Washington, DC: IEEE Computer Society http://dx.doi.org/10.1109/Grid.2012.28.

13

Development of Wearable Services with Edge Devices

Yuan-Yao Shih[1], Ai-Chun Pang[2], and Yuan-Yao Lou[2]

[1] *National Chung Cheng University, Department of Communications Engineering, Taipei City, Taiwan*
[2] *National Taiwan University, Graduate Institute of Networking and Multimedia and Department of Computer Science and Information Engineering, Taipei City, Taiwan*

13.1 Introduction

Wearable devices bring a new experience, which is more intuitive and closer to the living habits, of using technology to ordinary people. That explains why wearable devices grow explosively recently. In response to this trend, lots of major manufacturers, such as Google, Apple, Samsung, and FitBit, make considerable investments in the production of wearable devices. Cisco estimates that there will be 929 million wearable devices globally, growing nearly threefold from 325 million in 2016 [1]. Also, CCS Insight indicates that market for wearable technology will double over the next four years [2]. Those wear devices, including watches, wristbands, hearables, and eye-wear, are designed to be lightweight and power-saving; thus, their computing capability is limited. To eliminate the limitation on computing, it relies on a local-hub to do computing tasks for wearable devices.

The local-hub, or called coordinator, is mostly a smartphone that processes service requests transferred from wearable devices. The prerequisite of building the connection with the local-hub, wearable devices need to pair with the local-hub via Bluetooth low energy (BLE). After successful pairing, wearable devices can communicate with the local-hub through a specific application like Android Wear for Android installed on the local-hub [3]. By using the existing technology, wearable devices can connect to their local-hub, via BLE or Wi-Fi interface, depending on the different situations. When they are connected via BLE, two devices must be near to each other. Once they are not within the BLE networking coverage, wearable devices can only connect to the local-hub via Wi-Fi interface. This situation happens very often; for example, when people go jogging or swimming, they do not want to carry too many things. However, users still want to catch any incoming events and use the functions of

Fog and Fogonomics: Challenges and Practices of Fog Computing, Communication, Networking, Strategy, and Economics, First Edition. Edited by Yang Yang, Jianwei Huang, Tao Zhang, and Joe Weinman.
© 2020 John Wiley & Sons, Inc. Published 2020 by John Wiley & Sons, Inc.

the device with wearing the wearable device only. To address this shortcoming, many manufacturers utilize the Wi-Fi technology to enable the connection between wearable devices and local-hub. For Android, both wearable devices and the local-hub need to connect to the recognized access point (AP) to build connection [4]. Then, service requests and the raw data sent from wearable devices to the local-hub will go through the Google servers. Google servers know the exact location of the local-hub, so they redirect requests and data to it. After the local-hub receives, it processes requests with data and returns results to wearable devices. The path of sending results to wearable devices still needs to pass through Google servers. Consequently, users suffer from long response time waiting for results and reducing the battery life of wearable devices due to the long screen-on time. Figure 13.1 illustrates the process of building the connection between wearable devices and the local-hub.

Recently, a new computing paradigm and network architecture, fog computing arises. As shown in Figure 13.2, the main idea of fog computing is to provide computing, storage, and networking services on the edge of network [5]. Under the coworking between edge network and cloud server, it can provide latency-sensitive services for users [6–8]. Inspired by fog computing, we proposed a concept, named VLH, to eliminate the restrictions of using wearable devices [9]. The main idea of VLH is to utilize edge devices to serve wearable devices; hence, edge devices can be treated as a VLH. VLH can be set up on edge

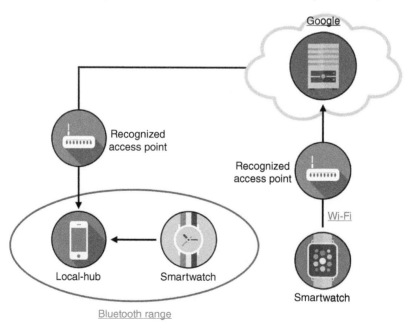

Figure 13.1 Building connection between wearable devices and the local-hub.

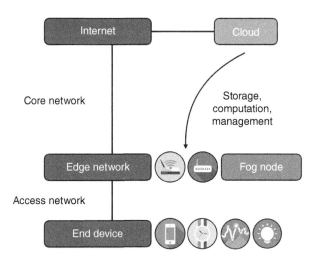

Figure 13.2 The network architecture of fog computing.

devices, such as Wi-Fi AP, such that wearable devices can connect to it directly via Wi-Fi interface. As mentioned earlier, wearable devices need to connect to the local-hub to complete wearable services. The VLH provides wearable services on the edge devices; thus, wearable devices can be served by edge devices instead of the local-hub. It means wearable devices can still be functional without connecting to the physical local-hub (PLH). Besides, the edge of the network is very close to the user, so service requests and the processing data are no longer needed to pass through the Internet to remote servers. Consequently, the long response time caused by the remote connection to the physical local-hub (RPLH) can be reduced. From another perspective, VLH is doing "Computation offloading" for wearable devices. Previous approaches to do computation offloading create virtual machines (VMs) for every user and migrates all the needed processing data and applications from the end-devices to VM to complete the offloading process. However, these approaches are based on the concept of cloud computing; the cloud server has much more computing capacity than edge devices. Due to the limitation of the computing capacity of the edge devices, the execution environment that serves wearable devices on edge devices must be lightweight. Therefore, VLH decomposes applications into several function modules and provides common function modules as a wearable service. Without running and managing a whole VM for each user, edge devices can serve many users under limited computing resources. This idea is similar to the execution model of cloud computing, serverless computing [10, 11], also referred to provide function as a service (FaaS). Figure 13.3 shows the network architecture of VLH. Since edge devices provide different kinds of wearable services for wearable devices, the service request can be completed at the edge of the network without traveling through the Internet.

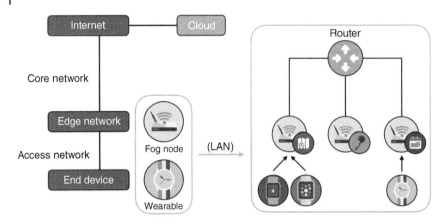

Figure 13.3 The network architecture of virtual local-hub.

In this chapter, first, we propose a system design based on the concept of VLH to utilize edge devices to provide low-latency services for wearable devices. Second, to make wearable devices be served by edge devices instead of the local-hub, we modify the system behavior of wearable devices. Since the system modifications are transparent to both users and developers, existing applications can naturally fit into our system without being rewritten. Third, considering the limited computing capacity of edge devices, we follow the idea of VLH to provide common and native function modules (e.g. localization and speech recognition) as a wearable service for users instead of running user-specific VM. Therefore, the execution environment is lightweight enough to be built on edge devices. Fourth, because native function modules are necessary for every wearable device, they can be shared and remotely accessed. This mechanism balances the workloads among edge devices and avoids edge devices being crushed by processing too many service requests at the same time. Finally, we conduct three experiments to analyze the performance of our approach. The results show that we reduce the execution time of wearable services by up to 60%, and the CPU usage of wearable devices is also reduced by up to 70% during the process of running wearable services.

13.2 Related Works

To implement a system based on the concept of VLH, the most critical issue is how to make wearable devices be served by edge device. In other words, how to make edge devices do computation offloading for wearable devices. Here in the following, we provide a comprehensive review of related works about computation offloading.

Due to the flourishing development of cloud computing, the common way to do computation offloading relies on the cloud server. This approach is often called mobile cloud computing (MCC). The main idea of MCC is to migrate computation to more capable computers (i.e. cloud servers) for mobile devices. The restrictions on mobile systems (e.g. battery life, network bandwidth, storage capacity, and processor performance) can be alleviated by MCC [12]. Then, there are two critical issues in the computation offloading. First, "How to do offloading," applications contain lots of different function modules, so it is difficult to decide which function modules are appropriate to be offloaded. Besides, the timing of triggering offloading process also needs to be evaluated carefully. Second, "What's the offloading granularity"; if offloading granularity is at the application level, we can simply migrate the whole application to the server and execute it on the VM. However, if offloading granularity is at the function level, we need to analyze and disassemble applications to evaluate every function. There exist two possible approaches to solve the above issues. First, ask the application developers to rewrite their applications to cooperate with the offloading system so that it can handle the offloading process precisely. Second, without the cooperation of developers, the offloading system does complex analysis on applications to control the offloading process. Thus, we categorize the related works into two main categories: offloading without/with application developer's effort.

13.2.1 Without Developer's Effort

References [13, 14] are at the application level of offloading granularity, and users need to install "Agent" on their device to complete the further operation. The former uses "Intent filter" to intercept the intent and transfers it to cloud server, then executes corresponding actions. The latter transfers the "Activity state" of applications to the cloud server to continue the ongoing process of applications. Reference [15], CloneCloud, is at the thread level of offloading granularity. Due to the fine-grained granularity, CloneCloud needs to do more complex analysis on applications to decide when to trigger the offloading process. CloneCloud designed a partitioning mechanism to decide which parts of applications need to be offloaded. With different execution conditions as input, the partitioning mechanism yields different partitions, resulting in a database. During the runtime, execution mechanism picks a partition matched with current execution condition to do offloading.

The above approaches can complete computation offloading without the help of application developers, which means developers do not need to rewrite applications in a particular rule or add any annotations in the code. However, it requires sophisticated analysis on the applications to find out which part of applications can be offloaded. On the server side, above approaches need to build and manage powerful VM on a resourceful server to support the execution part of offloading. Such requirements cannot match with our scenario because edge devices and wearable devices have limited computing capacity. Moreover, building and managing VM also lengthen the response time.

13.2.2 Require Developer's Effort

In this category, developers need to add annotations, such as "Remote", "Remoteable", and "Offloadable", on the functions needed to be offloaded. Thus, they are all at the function/method level of offloading granularity. Cuervo et al. propose a system, called MAUI [16], that enables energy-aware computation offloading. MAUI uses the data collected by three factors: device's energy consumption, program characteristics, and network characteristics, to decide the method should be offloaded or executed locally. Kosta et al. design an offloading scheme, called ThinkAir [17], providing the online decision on offloading according to current execution environment and past invocations. ThinkAir also has hardware, software, and network profilers as MAUI has, but it focuses more on the scalability issues. Huang et al. purposes the device-to-device (D2D) code computation offloading system, called Dust [18], for wearable computing. Dust designs task dissemination and task scheduler to speed up the execution time of applications.

Shi et al. claims that computation offloading is typically counterbalanced by communication costs and delays between users and servers [19]. Thus, their proposed system, called IC-Cloud, predicts the execution time of the program and the condition of network connectivity; moreover, the results of prediction are referred when the system is ready to start offloading process. They then propose an extension of IC-Cloud, called COSMOS [24], which can effectively allocate and schedule offloading service requests to resolve the contention for cloud resources.

The approaches in this category can complete computation offloading with the help of application developers. With prediction and profiling mechanism, the performance of computation offloading can be enhanced. However, it is hard to convince developers to rewrite their applications without any benefits for them. Moreover, the prediction and profiling mechanism might spend more time responding to users. Most important of all, those approaches also require powerful execution environment, which cannot be built on edge devices.

In summary, we can find that the granularity is various from the application, thread to function. VLH provides function as a wearable service on edge devices, so the offloading granularity of its system is at the function level. Then the shared service is the unique characteristic of VLH, and previous works do not have this feature. We propose our system design such that it can adapt existing applications so that developers do not need to rewrite the applications. The comparison results indicate our work is different from any other related works.

13.3 Problem Description

The limited computing capacity of wearable devices requires users to carry the local-hub to complete services. The existing mechanism, PLH, limits the connection distance between the wearable device and the local-hub. If users cannot carry the local-hub and the wearable device at the same time, the user experience might be severely affected due to the long response time caused by RPLH (the remote connection to the PLH via Wi-Fi). To address these shortcomings, we propose a system design based on the concept of VLH. With the proposed system and the concept of VLH, users can use wearable services without connecting to the local-hub. In addition, we design this system with two objectives and the details are in the following content.

(i) *Low-latency and shared wearable services*: Fog computing utilizes edge devices (i.e. AP and router) to do the computation for end devices. By providing wearable services on edge devices, the service request sent from wearable devices can be completed at the edge of the network without entering into the Internet. Therefore, the long response time caused by RPLH can be reduced. However, due to the limited computing capacity, we cannot setup unique execution environment for every user on edge devices. Following the proposed idea in VLH, we decompose applications into several function modules and provide common and native function modules (e.g. localization and speech recognition) as a wearable service. From this point of view, the execution environment on edge devices is lightweight enough to be built on edge devices, since we provide wearable services to users directly instead of running and managing lots of VMs for them. Furthermore, native wearable services are necessary for every wearable device, so they can be shared instead of being specific for users. The shared wearable services mean that they can be accessed by multiple users at the same time. To make wearable service to be remotely accessed, we add a controller in the system to handle the remote service request. With the remote wearable services provision, the workloads among edge devices can be balanced and ensured that the fog nodes are functional in advance.

(ii) *Adapting existing applications into our system*: By default, whenever service requests take place, wearable devices find the local-hub to process requests. To change this default and make edge devices serve wearable devices, we need to modify the system behavior of wearable devices. For the convenience of users and developers, that modification should be transparent

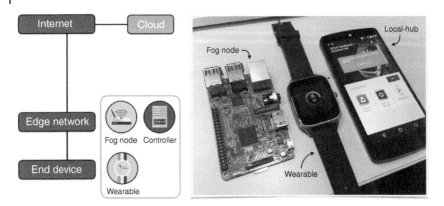

Figure 13.4 The proposed system architecture.

for them. If our system modification affects the user habits or requires developers to rewrite their applications, it might be hard to be accepted by popularity. It also means that the existing applications can fit into our modification on the system of wearable devices naturally. Besides, wearable devices are designed to be power-saving, so the modifications cannot bring additional system overhead to drain its battery.

We propose a system that possesses the design objectives mentioned above to solve the defect of using wearable devices. The high-level overview of the proposed system architecture is shown in Figure 13.4.

13.4 System Architecture

As shown in the left side of Figure 13.4, our system architecture consists of three major components: end device (wearable device), fog node (access point), and the controller. Fog nodes provide wearable services for end devices, and end devices connect to fog nodes via Wi-Fi interface. The right side of Figure 13.4 shows the physical hardware of fog nodes, wearable devices, and the local-hub. Our system serves wearable devices by fog nodes, so the local-hub does not involve into our system. We will go through the details of every component in the following content and explain what is the role they play in our system.

13.4.1 End Device

End devices in our system architecture are wearable devices, and we use smartwatches, ASUS ZenWatch 2, for the implementation. By default, whenever smartwatches execute applications and run specific services, they send the service request to the local-hub and wait for the results via BLE or Wi-Fi interface. However, in our system, we change this default by modifying the system behavior of smartwatches to make fog nodes serve them instead of the local-hub.

13.4.2 Fog Node

Fog nodes are edge devices in the network architecture of fog computing. Thus, we set up an access point via a development board, Raspberry Pi 3, to act as a fog node. Fog nodes in our system architecture are service providers; they process the service request sent from end devices and reply the results. According to the widespread usage of wearable devices and original functionality of them, we provide three kinds of wearable services for end devices in our system: localization, speech recognition, and retrieving Google calendar information. End devices connect to fog nodes via Wi-Fi interface and complete their services at the edge of the network without entering into the Internet. If a fog node does not provide particular service, it queries the controller to get the location of the fog node which possesses the service, moreover, then redirect the request and the data to it. After receiving the result, the fog node returns it to the end device. Also, fog nodes listen to the command of the controller to turn on/off the services when computing resources are enough/empty.

13.4.3 Controller

The controller records locations of wearable services. When fog nodes cannot process service requests and need to redirect them, the controller gives the correct location of the service provider to them. The controller also monitors the system status of fog nodes; it sends the command to fog nodes to turn on/off the service, depending on different conditions.

13.5 Methodology

To ensure that all the components cooperate with each other correctly, we modify the operating system of end devices and set up corresponding wearable services on fog nodes. Moreover, the controller manages fog nodes to balance the workloads among them. The core of the design is around with the end device since we need to provide precious information with correct data format at the same time to prevent our modifications crushing the system. Thus, the implementation of fog nodes and the controller needs to accommodate the system behavior of end devices. In this section, we present how we implement all the design on every component.

13.5.1 End Device

By default, wearable devices ask the local-hub to process the service request. However, in our system, the service request sent from wearable devices will be caught and completed by edge devices. To have this feature, we need to modify the system behavior of wearable devices, and we choose Android smartwatch to be our target. The operating system of Android smartwatch is Android Wear which is not open source, so we cannot modify its system code directly. Alternatively, we decide to intercept method calls of the system and modify its operation in advance. There are two open source projects that can help us to achieve the goal: Cydia [20] and Xposed [21]. To be noted, the installation of Cydia and Xposed requires the root access of devices. The first one, Cydia, is prevalent because it can jailbreak the iOS device, but regarding Android, its supporting OS version is only up to the 4.3. The second one, Xposed, updates regularly and can be functional on Android and Android Wear, so we use Xposed in our system implementation. Xposed is a framework for modules that can change the behavior of the system and apps without touching any APKs. It modifies the system behavior by "Hooking" method calls, and then we can inject our code before and after methods. This feature is flexible because it means that the modules of Xposed can work for different versions of operating systems and even ROMs without any changes. Also, users can decide when to activate specific modules or deactivate them when users think they are not useful. Whenever users modify (activate or deactivate) the status of modules, this modification becomes effective after rebooting.

The Xposed framework can help us modify the system behavior of Android smartwatch, however, using Xposed is not intuitive and easy for debugging. Before we use hooking mechanism to modify particular method, we need to trace the source code of operating system first. That is because we do not expect that our modifications crush the system or brick the device. After we fully understand the operating system, we know the kind of date that methods need with the specific data format. Mastering this information can prevent the system and device from harm by the hooking mechanism.

13.5.1.1 Localization

The location information is essential for wearable devices due to the demand for the fitness and the map navigation. There are two ways to retrieve the location information on Android smartwatches. Ask existing location providers (GPS and network) or use Google mobile services (GMS) to query the local-hub. Both ways need two important methods to complete location services, request updating of the location and get the last-known location. If devices require updating of the location information, the system will call another method to inform applications after we get the results. Thus, to make fog nodes help Android smartwatches get the location information, we hook

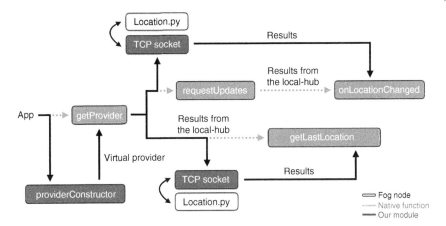

Figure 13.5 The process of retrieving the location information.

those methods and replace them with our module which gets the data about the physical location from fog nodes. After receiving the data, we encapsulate it in a specific format as arguments and call the method which informs the system that updating is completed with it. However, in some applications (e.g. Google Map), they will check whether there is any existing location provider first. In this case, if there is no existing location provider, smartwatches cannot get the location information. To solve this issue, we hook system methods to create a virtual GPS location provider when applications check and find if there is any location provider. Then, we hook the methods mentioned above to let fog nodes help smartwatches get the data they need. The process of retrieving the location information is illustrated in Figure 13.5.

Also, worth mentioning is that when smartwatches can only connect to the local-hub via Wi-Fi interface and they are using GMS to get location information, smartwatches will not get the response. Because when they cannot build connection via BLE, it means that they are not close to each other. Thus, it is useless to provide the local-hub's location to smartwatches. However, our system eliminates this restriction; thus, smartwatches can get the accurate location information even if the local-hub is not nearby.

13.5.1.2 Speech Recognition

The small screen of the smartwatches makes it difficult for the users to type on it quickly, so speech recognition becomes the necessary method to send instructions to applications or the system. The process of using speech recognition is shown in Figure 13.6. First, applications need to initialize an intent object with the predefined configuration of speech recognition. Moreover, then, call the method to start a new activity using the intent object as arguments. The next step is that the new activity will record voices, do recognition for users,

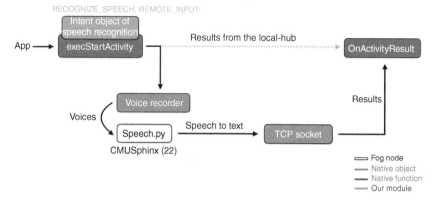

Figure 13.6 The process of speech recognition.

and inform the users of the results when the recognition process is completed. Similar to what we do in the localization part, to make fog nodes do speech recognition for smartwatches, we hook the method which starts a new activity with an intent object about speech recognition and replaces it by our module. Our module records the voices spoken by the users, sends it to fog nodes, and waits for the recognition process to be completed. After receiving the recognition results from fog nodes, we encapsulate it in a specific data format as arguments and call the method with it to inform the system that speech recognition is completed.

13.5.1.3 Retrieving Google Calendar Information

Checking personal information on smartwatches is more convenient than using hand-held for users to catch up incoming events. To retrieve Google calendar information, the applications need to do OAuth 2.0 authentication and get the credential first as shown in Figure 13.7. After authentication is completed,

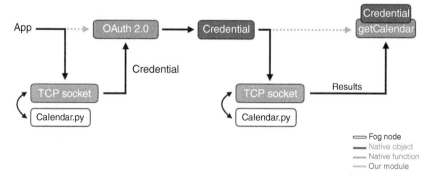

Figure 13.7 The process of retrieving Google calendar information.

applications initialize a service object with the credential and then ask Google servers for the calendar information. To make fog nodes retrieve Google calendar information for smartwatches, we hook and replace the method that authenticates with Google and gets credential by our module. Our module asks fog nodes to complete the authentication process for the smartwatch and reply to the credential. Next, we hook and interrupt another method that uses the credential to request Google for the calendar information. Fog nodes take the operation further and transfer the cached Google calendar information back to smartwatches. After receiving the results, we do similar encapsulation process for data and resume the process of the method call.

Besides the three modules mentioned above, we write another module to detect and automatically add the permissions to applications. Because some functionality needs the permissions specified in the manifest file of Android applications, we cannot assume that the applications already have those permissions. For example, the voice recorder needs the permission of microphone and socket connection needs the permission of Internet and Wi-Fi access.

13.5.2 Fog Node

Fog nodes are service providers that process the service requests sent from end devices in our system. According to the services we provide, we set up three socket servers to handle the different kinds of service requests. Before socket servers process the service request, fog nodes need to check a configuration file to get the status of every service. If the service is running, fog nodes process the request. However, if the service is not running, fog nodes request the controller to get the IP address of the correct service provider first. Then, fog nodes redirect the request and the data to it and wait for the results. Fog nodes also are managed by the controller; they always listen to the command sent from the controller to turn on/off the service. Thus, there is another socket server which keeps listening to the controller and maintains the configuration file which records the status of every service. All the socket servers are written using python program, and the relation charts are plotted as Figure 13.8.

The physical location information of every fog node is set manually when the system is built. If fog nodes are moved to another place, the location information can be updated by using the new IP address. We use an open source project of CMU, CMUSphinx [22], to do speech recognition. It features offline recognition so that whole process can be completed at the edge of the network instead of entering into the Internet. The personal Google calendar information is already cached on fog nodes, so when service requests come in, fog nodes do not need to do authentication and grab the data from Google server. Fog nodes only need to read the file and send the information back to the end device.

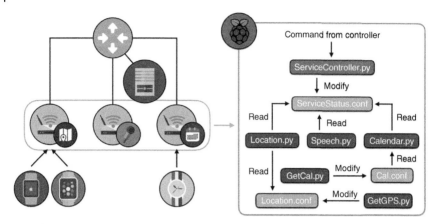

Figure 13.8 The execution environment of the fog node.

13.5.3 Controller

As shown in Figure 13.9, the controller helps fog nodes redirect service requests when they cannot process it, so we set up a socket server that responds to fog nodes with the IP address of the correct service provider. The controller also manages and monitors fog nodes; it runs a socket client that sends the command to fog nodes to turn on/off the service, depending on the computing capacity of fog nodes. At the same time, it maintains the record of the IP address

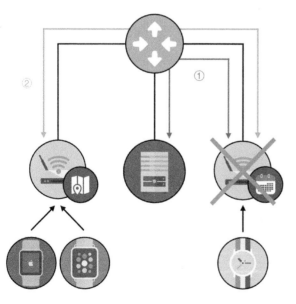

Figure 13.9 The mechanism of remote wearable services provision.

of correct service providers that are running the service. All the socket clients and servers are written using python program.

13.6 Performance Evaluation

In this section, we report the results of extensive experiments conducted to validate the proposed system and methodology and share the insight we get from the results.

13.6.1 Experiment Setup

As shown in Figure 13.10, we build a local area network (LAN) using TP-Link AC750 router. The wired interface of the router is 4 of 1000Mbps LAN ports connected by three fog nodes and one controller. Fog nodes are set up as AP, and end devices connect to it via Wi-Fi interface, and the wireless speed is around 20–30Mbps. Every component's hardware specification in our experiment is the following: end device (wearable device) is a ASUS smartwatch, ZenWatch 2, installed with Android Wear 1.5. The processor of the end device is quad-core 1.2GHz Cortex-A7 with 512MB RAM. By default, Android smartwatch needs a PLH to help them process service requests and reply results. The PLH is a mobile device, Nexus 5, installed with Android 6.0. The processor of the local-hub is quad-core 2.3GHz Krait 400 with 2GB RAM. The fog node is a development board, Raspberry Pi 3, installed with Ubuntu Mate. The processor of the fog node is quad-core 1.2GHz Broadcom BCM2837z with 1GB RAM. The fog node equips wireless LAN interface, so that it can be set up as an access point. The controller is a laptop, Lenovo B460, which is installed with Ubuntu 16.04. The processor of the controller is dual-core 2.26GHz Intel i3-350M with 2GB RAM.

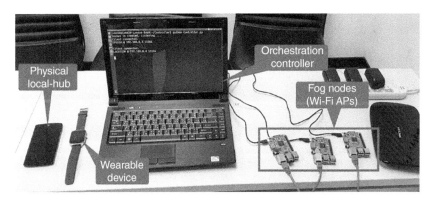

Figure 13.10 The prototype implementation of VLH.

The comparison methodologies are PLH, RPLH, and VLH. The main difference among them is the way of building connection from the smartwatch to the local-hub and how the service request is being completed. In PLH, because the smartwatch is close to the local-hub, they communicate to each other via BLE. But in RPLH, the smartwatch and the local-hub are not within the same BLE coverage, so the smartwatch only can build connect via Wi-Fi interface. The prerequisite of building connection in RPLH is that both smartwatch and local-hub need to connect to the recognized access point via Wi-Fi interface. Then, service requests and the data will enter into the Internet, pass through Google server, and reach to the local-hub. After local-hub completes services, results are replied to the smartwatch through the same path. In our solution, VLH, wearable devices are served by fog nodes instead of the local-hub. Thus, service requests and the data do not need to enter into the Internet because fog nodes are on the edge of the network. The performance metrics of all experiments is the execution time of different applications and the CPU usage of end devices and fog nodes.

13.6.2 Different Computation Loads

In Figure 13.11, we compare the execution time of different schemes under different length of sentences for speech recognition. We can see that longer the sentence, longer the execution time. This is because longer sentence means

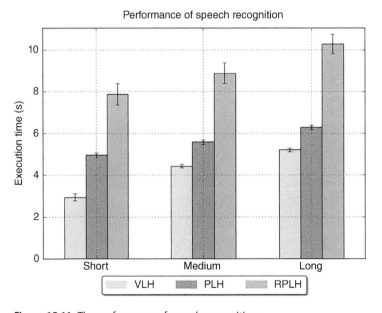

Figure 13.11 The performance of speech recognition.

more complex speech recognition. Also, regardless of the sentence length, RPLH performs worse than both PLH and VLH since RPLH requires the data to travel through the Internet while both PLH and VLH require only one-hop data transmission. In addition, VLH performs better than PLH. This is because the end-device connects to the fog node via Wi-Fi for VLH, while for PLH, the end-device connects to the phone via Bluetooth, and the bandwidth of Wi-Fi is larger than that of Bluetooth. This result demonstrates that the proposed VLH can achieve great performance.

In Figures 13.12 and 13.13, we compare the CPU usage of the end-device and fog node for different schemes and different lengths of sentences for speech recognition. First as shown in Figure 13.12, we can see that the CPU usages of PLH and RPLH are higher than the VLH, this is because the Google framework shows complex UI on the watch when executing speech recognition. Nevertheless, the CPU usages basically remain low regardless of the sentence length since all the complex computation is done at fog node or the smartphone, not on the watch. Then in Figure 13.13, we can see that the CPU usage at fog node becomes larger as the sentence length gets longer. This is because longer sentences require more complex computation for the speech recognition. Besides, the CPU usages at fog node all remain under 0.08. This means the computation resource at the fog node is quite enough to handle the complex speech recognition. The results show that the VLH does not cause large computation overhead on both the end-device and the fog node.

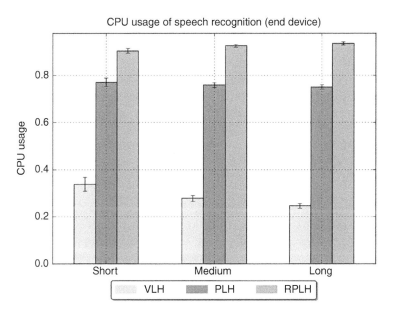

Figure 13.12 The CPU usage of speech recognition (end-device).

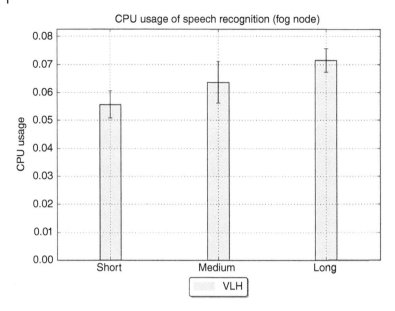

Figure 13.13 The CPU usage of speech recognition for (fog node).

13.6.3 Different Types of Applications

To evaluate the performance of different methodologies in different types of applications, we measure the execution time and the CPU usage of three types of applications, localization, speech recognition, and retrieving Google calendar. Because the wearable device we used in the experiment is installed with Android Wear 1.5, it does not have the standalone method call to authenticate with Google via OAuth 2.0. Therefore, retrieving Google calendar in VLH is implemented by using socket connection to communicate with fog nodes and get the results directly. For PLH and RPLH, we use the wearable data layer application programming interface (API) to send the request of retrieving Google calendar to the local-hub, and the local-hub replies the results via same API. RPLH cannot get the location information so the application of localization is only functional in VLH and PLH.

Localization and retrieving Google calendar do not require computing resource. Thus, as shown in Figure 13.14, the execution time of these two applications is extremely short. Figure 13.15 shows the comparison of the execution time of localization and retrieving Google calendar. In the part of localization, we can find that the execution time of VLH is longer than PLH because BLE has more advantage than Wi-Fi in small data transmission. The execution time of VLH in retrieving Google calendar is the shortest since VLH has already cached the calendar information. But in PLH and RPLH, whenever

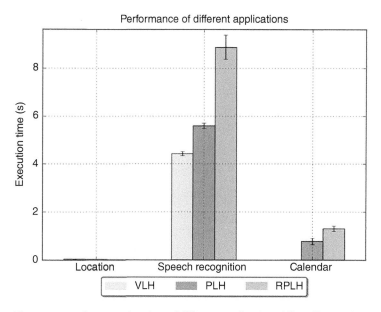

Figure 13.14 The execution time of different applications (all applications).

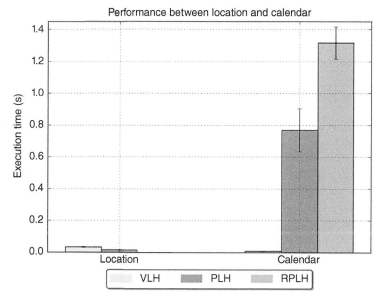

Figure 13.15 The execution time of different applications (localization and retrieving Google calendar).

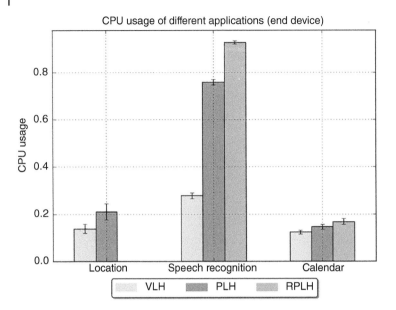

Figure 13.16 The CPU usage of different applications (end-device).

the local-hub receives the service request, it needs to do the authentication and then ask the server for the calendar information.

The CPU usage of three different applications is shown in Figure 13.16. As mentioned above, localization and retrieving Google calendar do not need computing resource, so the CPU usage between different methodologies is very close. But compared with speech recognition, the CPU usage of them is much smaller because speech recognition requires more computing resource. For the same reason, the CPU usage of the fog node in localization and retrieving Google calendar is extraordinarily low as shown in Figure 13.17. This experiment verifies that even in the different types of applications, VLH has great performance on the execution time and the CPU usage. In addition, it also verifies that the execution environment we built on fog nodes is very lightweight. Consequently, fog nodes can serve wearable devices with the limited computing capacity.

13.6.4 Remote Wearable Services Provision

Wearable services are shared and can be remote accessed to balance the workloads among fog nodes in the proposed system. In this experiment, we analyze the performance of this mechanism, remote wearable services provision (virtual local-hub-remote, VLHR). Figures 13.18 and 13.19 show that the execution time of different types of applications in VLHR is almost same as VLH, but

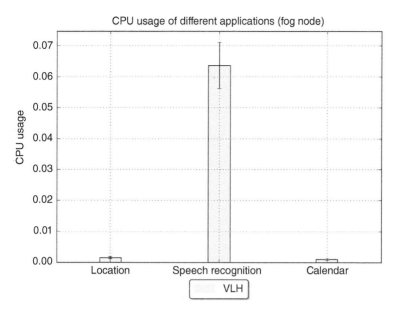

Figure 13.17 The CPU usage of different applications (fog node).

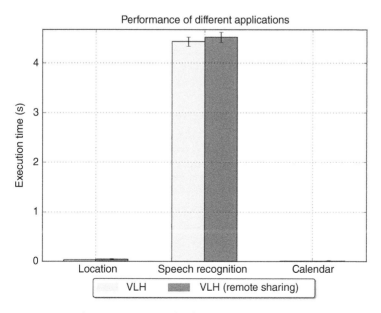

Figure 13.18 The execution time of different applications in VLHR (all applications).

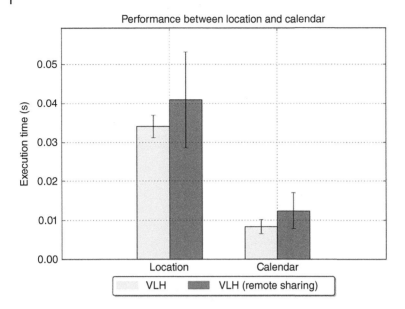

Figure 13.19 The execution time of different applications in VLHR (localization and retrieving Google calendar).

VLHR has larger standard deviation than VLH. This is because VLHR needs to query the controller and redirect requests and the data so it leads to more hops during the process.

Figure 13.20 shows that the CPU usage of end device will not be affected by the VLHR in different types of applications. Last, we verify that VLHR can balance the workloads among fog nodes by the result shown in Figure 13.21. The result shows that the CPU usage of the fog node is reduced by up to 90% after redirecting the service request of speech recognition to another fog node.

13.6.5 Estimation of Power Consumption

The most important factor that wearable services need to be offloaded is power-saving. Currently, wearable devices rely on the local-hub to process computing tasks, and this is also due to the power-saving issue. And VLH proposes the idea of offloading wearable services to access points under the network architecture of fog computing. In this experiment, to prove that offloading computing tasks is necessary for wearable devices, we estimate the power consumption of processing wearable services on the local side or remotely on fog nodes. Because we do not have the power model of ASUS ZenWatch 2, we use the power model of LG Urbane watch to do the estimation [23]. The processor of ZenWatch 2 and LG Urbane watch are the same, so we do not have to do the conversion of power model. On the other side,

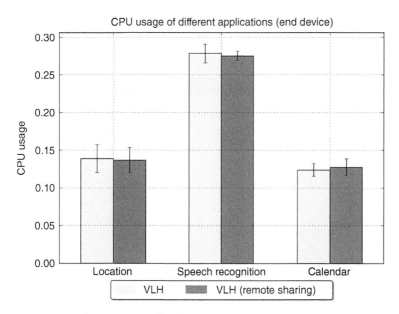

Figure 13.20 The CPU usage of different applications in VLHR (end-device).

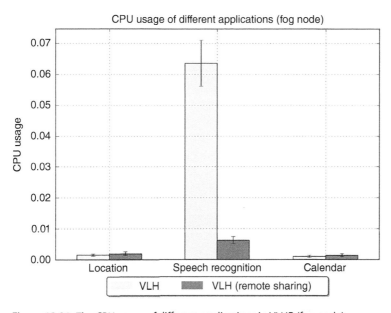

Figure 13.21 The CPU usage of different applications in VLHR (fog node).

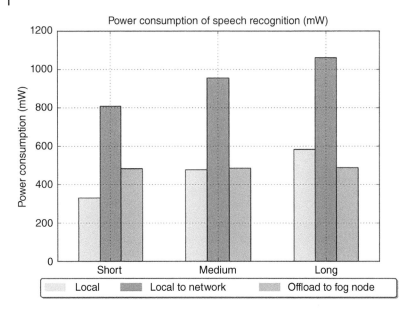

Figure 13.22 Power consumption of speech recognition.

the existing mechanism does not allow wearable devices complete wearable services all alone on the local side. Thus, we process wearable services on the development board, Raspberry Pi 3, and consider the processor difference between Urbane watch and development board to do the conversion of performance. For processing wearable services on the local side, we calculate the power consumption of CPU utilization and screen display. If results of wearable services are used in other network services, the power consumption needs to add the network connection. For offloading wearable services to the fog node, transferring data and screen display dominate the power consumption. Figure 13.22 shows the results of power consumption of processing speech recognition in the local side or remotely on the fog node. We can find that if users use speech recognition only to trigger some local system commands, it consumes less power than offloading to fog nodes. However, when speech recognition is used in the map navigation, messenger, or others applications which require the network connection, it consumes more power than offloading to fog nodes.

13.7 Discussion

Here, to implement the idea of VLH, we use a development board to act as an access point and modify the system behavior of wearable devices.

However, our system architecture encounters two deployability issues. First, the system modifications on wearable devices need the root access. That could be unacceptable for some users because root access can manipulate the whole system without any restriction. Second, we build the execution environment to process service request sent from wearable devices on the access point. Although there are some open source access points that allow programmers to develop on it, the enterprise access points have closed system so that we cannot deploy our execution environment. The aforementioned deplorability issues can be solved once fog computing is ubiquitous; there are several scenarios that might come up after this. One possible working scenario is described as follows: following the trend of fog computing, manufacturers of the access point can release some APIs for developers to utilize the computing resources. For example, just like FaaS platform, developers can upload the program to access points and use REST API to execute that program in other programs. In this way, end devices can complete the local service request by utilizing the computing resources of access points without endangering its system. On the other side, by default, wearable devices cannot utilize other computing resources except the local-hub. However, with the core concept of fog computing, the operating system of wearable devices will release application APIs for developers to choose the offloading target. Consequently, wearable devices can utilize other computing resources primitively instead of being rooted. In such a scenario, the deplorability issues are solved by the cooperation between manufacturers of access points and wearable devices. Therefore, the idea of VLH can be realized and deployed in our living life.

13.8 Conclusion

In this chapter, we propose a system design based on the concept of VLH to overcome the limitations of using wearable devices. The proposed system serves wearable devices by edge devices instead of the local-hub. Wearable services are deployed on the edge devices so that services are directly completed on the edge of the network instead of entering into the Internet. To achieve this feature, our system modifies the system behavior of wearable devices. Most importantly, these system modifications are transparent for users and application developers so that existing applications can fit into our system naturally without being rewritten. However, considering that the computing capacity of edge devices are limited, the execution environment must be lightweight on edge devices. Inspired by the proposed idea of VLH, we decompose applications into several function modules and provide common and native function modules as a wearable service. Therefore, edge devices provide wearable services directly to users instead of managing VMs for them. Also, the native wearable services are necessary for every wearable device so that it can be shared and remotely accessed. With the remote wearable services provision, the workload among edge devices can be balanced. We conduct extensive experiments based on the common usage of wearable devices to evaluate the performance of the proposed system. The experiment results indicate that the proposed system can reduce the execution time of wearable services by up to 60%. Also, the CPU usage is reduced by up to 70% during the process of running wearable services.

References

1 Cisco Mobile VNI (2018). CISCO white paper: visual networking index global mobile data traffic forecast update, 2016–2021. www.cisco.com/c/en/us/solutions/collateral/service-provider/visual-networking-index-vni/mobile-white-paper-c11-520862.html (accessed 15 April 2018).

2 Marko Maslakovic (2017). CCS Insight: wearable tech market to double over the next four years. gadgetsandwearables.com/2017/03/16/ccs-insight-forecast-2021 (accessed 15 April 2018).

3 Google (2018). Google connect your watch with your phone – android wear help. https://support.google.com/androidwear/answer/6056630?hl=en&ref_topic=6056389 (accessed 15 April 2018).

4 Google (2018). Google sending and syncing data – android developers. developer.android.com/training/wearables/data-layer/index.html (accessed 15 April 2018).

5 Bonomi, F., Milito, R., Zhu, J., and Addepalli, S. (2012). Fog computing and its role in the internet of things. In: *ACM Mobile Cloud Computing (MCC)*, 13–16. New York, NY: ACM.

6 Chiang, M. and Zhang, T. (2016). Fog and IoT: an overview on research opportunities. *IEEE Internet of Things Journal* 3: 854–864.

7 Yannuzzi, M., Milito, R., Serral-Gracia, R. et al. (2014). Key ingredients in an IoT recipe: fog computing, cloud computing, and more fog computing. In: *2014 IEEE 19th International Workshop on Computer-Aided Modeling Analysis and Design of Communication Links and Networks (CAMAD), Athens, Greece* (1–3 December 2014).

8 Cisco (2015). Fog computing and the internet of things: extend the cloud to where the things are. CISCO white paper. http://www.cisco.com/c/dam/en_us/solutions/trends/iot/docs/computing-overview.pdf (accessed 15 April 2018).

9 Lin, H.-P., Shih, Y.-Y., Pang, A.-C., and Lou, Y.-Y. (2016). A virtual local-hub solution with function module sharing for wearable devices. In: *ACM MSWiM 16' Proceedings of the 19th International Conferences on Modeling Analysis and Simulation of Wireless and Mobile Systems, Malta*, (13–17 November 2016).

10 Kumar, K., Liu, J., Lu, Y.-H., and Bhargava, B. (2013). A survey of computation offloading for mobile systems. *Mobile Networks and Applications* 18: 129–140.

11 Yan, M., Castro, P., Cheng, P., and Ishakian, V. (2016). Building a chatbot with serverless computing. In: *MOTA'16 Proceedings of the 1st International Workshop on Mashups of Things and APIs, Trento, Italy* (12–16 December 2016).

12 Hendrickson, S., Sturdevant, S., Harter, T., and Venkataramani, V. (2016). Serverless computation with openLambda. In: *HotCloud'16 Proceedings of the 8th USENIX Workshop on Hot Topics in Cloud Computing, Denver, CO* (20–21 June 2016).

13 Ghorpade, S., Chavan, N., Gokhale, A., and Sapkal, D. (2013). A framework for executing android applications on the cloud. In: *2013 International Conference on Advances in Computing, Communications and Informatics (ICACCI), Mysore, India* (22–25 August 2013).

14 Hung, S.-H., Shih, C.-S., Shieh, J.-P. et al. (2012). Executing mobile applications on the cloud: framework and issues. *Computers and Mathematics with Applications* 63: 573–587.

15 Chun, B.-G., Ihm, S., Maniatis, P. et al. (2011). CloneCloud: elastic execution between mobile device and cloud. In: *ACM EuroSys'11 Proceedings of the Sixth Conference on Computer Systems*, 301–314. New York, NY: ACM.

16 Cuervo, E., Balasubramanian, A., Cho, D.-k. et al. (2010). MAUI: making smartphones last longer with code offload. In: *ACM MobiSys'10 Proceedings of the 8th International Conference on Mobile Systems, Applications, and Services*, 49–62. New York, NY: ACM.

17 Kosta, S., Aucinas, A., Hui, P. et al. (2012). ThinkAir: dynamic resource allocation and parallel execution in the cloud for mobile code offloading. IEEE Computer Communications (INFOCOM), Orlando, FL (25–30 March 2012).

18 Huang, D., Yang, L., and Zhang, S. (2015). Dust: real-time code offloading system for wearable computing. In: *2015 IEEE Global Communications Conference (GLOBECOM), San Diego, CA* (6–10 December 2015).

19 Shi, C., Pandurangan, P., Ni, K. et al. (2014). IC-Cloud: Computation Offloading to an Intermittently-Connected Cloud. Georgia Institute of Technology, Tech. Rep. Number: GT-CS-13-01.

20 SaurikIT, LLC (2014). Cydia substrate. http://www.cydiasubstrate.com (accessed 15 April 2018).

21 Rovo89 and Tungstwenty (2017). Xposed framework. repo.xposed.info/module/de.robv.android.xposed.installer (accessed 15 April 2018).

22 Carnegie Mellon University (2015). CMU Sphinx. cmusphinx.github.io (accessed 15 April 2018).

23 Liu, X. and Qian, F. (2016). Poster: measuring and optimizing android smartwatch energy consumption. In: *Mobicom'16 Proceedings of the 22nd*

Annual International Conference on Mobile Computing and Networking, 421–423. New York, NY: ACM.

24 Shi, C., Habak, K., Pandurangan, P. et al. (2014). COSMOS: computation offloading as a service for mobile devices. In: *ACM MobiHoc'14 Proceedings of the 15th ACM International Symposium on Mobile Ad Hoc Networking and Computing,* 287–296. New York, NY: ACM.

14

Security and Privacy Issues and Solutions for Fog

Mithun Mukherjee[1], Mohamed Amine Ferrag[2], Leandros Maglaras[3], Abdelouahid Derhab[4], and Mohammad Aazam[5]

[1] *Guangdong Provincial Key Laboratory of Petrochemical Equipment Fault Diagnosis, Guangdong University of Petrochemical Technology, Maoming, China*
[2] *LabSTIC Laboratory, Department of Computer Science, Guelma University, Guelma, Algeria*
[3] *School of Computer Science and Informatics, Cyber Technology Institute, De Montfort University, Leicester, UK*
[4] *Center of Excellence in Information Assurance (CoEIA), King Saud University, Riyadh, Saudi Arabia*
[5] *Carnegie Mellon University, USA*

14.1 Introduction

With the advancement in computing and wireless technologies, the world has witnessed a growing number of connected devices to the Internet at an unprecedented rate. International Data Corporation expected that sensor-enabled objects connected to network will rise to 30 billion by 2020 and the number of connected devices will increase ranging from 50 billion to 1 trillion consisting of 500 million sensors in US factories, 212 billion available sensors, 110 million connected cars with 5.5 billion sensors, 1.2 million connected homes with 200 million sensors [1]. To meet the massive amount of data processing, cloud computing [2–4] is viewed as an attractive choice while providing a cost-effective solution with enough storage and computing resources in cloud data centers.

14.1.1 Major Limitations in Traditional Cloud Computing

Cloud computing suffers from substantial yet unsolved challenges such as large end-to-end delay, traffic congestion, lack of mobility, location awareness, and communication cost. Some of these issues are caused mainly due to large physical distance between cloud service provider's data centers (DCs) [5] (some of the examples are Amazon Web Services (AWS), Google, ALTUS, Anacondaweb, Apple, Facebook, Matrix, Microsoft, TATA, China Telecom,

Fog and Fogonomics: Challenges and Practices of Fog Computing, Communication, Networking, Strategy, and Economics, First Edition. Edited by Yang Yang, Jianwei Huang, Tao Zhang, and Joe Weinman.

China Unicom, AT&T, and Bell) and end user. In addition, since these DCs need to operate round the clock, the carbon footprint is an important issue. According to the Natural Resources Defense Council, in 2013, US data centers in total used 91 billion kilowatt-hours (kWh) of electrical energy, and the DC electricity consumption is projected to increase to roughly 140 billion kWh annually by 2020, the equivalent annual output of 50 power plants [6]. This will cost American businesses $13 billion annually in electricity bills and emit nearly 100 million metric tons of carbon pollution/yr. Although some companies such as Apple are moving toward more environmental friendly 100% renewable DCs [7, 8] with the wind, solar, and geothermal energy, the carbon emission from DCs will dominate on global carbon footprint [9]. In addition, DCs in most of the cloud service providers (e.g. AWS, Google cloud) are geographically far apart from each other (see Figure 14.1(a)), resulting in large DC to end user delay and quality-of-service (QoS) degradation. As a result, the delay-critical service provisioning for the Internet of Things (IoT) applications becomes a challenging issue.

14.1.2   Fog Computing: An Edge Computing Paradigm

Edge computing, envisioned to provide context-aware storage and distributed computing, facilities storage and computing capabilities toward the network edge. Several leading companies, such as Google (see Figure 14.1(b)), represent their infrastructure closer to the end users. The popular contents are temporarily cached in the edge nodes, thereby improving the end-to-end latency in content delivery to the end users.

Cloudlets [10] is one of the first edge computing concepts that brings computing and resource storage closer to the edge. The resource-rich computers with high computation power are placed in a strategic location within end user's vicinity [11]. Similar to the Wi-Fi concept, cloudlets provide the cloud service to the mobile users instead of proving Internet connectivity as in Wi-Fi [12]. The mobile users can find their preferable cloudlets. It is important to note that the cloudlets exist as a standalone environment since virtual machine (VM) provisioning can be supported in cloudlets without the intervention of the cloud computing.

In 2014, the Industry Specification Group (ISG) within European Telecommunication Standards Institute (ETSI) integrated edge computing into the mobile network architecture, outlined as Mobile Edge Computing [13]. The purpose of the ISG is to create a "standardized, open environment which will allow the efficient and seamless integration of applications from vendors, service providers, and third-parties across multivendor Mobile-edge Computing platforms" [14]. This standardization effort is driven by several leading mobile operators such as DOCOMO, Vodafone, and TELECOM Italia and manufacturers such as IBM, Nokia, Huawei, and Intel [15]. In March 2017, the

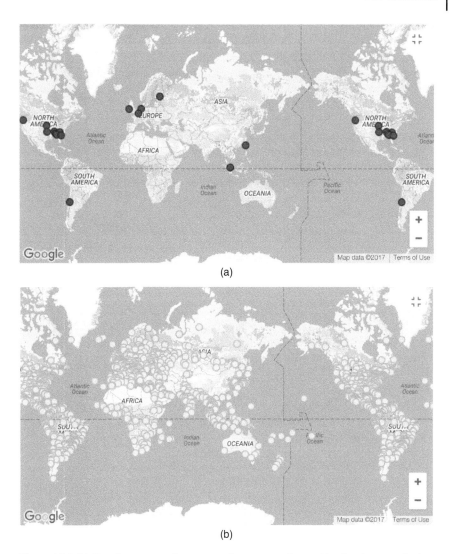

(a)

(b)

Figure 14.1 (a) Google operates data centers for computation and backend storage in the Americas, Europe, and Asia. (b) With the edge nodes (called Google Global Cache), network operators and Internet service providers deploy Google-supplied servers inside their network. Static content that is very popular with the local host's user base, including YouTube and Google Play, is temporarily cached on edge nodes [accessed on 20 February 2018]. Source: https://www.google.com/maps/.

ETSI has expanded the scope of MEC and after that replaced the term *Mobile* by *Multiaccess*. The edges of nonmobile networks are also being considered in multiaccess edge computing (MEC) [14].

From the aforementioned discussion, it is observed that cloudlets, MEC, and fog computing focus on edge computing; however, there is a significant difference between these technologies. Note that the ETSI is the main driving force for MEC [14, 15]; however, fog and cloudlets are driven by research and development, although OpenFog consortium aims at the development of fog architecture and standardization details. Moreover, cloudlets use only VM for virtualization; however, both fog and MEC can use other technologies other than VMs for virtualization [11]. Another main difference is that cloudlet mainly focuses on mobile offloading, and MEC aims to handle the applications that are better suited in either mobile or nonmobile edge networks; however, fog computing overlaps between edge and cloud while satisfying a large number of latency-sensitive applications for resource-constraint end devices [11, 16, 17]. Finally, until today, MEC works only in standalone mode, while cloudlets can function in standalone and can connect to the cloud, although there is a lack of detailed discussion in literature how to interact between cloud and cloudlets, fog computing needs the support of the cloud for the tasks that might not be performed in the resource-constraint fog layer. More details on various IoT-cloud middleware technologies and the difference among them can be found in [18]. Table 14.1 summarizes the basic differences between fog, cloudlets, and MEC.

To reduce the burden of DCs in traditional cloud computing, fog computing emerges as an alternative solution to support geographically distributed, latency sensitive, and QoS-aware IoT applications. *Fog computing* was first initiated by Cisco to extend the cloud computing to the edge of a network [19]. The term *"fog"* is used simply because *"fog is a cloud close to ground,"* [19], i.e. From cOre to edGe computing [20] enabling refined and better applications

Table 14.1 Basic difference between edge computing technologies.

Features	Cloudlets	MEC	Fog computing
Applications	Mobile offloading	Focus on the applications that are better suited for both mobile and nonmobile network edge	Support a wider range of latency-sensitive applications for resource-constraint end devices
Virtualization	Only depends on VMs	Can use other technologies apart from VMs	Other virtualization technologies can be used
Operational mode	Can work in standalone mode	Can work in standalone and can connect to the cloud	Cannot work in standalone, need the support of the cloud

or services. Fog computing is a highly virtualized platform [21] that provides computing, storage, and networking services between end user (EU) and DC of the traditional cloud computing.

Definition 14.1   According to [22] "Fog computing" term is defined as "*a scenario where a huge number of heterogeneous (wireless and sometimes autonomous) ubiquitous and decentralized devices communicate and potentially cooperate among them and with the network to perform storage and processing tasks without the intervention of third parties. These tasks can be for supporting basic network functions or new services and applications that run in a sandboxed environment. Users leasing part of their devices to host these services get incentives for doing so.*"

In another way, fog computing is defined by the OpenFog Consortium [23] as "*a system-level horizontal architecture that distributes resources and services of computing, storage, control and networking anywhere along the continuum from cloud to Things.*"

14.1.3   A Three-Tier Fog Computing Architecture

The three-tier architecture (see Figure 14.2) is one of the basic and widely used architectures in fog computing. The basic features of the tiers are discussed as follows:

- *Tier 1 – Things/end devices*: This tier consists of IoT-enabled devices including sensor nodes, EU's smart hand-held devices (e.g. smartphones, tablets, smart cards, smart vehicles, and smartwatch), and others. These end devices are often termed as terminal nodes (TNs). It is assumed that these TNs are equipped with global positioning system (GPS).
- *Tier 2 – Fog*: This tier also termed as fog computing layer. The fog nodes in this layer are comprised of network devices such as a router, gateway, switch, and access points (APs). These fog nodes can collaboratively share storage and computing facilities.
- *Tier 3 – Cloud*: Traditional cloud servers and cloud DC reside in the topmost tier. This tier has sufficient storage and computing resources.

The core benefit of the three-tier architecture is better management of the service. The services that are time-sensitive are dealt by fog computing nodes, which are located closer to the users. Similarly, data aggregation, data filtration, data trimming-related tasks have to be performed before the data is sent to the core network, thenceforth, necessitating the middleware fog in the architecture. Moreover, location-sensitive and privacy-aware services also require several types of data and information to be handled locally, instead of sending the data to the public network. This also emphasizes the importance of

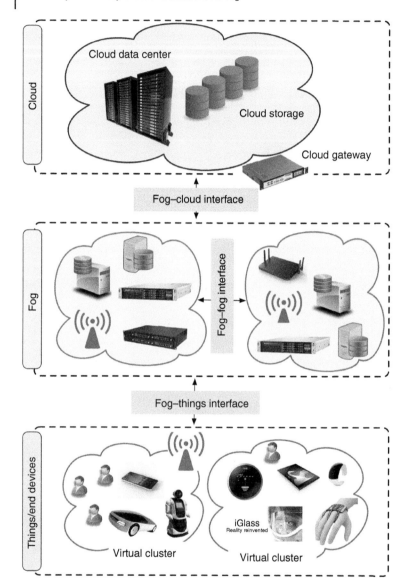

Figure 14.2 Three-tier fog computing architecture.

three-tier architecture. Nevertheless, Figure 14.3 illustrates several hierarchical fog deployment models (including N-tier fog deployment) suggested to various use cases for fog computing.

The fog and cloud providers are *honest-but curious* [25]. In many cases, the fog, cloud, and users reside in different trusted domain. This brings a huge

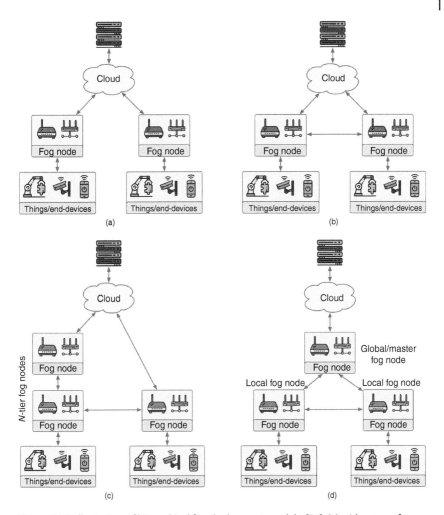

Figure 14.3 Illustration of hierarchical fog deployment models [24]: (a) without any fog cooperation; (b) fog node cooperation; (c) multi-tier (say, N-tier) fog node deployment; and (d) Global and local fog hierarchy.

threat to the data and IoT services. Although, the fog nodes are generally on agreed security protocols, they may snoop the personal information if there is any lack of proper authentication and access control [26]. As a result, the IoT services become vulnerable to the various attacks, such as forgery, data tampering, jamming, denial-of-service (DoS), man-in-the-middle, and impersonation. Moreover, the resource utilization can also be compromised. Basically, the decentralization and low-latency service provisioning create a barrier to the secure authentication. In addition, the user mobility demands frequent joining/exit to/from the fog cluster that consists of several fog nodes. As a result, the key management becomes a challenging issue. Apart from these, it is very difficult to collect behavior-related information of all the fog users; therefore, the trustworthiness among all fog nodes in the network is a critical issue.

14.2 Security and Privacy Challenges Posed by Fog Computing

We briefly summarize the following main factors that cause the security and privacy challenges in fog computing as follows:

1) *Proximity of the fog nodes*: In general, the service request from the end users is handled by nearby fog nodes, thus, the approximate physical location can be inferred by the fog nodes. Although the end user can hide his identity while sending the data to the closest fog nodes, the controlling authority can easily de-anonymize the user's location by simply observing the uniqueness of the data type. Since fog clients offload its tasks to nearest fog nodes, location, trajectory, and even mobility habits can be revealed from an adversary. User habits can also be revealed from an adversary by analyzing his/her usage habits of fog services, e.g. smart grid. As shown in [27] smart meters' readings can disclose information about the time that the house is empty or even the TV programs that the EU prefers to watch. As a result, a privacy concern arises to prevent the location of the end users from the higher layers that use intermediate fog nodes.
2) *Stringent requirements in fog computing*: Fog nodes need to handle a large number of IoT devices. In several cases, the user's data are exposed due to untrusted parties residing on fog nodes. To hide the identity, the end user may need to connect with a distant third-party for several reasons (for example, authentication and access control); however, the latency and communication overhead may restrict the design goal of the fog computing, such as low latency and reduced communication overhead. As a result, the identity of a user is vulnerable to the fog nodes.

In fact, the way of processing and storage requirements can be offloaded to fog nodes, however, the security requirements cannot be offloaded. Even IoT devices need to implement the minimum security requirements at their premises. The fog nodes interact with each other when they need to effectively manage network resources or to manage network itself. They may even operate in distributed manner to perform a specific task. Due to lack of centralized and privileged control in security management as in cloud computing, access control becomes a challenging issue in fog computing with heterogeneous and service requirements. In fact, the on-board sensor nodes collect the sensitive data from their surroundings. Therefore, fog computing imposes serious security threats from the malicious attackers.

In general, an IoT device can initiate communication with any of the fog nodes in the fog network, requesting for a processing or storage requirement. In fact, the IoT device may not even be aware of the existence of the fog network; therefore, messages sent by such a device cannot be secured by using symmetric cryptographic techniques. Alternatively, asymmetric key cryptography has its set of challenges that are unique to IoT environment. Maintaining the PKI that is required to facilitate secure communication is one of the major challenges. At the same time, minimizing the message overhead becomes an important issue from the context of end-devices.

Due to the fog computing architecture, several new privacy challenges are introduced. At the same time, the existing privacy-preserving schemes are not well suited for the fog computing. Basically, fog computing relies on the computational power of distributed nodes for reducing the total burden on the cloud data center. Privacy preservation is more challenging in fog computing since these fog nodes that are in vicinity with EUs may collect sensitive data concerning the identity, usage of utilities, e.g. smart grid or location of end users compared to the remote cloud server that lies in the core network. Moreover, since fog nodes are scattered in large areas, centralized control is becoming difficult. The compromise of a poorly secured edge node can be the entry point for an intruder to the network. The intruder once inside the network can mine and steal user's private data that are exchanged among entities.

14.3 Existing Research on Security and Privacy Issues in Fog Computing

Table 14.2 presents existing research in security and privacy for fog computing published in 2018. The details about the research published before 2018 can be found in [34].

14.3.1 Privacy-preserving

To resist attack from the inside of cloud server, Wang et al. [30] proposed a three-layer privacy-preserving cloud storage scheme, namely, three-Layer privacy preserving cloud storage scheme (TLS), which is based on fog computing model and Hash-Solomon code. The TLS scheme considers three layers, namely, cloud server, fog server, and local machine. Using Hash-Solomon code, the TLS can produce a portion of redundant data blocks that is used in the decoding procedure. Therefore, Yang et al. [31] proposed a privacy-preserving data aggregation scheme for fog computing. Based on machine learning and data aggregation method, the Yang et al.'s scheme can satisfy differential privacy

Table 14.2 Existing research in security and privacy for fog computing.

Authors	Year	Network model	Security model	Countermeasures	Performances (+) and limitations (−)
Wang et al. [28]	2018	Consists of four different entities, namely, system manager, terminal devices, fog node bridges, and public cloud server	− Secure aggregation − Identity privacy − Authentication	− Elliptic curve public-key cryptography − Castagnos−Laguillaumie cryptosystem	+ Secure against ciphertext-only attacks − Attacks model is limited
Zhang et al. [29]	2018	There are five parties, namely, cloud service provider, fog nodes, data owner, end user, and key authority	− Access control	− Access tree − Bilinear maps	+ Supporting outsourcing capability and attribute update − Privacy-preserving is not considered
Wang et al. [30]	2018	Three layers, namely, cloud server, fog server, and local machine	− Data privacy	− Hash-Solomon code algorithm	+ Can protect the fragmentary information − Authentication is not considered
Yang et al. [31]	2018	Four entities, namely, sensors, fog nodes, the fog center, and a cloud server	− Differential privacy	− Machine learning	+ Ensure multifunctional aggregation + Guarantee the privacy of the collected user data − Location privacy is not considered
Lyu et al. [32]	2018	Fog-enabled aggregation in smart grid	− Differential privacy	− Stream cipher and public-key cryptography − Homomorphic encryption	+ Guarantee the privacy of the collected data − Attacks model is limited
Yu et al. [33]	2018	Three layers, namely, cloud server, fog server, and local machine	− Fine-grained access control	− Leakage-resilient functional encryption	+ Guarantee security against side channel attacks − Mobility model is not considered

for sensory data. Note that the differential privacy can be preserved by the Stream cipher and public-key cryptography, as the Lyu et al.'s scheme [32].

14.3.2 Authentication

Based on the data aggregation technique, Wang et al. [28] proposed the concept of anonymous and secure aggregation scheme, named, ASAS, for fog-based public cloud computing. The ASAS considers four different entities, namely, system manager, terminal devices, fog node bridges, and public cloud server. The system manager generates the system parameters and helps the other entities to generate the private key/public key pairs. The terminal devices are user or near-user entities on the edge of the network. The fog node bridges are used as a multitude for storage, for communication, and for control, configuration, measurement, and management. Therefore, the ASAS considers that terminal devices and fog node bridges trust the public cloud server. In addition, the ASAS can preserve the following security requirements, anonymity, indistinguishability of the ciphertexts, and revocation. For more details about the authentication protocols for the IoTs, we refer the reader to work with [35].

14.3.3 Access Control

Due to lack of centralized and privileged control in security management as in cloud computing, access control becomes a challenging issue in fog computing with heterogeneous and service requirements. In fact, the on-board sensor nodes collect the sensitive data from their surroundings. Therefore, secure access control is an important issue in fog computing. According to Zhang et al. [26], there are seven different models to design access control, namely,

1) Discretionary access control model, which is used in environments that do not need a high level of security, such as the UNIX operating system.
2) Mandatory access control model, which is adapted for a distributed system.
3) Role-based access control model, which is more adapted to use in the fog/cloud computing environment.
4) Attribute-based access control model, which is proposed to satisfy the security and flexibility of cloud computing.
5) Usage-control-based access control model, which is adapted to manage sessions used by users.
6) Reference monitoring access control model, which is unsuitable for dynamic distributed systems such as fog computing according to high computation cost.
7) Proxy re-encryption model, which is proposed to provide a secure distributed storage system.

These different models can be used for fog computing but considering some requirements, such as nontransferability and chosen-ciphertext safeness. To supporting outsourcing capability and attribute update, Zhang et al. [29] proposed an access control scheme, named, CP-ABE, for fog computing. The CP-ABE scheme considers five parties, namely, cloud service provider, fog nodes, data owner, end user, and key authority. In addition, the CP-ABE scheme incurs less computation cost for key generation, encryption, decryption, and attribute update. Therefore, the Yu et al.'s scheme [33] uses the leakage-resilient functional encryptions in order to provide access control in fog computing secure against side-channel attacks.

14.3.4   Malicious attacks

As illustrated in Figure 14.4, the malicious attack poses a significant threat to fog computing. Based on the articles [36] and [37], we classify the malicious attacks to fog computing into the following categories:

- *Attacks against the Network infrastructure*: The attacks under this category target the components of the communication infrastructure of the fog network. We can list the following attacks:
  - *Denial-of-service (DoS)*: As many of the IoT devices are not mutually authenticated, it becomes easy to launch a DoS attack. The compromised IoT devices may request infinite processing/storage resources to a fog node and prevent the legitimate devices from accessing the fog services.

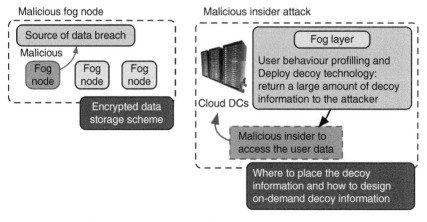

Figure 14.4 Malicious attacker steals the end user's private key and illegitimately accesses the user data. How to identify the malicious insider and reduce the amount of stolen information are major research challenges in fog computing. Based on the user behavior, proliferation decoy technology is deployed using fog computing. The decoy information returns a massive amount of garbage data which is assumed to be true user's data to the attacker.

- *Man-in-the-middle*: An adversary takes control of a section of the network and then launches attacks such as: eavesdropping and/or traffic injection.
- *Rogue gateway*: Due to the open nature of fog networks, adversaries might deploy their own gateways. In this way, the rogue gateway pretends to be a legitimate one and launches some attacks such as: man-in-the-middle attack.
- *Attacks against the edge data center*: As the edge data center hosts many management services, it can be targeted by the following attacks:
  - *Data leakage*: An edge data center mainly stores and processes information from the entities that are located in its vicinity. Some vulnerabilities such as insecure application programming interface (API), misconfiguration, or malicious insider can be exploited to leak data.
  - *Privilege escalation*: This attack can be launched by external adversaries that exploit some vulnerabilities at the edge data center, or by insider adversaries who decided to act maliciously.
  - *Service manipulation*: An external adversary takes control of certain parts of the edge data center through privilege escalation, or a legitimate administrator acts maliciously. In this way, it is possible to manipulate the services of the edge data center.
  - *Rogue data center*: Under this attack, external and internal adversaries take control of the entire edge data center.
- *Attacks against the core infrastructure*: In this category, we can find the following attacks:
  - *Illegal data access*: The stored and processed data can be accessed by unauthorized parties or honest but curious adversaries.
  - *Service manipulation*: An internal adversary with sufficient privileges can instantiate rogue services to generate bogus information.
  - *Rogue infrastructure*: This attack takes control of some elements of the core infrastructure by exploiting some vulnerabilities, and consequently it can manipulate its services.
- *Attacks against virtualization infrastructure*: Like any physical host, the virtual machines might be controlled by malicious adversaries who are trying to misuse the resources provided to those machines. In this category, we can find the following attacks:
  - *Denial of service (DoS)*: A malicious virtual machine can try to run out the computational and storage resources of the host where it is running.
  - *Misuse of resources*: A malicious virtual machine can execute various malicious programs to target remote entities. For example, searching vulnerable IoT, or hosting botnet servers.
  - *Data leakage*: Most of virtualization infrastructures at the edge data center implement APIs that provide information about the physical and logical environment. If these APIs are not secure, a malicious virtual machine can obtain information about the execution environment.

- *Privilege escalation*: Malicious virtual machines might exploit vulnerabilities of their hosts to perform privilege escalation and manipulate other virtual machines.
- *VM manipulation*: A host that is controlled by an external or internal adversary can launch different attacks against its hosted virtual machines such as data leakage or manipulation of the computational tasks.
- *Attacks launched by user devices*: The devices that are controlled by the users might try to disrupt the services of the fog. Under this category, we can find the following attacks:
  - *Data injection*: A compromised device can generate fake information when queried (e.g. sensors or vehicles reporting wrong values).
  - *Service manipulation*: A device might take part in providing services (e.g. crowdsourcing services). If the device is compromised or under the control of an adversary, it can manipulate the outcome of the service.
- *Web-based attacks*: Some attacks such as SQL injection, cross-site scripting, cross-site request forgery, session/account hijacking, and insecure direct object references can target vulnerable web applications, which are hosted by the edge data center, and hence launch various attacks such as data leakage and installation of malicious applications.
- *Malware-based attacks*: An adversary can infect the host or the virtual machines with different types of malware such as logical bomb, Trojans, worms, spyware, and ransomware.

In order to defend against the abovementioned attacks, it is required to deploy preventive and detective countermeasures. The preventive countermeasures consist in securing the systems from vulnerabilities such as insecure APIs and web-based vulnerabilities. It is also important to deploy policy enforcement access mechanisms against illegal data accesses, regular software updates, and anti-malware. As for the security of virtual machines, the countermeasures that can be adopted are isolation policies, hypervisor hardening, separation of roles and VMs, and networking abstraction [38]. There is also a vital need to implement an intrusion detection system (IDS) at the different domains of the fog networks. The IDS should monitor the behavior of the hosts, the VMs, and the network traffic.

14.4 Open Questions and Research Challenges

Unlike computation and storage offloading to the fog layer, the security requirements cannot be offloaded to the fog. The secure communication between EU's devices and fog nodes is an equally important issue. Compared to cloud computing, fog computing is more vulnerable to attacks due to its distributed nature. Cloud computing is heavily protected by cloud operators, all of the security solutions cannot be easily extended to fog computing since fog computing is more close toward network edge. Nevertheless, as EU devices at the bottom layer of fog computing do not have enough resources, it is equally essential to ensure many security challenges. A few articles [2–4, 34, 39] discussed the security and privacy issues in fog computing, however, still they are in the very early stage. In the following, we discuss main research issues related to security and privacy concerns in fog computing.

14.4.1 Trust

The devices in a fog network must ensure a certain level of trust among each other. Although authentication establishes an initial secure agreement between EU devices and fog nodes, the devices are vulnerable to malicious attacks and malfunctions. Therefore, the trust based on previous relation or interaction play an important role in fog network [40–42]. The fog node must ensure that the service requests are generated from genuine end devices. At the same time, there must be a mechanism to measure the trust of a fog service as well as what are the primary attributes that define the trust of a fog service.

Mainly, a two-way trust relationship is desirable to develop trusted interaction between fog nodes and end devices. Opinion-based model is helpful to choose a fog service. However, the reliability will become an essential factor to be considered. Service-level agreement (SLA) between a cloud service and EU is limited if the service is processed in the fog layer, a professional and licensed third-party should monitor SLA verification for the EUs and small organizations who lack in technical capability. Although a fog service provider offers attributes to measure the trust of service, at the same time, verification and monitoring of these attributes are not yet studied in detail.

14.4.2 Privacy preservation

Privacy preservation becomes an essential issue in fog computing since the fog nodes collect sensitive EU's data such as location, identity, and resource utilization. Due to the distributed nature, centralized privacy control becomes difficult in fog computing compared to traditional cloud computing. The intruder can steal the users' private data as well as the increased communication between the EUs also leads to privacy leakage. It is expected that the fog network must ensure the location and identify privacy issues [43–45] of the EUs in fog computing paradigm.

14.4.3 Authentication

To avail the fog service, a device needs to be authenticated. This will prevent the unauthorized end users from accessing the fog services. A certifying

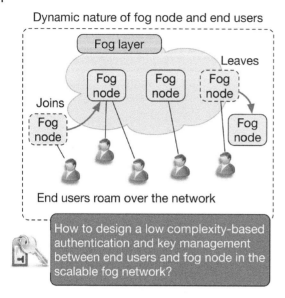

Dynamic nature of fog node and end users

Figure 14.5 The end users roam randomly over the network. Besides, fog nodes also frequently join and leave the fog layer. Thus, the mutual authentication EU and fog node is a challenging issue. How an end user is able to mutually authenticate with new fog node that joins the network without any significant increase in overheads becomes an important issue.

authority (CA) model is used to allow fog nodes to restrict service requests from malicious nodes. At the same time, it will also prevent any devices from becoming a part of the fog network. Compared to cloud computing, mobility issues and dynamic nature of fog nodes (see Figure 14.5) are inherent challenges in fog computing [46]. Thus, how to ensure uninterrupted services among fog nodes and end devices in newly formed fog layer become a critical issue. From a resource-constraint EU perspective, re-authentication and re-registration phase must be straightforward with less overhead.

14.4.4 Malicious Attacks and Intrusion Detection

Malicious attacks are widespread in fog computing environment without proper security measurements. DoS attack is one of them. Due to the large network size, lack of device authentication leads to malicious attacks [47]. Nevertheless, the complex authentication on service requests restricts the real-time QoS guarantee. At the same time, spoofing of addresses becomes more accessible due to large address space. Apart from the attacks from outside, a malicious insider to the cloud provider becomes a severe issue since the EU entirely rely on cloud service provider. The malicious insider can easily be compromised by the security issues in fog computing. Proper intrusion detection [48, 49], misbehavior detection of EUs as well as fog nodes, real-time notification, false alarm control, and fast prevention are major security-related issues in fog computing.

The security architecture for the federated cloud environments [50] needs to be deployed for every trust domain. The reason is that due to the distributed

Figure 14.6 Fog computing consists of a massive number of fog nodes as infrastructure, retrieving the *log data* from these fog nodes becomes very difficult.

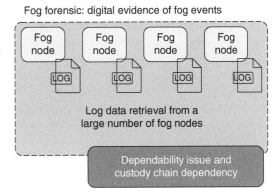

Fog forensic: digital evidence of fog events

nature of the fog networks, many IDS solutions might exist at different trust domains. In fact, the main challenges are how to make the different IDSs collaborate and exchange information in an interoperable way.

14.4.5 Cross-border Issues and Fog Forensic

Followed by cloud forensic, fog forensic is defined as the application of digital forensics in fog computing. As observed by Wang et al. [51], fog forensics that have some steps similar to cloud forensic, however, is not a part of cloud forensics. Although some challenges in fog forensics are same as cloud forensics (e.g. cyber-physical systems (CPS) and custody chain dependency, and integrity preservation), many challenges are more significant in fog forensics compared to cloud forensics. For example, since fog computing consists of a massive number of fog nodes as infrastructure, retrieving the *log data* from these fog nodes becomes very difficult (see Figure 14.6). Although the cross-border issues are less significant as compared to cloud computing due to the distributed nature of fog computing, the dependability issue becomes more crucial with a large number of fog nodes. Thus, fog forensics still require international legislation and jurisdictions [52, 53] and application level logging [54]. Therefore, it is still an important task to overcome cross-border legislation challenges from the security and privacy aspects in fog computing.

14.5 Summary

Fog computing paradigm extends a substantial amount of data storage, computing, communication, and networking of cloud computing toward the edge of the networks while offloading the cloud data centers and reducing service latency to the end users. Due to close integration with the front-end intelligence-enabled end devices, fog computing enhances the overall system efficiency, after that improving the performance of critical **CPS**. Fog computing is also a virtualized service computing architecture, which may provide computation, storage, and networking services between terminal devices and back-end cloud services. However, the characteristics of fog computing give rise to new security and privacy challenges. All of the privacy and security solutions proposed for cloud computing are not suitable for fog computing due to several distinct characteristics such as the decentralized and distributed architecture of fog computing as well as a wider scale of fog devices at the edge of the network.

In this chapter, we have presented an overview of the primary security and privacy issues in fog computing. Afterward, this chapter surveys the state-of-the-art solutions that deal with fog computing-related security and privacy challenges. Then, we discuss the major attacks on fog-based IoT applications. We provide a side-by-side comparison of the state-of-the-art methods toward secure and privacy-preserving fog-based IoT applications. This chapter aims to summarize all up-to-date research contributions and to outline future research directions that researcher can follow in order to address different security and privacy preservation challenges in fog computing.

Exercises

1 What are the various factors in fog computing architecture responsible for security and privacy vulnerabilities in fog computing?

2 How trust can be compromised in fog computing?

3 Discuss about different attacks in fog computing and possible solutions.

4 Why privacy preserving schemes designed for cloud computing cannot be directly applied to the fog computing paradigm?

5 Discuss about authentication protocols in fog computing.

6 What are the different malicious attacks that can target fog computing architecture?

References

1 IDC (2016). Worldwide internet of things forecast update 2015–2019. Doc #US40983216.
2 Bilal, K., Malik, S.U.R., Khan, S.U., and Zomaya, A.Y. (2014). Trends and challenges in cloud datacenters. *IEEE Cloud Computing* 1 (1): 10–20.

3 Buyya, R., Yeo, C.S., and Venugopal, S. (2008). Market-oriented Cloud computing: vision, hype, and reality for delivering IT services as computing utilities. IEEE 10th International Conference on High Performance Computing and Communications, September 2008, pp. 5–13.

4 Rimal, B.P., Choi, E., and Lumb, I. (2009). A taxonomy and survey of cloud computing systems. Proceedings of the 5th International Joint Conference on INC, IMS and IDC, August 2009, pp. 44–51.

5 DATACENTERS.com. Data Center Companies. https://www.datacenters .com/directory/companies (accessed 20 September 2017).

6 NRDC (2015). Americas Data Centers Consuming and wasting growing amounts energy. https://www.nrdc.org/resources/americas-data-centers-consuming-and-wasting-growing-amounts-energy (accessed 20 September 2017).

7 DataCenterKnowledge (2013). Apple hits 100% renewable energy in its data centers. https://www.datacenterknowledge.com/archives/2013/03/22/apple-hits-100-renewable-energy-in-its-data-centers/ (accessed 20 September 2017).

8 Newsroom (2017). Apple's next US data center will be built in iowa. https://www.apple.com/newsroom/2017/08/apples-next-us-data-center-will-be-built-in-iowa/ (accessed 20 September 2017).

9 The carbon footprint of the Internet. https://www.commmade.com/blog/carbon-footprint-of-internet/ (accessed 12 September 2017).

10 Satyanarayanan, M., Bahl, P., Caceres, R., and Davies, N. (2009). The case for VM-based Cloudlets in mobile computing. *IEEE Pervasive Computing* 8 (4): 14–23.

11 Mouradian, C., Naboulsi, D., Yangui, S. et al. (2018). A comprehensive survey on fog computing: state-of-the-art and research challenges. *IEEE Communication Surveys and Tutorials* 20 (1): 416–464.

12 Barbarossa, S., Sardellitti, S., and Lorenzo, P.D. (2014). Communicating while computing: distributed mobile cloud computing over 5G heterogeneous networks. *IEEE Signal Processing Magazine* 31 (6): 45–55.

13 Hu, Y.C., Patel, M., Sabella, D. et al. (2015). *Mobile Edge Computing: A Key Technology Towards 5G*, 1e, vol. 11. ETSI White Paper.

14 ETSI. Multi-acess Edge computing. http://www.etsi.org/technologies-clusters/technologies/multi-access-edge-computing (accessed 20 January 2018).

15 Mach, P. and Becvar, Z. (2017). Mobile edge computing: a survey on architecture and computation offloading. *IEEE Communications Surveys & Tutorials* 19 (3): 1628–1656.

16 Mukherjee, M., Shu, L., and Wang, D. (2018). Survey of fog computing: fundamental, network applications, and research challenges. *IEEE Communication Surveys and Tutorials* 20 (3): 1826–1857.

17 Aazam, M., Zeadally, S., and Harras, K.A. (2018). Fog computing architecture, evaluation, and future research directions. *IEEE Communications Magazine* 56 (5): 46–52.

18 Aazam, M., Zeadally, S., and Harras, K.A. (2018). Offloading in fog computing for IOT: review, enabling technologies, and research opportunities. *Future Generation Computer Systems* 87: 278–289.

19 Bonomi, F., Milito, R., Zhu, J., and Addepalli, S. (2012). Fog computing and its role in the internet of things. Proceedings of the First Edition of the MCC Workshop on Mobile Cloud Computing, January 2012, pp. 13–16.

20 Cisco (2014). *Cisco Delivers Vision of Fog Computing to Accelerate Value from Billions of Connected Devices*. Press Release. https://newsroom.cisco .com/press-release-content?type=webcontent&articleId=1334100 (accessed 20 September 2017).

21 Aazam, M. and Huh, E.N. (2016). Fog computing: the cloud-IoT/IoE middleware paradigm. *IEEE Potentials* 35 (3): 40–44.

22 Vaquero, L.M. and Rodero-Merino, L. (2014). Finding your way in the fog. *ACM SIGCOMM Computer Communication Review* 44 (5): 27–32.

23 Definition of fog computing. https://www.openfogconsortium.org/ resources/#definition-of-fog-computing (accessed 12 May 2017).

24 OpenFog Consortium. https://www.openfogconsortium.org (accessed 20 September 2017).

25 Ni, J., Zhang, K., Lin, X., and Shen, X.S. (2018). Securing fog computing for internet of things applications: challenges and solutions. *IEEE Communication Surveys and Tutorials* 20 (1): 601–628.

26 Zhang, P., Liu, J.K., Yu, F.R. et al. (2018). A survey on access control in fog computing. *IEEE Communications Magazine* 56 (2): 144–149.

27 Hong, Y., Liu, W.M., and Wang, L. (2017). Privacy preserving smart meter streaming against information leakage of appliance status. *IEEE Transactions on Information Forensics and Security* 12 (9): 2227–2241.

28 Wang, H., Wang, Z., and Domingo-Ferrer, J. (2018). Anonymous and secure aggregation scheme in fog-based public cloud computing. *Future Generation Computer Systems* 78: 712–719.

29 Zhang, P., Chen, Z., Liu, J.K. et al. (2018). An efficient access control scheme with outsourcing capability and attribute update for fog computing. *Future Generation Computer Systems* 78: 753–762.

30 Wang, T., Zhou, J., Chen, X. et al. (2018). A three-layer privacy preserving cloud storage scheme based on computational intelligence in fog computing. *IEEE Transactions on Emerging Topics in Computational Intelligence* 2 (1): 312.

31 Yang, M., Zhu, T., Liu, B. et al. (2018). Machine learning differential privacy with multifunctional aggregation in a fog computing architecture. *IEEE Access* 6: 17119–17129.

32 Lyu, L., Nandakumar, K., Rubinstein, B. et al. (2018). PPFA: privacy preserving fog-enabled aggregation in smart grid. *IEEE Transactions on Industrial Informatics* 14 (8): 3733–3744.

33 Yu, Z., Au, M.H., Xu, Q. et al. (2018). Towards leakage-resilient fine-grained access control in fog computing. *Future Generation Computer Systems* 78: 763–777.

34 Mukherjee, M., Matam, R., Shu, L. et al. (2017). Security and privacy in fog computing: challenges. *IEEE Access* 5: 19293–19304.

35 Ferrag, M.A., Maglaras, L.A., Janicke, H. et al. (2017). Authentication protocols for internet of things: a comprehensive survey. *Security and Communication Networks* 2017: 1–41.

36 Roman, R., Lopez, J., and Mambo, M. (2018). Mobile edge computing, fog et al.: a survey and analysis of security threats and challenges. *Future Generation Computer Systems* 78: 680–698.

37 Khan, S., Parkinson, S., and Qin, Y. (2017). Fog computing security: a review of current applications and security solutions. *Journal of Cloud Computing* 6 (1): 1–9.

38 Pék, G., Buttyán, L., and Bencsáth, B. (2013). A survey of security issues in hardware virtualization. *ACM Computing Surveys (CSUR)* 45 (3): 40.

39 Yi, S., Qin, Z., and Li, Q. (2015). Security and privacy issues of fog computing: a survey. Proceedings of the 10th International Conference Wireless Algorithms, Systems, and Applications: (WASA), August 2015, pp. 685–695.

40 Atzori, L., Iera, A., Morabito, G., and Nitti, M. (2012). The Social Internet of Things (SIoT) – when social networks meet the internet of things: concept, architecture and network characterization. *Computer Networks* 56 (16): 3594–3608.

41 Farris, I., Girau, R., Militano, L. et al. (2015). Social virtual objects in the edge cloud. *IEEE Cloud Computing* 2 (6): 20–28.

42 Nitti, M., Girau, R., and Atzori, L. (2014). Trustworthiness management in the social internet of things. *IEEE Transactions on Knowledge and Data Engineering* 26 (5): 1253–1266.

43 Ni, J., Zhang, A., Lin, X., and Shen, X.S. (2017). Security, privacy, and fairness in fog-based vehicular crowdsensing. *IEEE Communications Magazine* 55 (6): 146–152.

44 Koo, D. and Hur, J. (2018). Privacy-preserving deduplication of encrypted data with dynamic ownership management in fog computing. *Future Generation Computer Systems* 78 (Pt. 2): 739–752.

45 Motlagh, N.H., Bagaa, M., and Taleb, T. (2017). UAV-based IoT platform: a crowd surveillance use case. *IEEE Communications Magazine* 55 (2): 128–134.

46 Peng, M., Yan, S., Zhang, K., and Wang, C. (2016). Fog-computing-based radio access networks: issues and challenges. *IEEE Network* 30 (4): 46–53.

47 Yaseen, Q., Albalas, F., Jararwah, Y., and Al-Ayyoub, M. (2017). Leveraging fog computing and software defined systems for selective forwarding attacks detection in mobile wireless sensor networks. *Transactions on Emerging Telecommunications Technologies* 29 (4): 1–13.

48 Anwar, S., Mohamad Zain, J., Zolkipli, M.F. et al. (2017). From intrusion detection to an intrusion response system: fundamentals, requirements, and future directions. *MDPI Algorithms* 10 (2): 1–24.

49 Yaseen, Q., AlBalas, F., Jararweh, Y., and Al-Ayyoub, M. (2016). A fog computing based system for selective forwarding detection in mobile wireless sensor networks. 2016 IEEE 1st International Workshops on Foundations and Applications of Self* Systems (FAS*W), September 2016, pp. 256–262.

50 Luo, W., Xu, L., Zhan, Z. et al. (2014). Federated cloud security architecture for secure and agile clouds. In: *High Performance Cloud Auditing and Applications* (ed. K. Han, B.Y. Choi, and S. Song), 169–188. New York: Springer. https://link.springer.com/chapter/10.1007/978-1-4614-3296-8_7.

51 Wang, Y., Uehara, T., and Sasaki, R. (2015). Fog computing: issues and challenges in security and forensics. Proceedings of IEEE 39th Annual Computer Software and Applications Conference, vol. 3, July 2015, pp. 53–59.

52 Biggs, S. and Vidalis, S. (2009). Cloud computing: the impact on digital forensic investigations. Proceedings of IEEE International Conference for Internet Technology and Secured Transactions (ICITST), November 2009, pp. 1–6.

53 Wolthusen, S.D. (2009). Overcast: forensic discovery in cloud environments. Proceedings IEEE 5th International Conference on IT Security Incident Management and IT Forensics, September 2009, pp. 3–9.

54 Marty, R. (2011). Cloud application logging for forensics. Proceedings of ACM Symposium on Applied Computing (SAC), March 2011, pp. 178–184.

Index

Fog and Fogonomics: Challenges and Practices of Fog Computing, Communication, Networking, Strategy, and Economics, First Edition. Edited by Yang Yang, Jianwei Huang, Tao Zhang, and Joe Weinman.
© 2020 John Wiley & Sons, Inc. Published 2020 by John Wiley & Sons, Inc.